WORKFORCE
CROSS TRAINING

WORKFORCE CROSS TRAINING

Edited by
David A. Nembhard

CRC Press
Taylor & Francis Group
Boca Raton London New York

CRC Press is an imprint of the
Taylor & Francis Group, an **informa** business

CRC Press
Taylor & Francis Group
6000 Broken Sound Parkway NW, Suite 300
Boca Raton, FL 33487-2742

First issued in paperback 2019

ISBN-13: 978-0-8493-3632-4 (hbk)
ISBN-13: 978-0-367-38918-5 (pbk)

Library of Congress Cataloging-in-Publication Data

Nembhard, David A.
    Workforce cross training / David A. Nembhard.
      p. cm.
    Includes bibliographical references and index.
    ISBN-13: 978-0-8493-3632-4 (alk. paper)
    ISBN-10: 0-8493-3632-5 (alk. paper)
    1. Employees--Training of. I. Title.

HF5549.5.T7N42 2007
658.3'124--dc22
                                        2006035220

Visit the Taylor & Francis Web site at
http://www.taylorandfrancis.com

and the CRC Press Web site at
http://www.crcpress.com

Table of Contents

Preface

The idea for this project began many years ago as I started to conduct research in the area of workforce cross training. Several industrial managers and researchers began contacting me to ask about the body of knowledge in this area. Of course, many organizations have current or planned programs for cross training employees, yet few clear guidelines exist for designing these programs effectively with a focus on productivity and performance. For those who inquired, I was able to point them to a few research papers, but in the process I realized that there was a dearth of material that aggregated what was currently known, synthesized best practices, or even gave a clear indication of what was well known or not well known in this area. This book is intended to be a modest step toward that end.

This book integrates academic work on workforce cross training, current practices, and discussion of future needs and opportunities. It is not intended to be a comprehensive clearinghouse of all the work that has been done in the field. Rather, I hope the descriptions of best practices, effective research models, and results will be of benefit to both the interested researcher and the practitioner. It is through the gracious participation of the contributing authors that this project has been possible. For this I offer my heartfelt thanks. I believe that their varied viewpoints, approaches, and skill sets have resulted in a wide-ranging discussion of workforce cross-training technology. I hope that this book can serve as one of perhaps a number of starting points, where we can progress toward a better understanding of some of the how, why, when, who, and what that are involved in managing and improving workforce cross-training systems.

Acknowledgment

I would like to thank all of those who were involved with this project, first and foremost, the contributing authors, whose efforts are greatly appreciated, and the team at Taylor & Francis Group, Cindy Carelli and Jessica Vakili, for helping to get this project rolling, understanding delays, and keeping me on schedule. I also thank the National Science Foundation for its support of my research in this area, and specifically portions of my research that are summarized herein.

The Editor

David A. Nembhard, Ph.D., is Associate Professor and Harold and Inge Marcus Career Professor of Industrial and Manufacturing Engineering at The Pennsylvania State University, University Park. He received his Ph.D. from the University of Michigan, Ann Arbor, in 1994 in industrial and operations engineering. He also holds B.S.E. and M.S.E. degrees in systems and control engineering from Case Western Reserve University, Cleveland, Ohio. He has held appointments in industrial engineering at The University of Wisconsin–Madison, operations management at Auburn University, Auburn, Alabama, and a position as a visiting scholar of industrial engineering at Ecole Centrale Paris.

Dr. Nembhard's research interests include: workforce cross training, workforce learning and forgetting, knowledge worker measurement, routing, scheduling, and HAZMAT transport. His research has been funded by the National Science Foundation and various industrial sponsors. He has published in journals including: *IEEE Transactions on Engineering Management, Journal of Operations Management, The Engineering Economist, Management Science, Human Factors,* the *International Journal of Production Research,* the *European Journal of Operational Research,* the *Journal of Transportation Engineering,* and the *International Journal of Industrial Ergonomics.* He is a member of INFORMS and a senior member of the Institute of Industrial Engineers (IIE).

Contributors

O. Zeynep Aksin
Graduate School of Business
Koç University
Sariyer, Istanbul, Turkey

Sigrún Andradóttir
School of Industrial and Systems
 Engineering
Georgia Institute of Technology
Atlanta, Georgia

Nilay Tanik Argon
Department of Statistics and
 Operations Research
The University of North Carolina
 at Chapel Hill
Chapel Hill, North Carolina

Jos. A. C. Bokhorst
Faculty of Management and
 Organization
University of Groningen
Groningen, The Netherlands

Viviana I. Cesaní
Department of Industrial
 Engineering
University of Puerto
 Rico-Mayagüez
Mayagüez, Puerto Rico

Albert Corominas
IOC Research Institute
Universitat Politècnica de
 Catalunya
Barcelona, Spain

Stephen M. Fiore
Institute for Simulation & Training
University of Central Florida
Orlando, Florida

Fikri Karaesmen
Department of Industrial
 Engineering
Koç University
Sariyer, Istanbul, Turkey

David A. Nembhard
Harold and Inge Marcus
 Department of Industrial
 Engineering
The Pennsylvania State University
University Park, Pennsylvania

Bryan A. Norman
Department of Industrial
 Engineering
University of Pittsburgh
Pittsburgh, Pennsylvania

E. Lerzan Örmeci
Department of Industrial
 Engineering
Koç University
Sariyer, Istanbul, Turkey

Rafael Pastor
IOC Research Institute
Universitat Politècnica de
 Catalunya
Barcelona, Spain

Karndee Prichanont
Department of Industrial
  Engineering
Thammasat University
Bangkok, Thailand

Ruwen Qin
Harold and Inge Marcus
  Department of Industrial
  Engineering
The Pennsylvania
  State University
University Park, Pennsylvania

Ling Rothrock
Harold and Inge Marcus
  Department of Industrial
  Engineering
The Pennsylvania
  State University
University Park, Pennsylvania

Eduardo Salas
Institute for Simulation & Training
University of Central Florida
Orlando, Florida

Jannes Slomp
Faculty of Management and
  Organization
University of Groningen
Groningen, The Netherlands

Kevin C. Stagl
Institute for Simulation & Training
University of Central Florida
Orlando, Florida

Hari Thiruvengada
Harold and Inge Marcus
  Department of Industrial
  Engineering
The Pennsylvania State University
University Park, Pennsylvania

section I

Workforce operations

chapter 1

Design and operation of a cross-trained workforce

Jos A. C. Bokhorst and Jannes Slomp

Contents

1.1   Introduction

Due to increased global competition, firms are under constant pressure to cut their costs, while having to improve their delivery speed, quality, flexibility, and dependability. It has become clear that improvements should focus not only on the efficiency and effectiveness of (technical) processes but also on the workers involved in these processes. Workers increasingly need to be flexible — able to do several tasks and assume tasks or help other workers with their tasks — while remaining efficient and motivated. Acknowledging the value of the workforce and carefully considering the design and operation of a firm's workforce can significantly contribute to the improvement of the objectives of the firm. "Human capital" is a key success factor nowadays, and firms must try to achieve a fit between their own goals and the goals of their workforce. Training workers and using the acquired skills effectively is one of the ways both goals can be achieved.

In essence, a cross-trained workforce consists of (one or more teams of) workers who have (partly) overlapping skills or tasks they are able to perform. Research on cross training within the field of Operations Management often entails comparing the performance of teams having alternative numbers and/or distributions of skills, where the focus then is more on performance implications of the result of training — the qualifications — than on the process of training. Furthermore, even though a cross training may be regarded as the result of training someone for a skill already mastered by someone else (an overlapping skill), the terms *training* and *cross training* are more often used interchangeably. In this chapter, we also focus on the result of training instead of on the process, and we do not particularly distinguish a *cross training* from any other *training* or qualification.

This chapter focuses on the development of effective cross-training policies and labor assignment rules. These issues play a large role in Dual Resource Constrained (DRC) systems. In DRC systems, two resources are considered to be constraining factors for the level of output. In this chapter, we consider labor and machines to be the two constraining resources of our concern. This type of DRC system is also called a labor and machine-limited system, as opposed to a machine-limited system in which only machines are the constraining factor. In a labor and machine-limited system, jobs can only be processed if both a machine and a skilled worker are available. A key characteristic of labor and machine-limited systems is that the number of workers is less than the number of machines, which implies that (some) workers need to be multifunctional and worker transfer between machines is necessary. For smooth operation of these systems, attention should be given to cross training and labor assignment rules.

Table 1.1 An Example of a Skill Matrix

Machines workers	1	2	3	4	5	6	7	8	9	10
A		X		X		X				
B	X	X								X
C	X		X				X			
D		X		X	X				X	
E					X	X				X
F		X	X	X			X	X		
G		X					X			

Note: X denotes a skill.

Cross-training policy is defined as a set of rules for determining the extent of training and the distribution of workers' skills in a team. In practice, many firms use a skill matrix (also known as a worker–machine matrix or worker–task matrix) to display the current set of skills available in a team. Table 1.1 shows an example of a skill matrix, where each column represents a machine (a total of 10 machines) and each row represents a worker (a total of 7 workers in the team). An X in the matrix represents a worker skill. From the matrix, it is easy to see which skills a particular worker has mastered by looking at the corresponding row. For instance, worker A is skilled for machines 2, 4, and 6 in the example in Table 1.1. Similarly, by looking at a specific column, it can easily be seen which workers master the corresponding machine (for instance, machine 9 can only be operated by worker D).

Applying a cross-training policy to a manufacturing team results in an optimized skill matrix indicating which workers should be trained for which machines. We call the resulting changed skill matrix a *cross-training configuration*, which represents a cross-trained workforce. When having a cross-trained workforce, labor assignment rules must be set to properly assign workers to machines or tasks for which they are skilled.

Having a cross-trained workforce may support an organization's strategy, if carefully designed and operated. The extent and distribution of cross training impacts the performance of the workforce, as well as the assignment rules that are chosen to assign skilled workers to machines or tasks. Hopp and Van Oyen (2004) developed a strategic assessment framework that structures the key direct and indirect mechanisms by which a cross-trained workforce can support organizational strategy. They state that the *cross-training skill pattern* (cross-training configuration) and the *worker coordination policy* (labor assignment rule) may impact labor productivity, responsiveness, internal/external quality, and the offerings of products/services, which directly impact strategic objectives, such as cost, time, quality, and variety. Furthermore, issues with respect to team structure, such as collaboration, authority, communication, incentives, etc., indirectly impact the strategic objectives.

This chapter embraces an Operations Management viewpoint on cross training and labor assignment. Other aspects (i.e., human factors) are referred to and play a role in choices and considerations but are not dealt with in the research-based parts of the chapter. Time and costs are the main strategic objectives we consider.

Section 1.2 deals with the development of effective cross-training policies. Developing cross-training policies involves deciding which strategic goals to support, which aspects are important to include, and also defining decision rules to specify how these aspects will be addressed. Among the aspects considered are the extent of cross training, chaining, the level and distribution of multifunctionality and redundancy, and collective responsibility. These terms will be explained later in the chapter. We show how an Integer Goal Programming model can support making effective cross-training decisions. We evaluate cross-training decisions, or policies, by means of a simulation study. Finally, we show the applicability of an Integer Programming Model that — besides operational performance — focuses more on the training costs in an industrial setting.

Section 1.3 deals with labor assignment, which is addressing the question of which rules should be designed to assign workers to tasks for maximum performance. Here, the qualifications of workers or the tasks they are able to perform are fixed, but different labor assignment rules are designed that may alter the deployment of these qualifications. We first focus on industrial practice with respect to worker assignment. We then review the literature on labor assignment and worker differences. Previous studies on labor assignment mostly study the "when-rule" and "where-rule," which decide when a worker is eligible for transfer and where he/she should be transferred to, respectively. Furthermore, most studies consider a homogeneous workforce, or workers who have the same characteristics. In Section 1.3, we also draw attention to the "who-rule." This rule has gained only limited attention in literature so far. Finally, by means of several simulation experiments, we investigate labor assignment rules in systems with worker differences.

Section 1.4 discusses future research issues. The section extends the issues dealt with in Section 1.2 and Section 1.3 but also covers a broader range of (Operations Management) issues related to cross training and worker assignment that need to be addressed.

1.2 Development of cross-training policies

By developing a cross-training policy, an organization strives to design a cross-trained workforce that will support its strategy. Developing cross-training policies involves deciding which performance measures should be targeted, which aspects are important to include in light of this performance, and also defining decision rules to specify how these aspects will be addressed. The set of aspects to be included or the relative importance to be given to the aspects may depend on the specific strategy of the organization and the context.

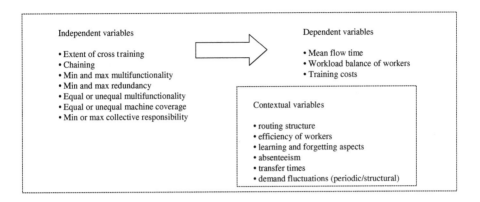

Figure 1.1 A model on developing and evaluating cross-training policies.

The following are five important aspects to consider when developing a cross-training policy: the extent of cross training, the concept of *chaining*, multifunctionality, machine coverage, and collective responsibility. Applying a cross-training policy to an existing workforce results in recommendations as to which workers to train for which tasks. This resulting cross-training configuration may be evaluated by using simulation. Figure 1.1 represents the theoretical model with the independent variables (aspects to consider), dependent variables (performance measures), and contextual variables.

Section 1.2.1 deals with performance measures and the contextual variables. The aspects to be considered in the development of a cross-training policy will be discussed consecutively in Section 1.2.2. Section 1.2.3 shows how an Integer Goal Programming model can lend support in making effective cross-training decisions and evaluates cross-training policies in a specific situation by means of simulation. Section 1.2.4 shows the applicability of an Integer Programming Model that focuses on the tradeoff between training costs and operational performance in an industrial setting. Finally, Section 1.2.5 summarizes the above discussions.

1.2.1 Performance measures and contextual factors

As mentioned in the Introduction, we embrace an Operations Management viewpoint and mainly focus on improving time (mean flow time of jobs) and costs (training costs and operational costs). We also consider the workload balance of workers to be an important measure from a Human Resource Management point of view. Of course, there are other reasons to perform cross training — both from a worker's point of view and a firm's point of view. A worker, for example, may be motivated more if his/her desire to be all round is fulfilled. A firm, for instance, may require extra training to decrease the frequency of entity handoffs in order to enable the workforce to develop broad capabilities that provide better ways of meeting customer needs (Hopp and VanOyen, 2004).

Several factors may be considered contextual variables. By this, we mean that these factors differ depending on the context of the workforce and may have an impact on (the relation between) independent and dependent variables. With respect to jobs, the routing of the jobs (i.e., routing structure: parallel, serial, job shop) has an impact on how to cross train workers (see Bokhorst et al., 2004b, and Section 1.2.3.2). Also, the complexity of tasks required to process the job or the variance in complexity between different tasks may impact cross-training decisions. The complexity of a task is most likely related to the amount of training costs and the specific learning and forgetting effects. That is, complex tasks may require more intensive training programs and/or a longer period of on-the-job learning than simple tasks, leading to higher investments for a company. Further, complexity significantly affects learning/forgetting parameters (of manual tasks), and the effects depend on the experience of workers (Nembhard, 2000). With respect to workers, several factors may or may not be included, e.g., the efficiency of workers, learning and forgetting aspects, absenteeism, and transfer times. Finally, with respect to the demand for machines, periodic and/or structural fluctuations may be considered.

We are interested in contexts where cross training leads to operational advantages, but at the same time involves significant training costs. In these situations, real tradeoffs have to be made. Section 1.2.3 focuses more on operational advantages and includes routing structure as a contextual variable and absenteeism and periodic demand fluctuations as given context. Section 1.2.4 presents a model that pays more attention to training costs and includes static efficiency differences between workers and several absenteeism scenarios as given context.

1.2.2   Important aspects to consider in developing cross-training policies

This subsection describes important aspects to consider when developing a cross-training policy. Throughout the discussion of important aspects, we will make use of an illustrative example. A cross-training configuration, which shows the distribution of skills within the workforce, can be represented by a worker–machine matrix or a bipartite graph, for instance. For the illustrative example, Figure 1.2 shows an initial cross-training configuration, representing four workers (A to D) and seven machines (I to VII) connected by worker skills. The graph is bipartite, since it can be partitioned into two disjoint subsets of vertices (i.e., workers and machines) such that each edge connects a worker to a machine. In this representation, the edges represent worker skills. Worker A is trained for machines I and II; worker B is able to operate machines III, IV, and V; worker C is qualified to operate machines V and VI; and worker D is trained for machines V and VII. The machine loads, which indicate the percentage of time in which the machines have to be used by the workers, are shown in brackets in Figure 1.2.

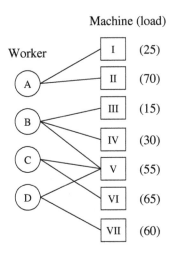

Figure 1.2 Initial cross-training configuration of the illustrative example.

1.2.2.1 Extent of cross training

A first aspect to consider when developing a cross-training policy is the extent of cross training required in the team. By the extent of cross training, we mean the number of (additional) cross trainings that are needed in the manufacturing team. A complete bipartite graph (i.e., each worker is connected to every machine) represents full flexibility. In case the bipartite graph is not complete, as in Figure 1.2, there is limited flexibility. Figure 1.2 shows limited flexibility, with 9 out of a possible 28 skills with full flexibility. Although increases in cross training can positively affect system performance, several papers have shown a diminishing positive effect of a stepwise increase of the level of labor flexibility (Park and Bobrowski, 1989; Malhotra et al., 1993; Fry et al., 1995; Campbell, 1999; Molleman and Slomp, 1999). Most of the positive effects can be achieved without going to the extreme of full flexibility.

Full flexibility is not needed, nor is it desirable in practical situations. Since it requires training of all workers for all machines, it can be very costly. Further, Kher and Malhotra (1994) showed that higher levels of labor flexibility lead to more labor transfers, resulting in considerable losses in productivity. This productivity loss results from, among other factors, the time required for orientation at new workstations, to access information about the job to be performed at the new machine, and to learn or relearn the setup procedures. This is especially the case if the firm applies a centralized assignment rule (i.e., a worker reassignment is considered after completion of each job). The effect is less in the case of a decentralized rule, i.e., where a worker reassignment is considered only when the job queue is empty. In both cases, however, productivity loss due to an increase in the number of worker transfers is an argument to limit the level of labor flexibility.

There are also several social arguments for limiting labor flexibility in manufacturing cells (see, e.g., Van den Beukel and Molleman, 1998). High levels of labor flexibility may impair social identity because the different jobs in a team/cell will be more similar. This may cause motivational deficits (Fazakerley, 1976). With respect to their abilities, people may prefer diversity within the team/cell. Being a specialist enhances feelings of being unique and indispensable and makes the contribution to group perfor- mance visible (Clark, 1993). In addition, studies pertaining to diversity reveal that creativity and motivation are greater in teams whose members have different, but somewhat overlapping, skills (e.g., Jackson, 1996). High levels of labor flexibility may also cause social loafing and, for example, cause a situation in which no one is willing to do the dirty work (Wilke and Meertens, 1994). Cross training may also lead to perceived lowering of status differentials within teams, which may result in negative attitudes, particularly among the higher-status team members who oppose learning and performing the lower status jobs (Carnall, 1982; Cordery et al., 1993; Hut and Molleman, 1998).

In a cross-training policy, a value may be set for the *ideal* number of cross trainings in the desired cross-training configuration. Another option is to minimize the number of additional trainings or training costs, or to minimize the deviation of a budget for training set by a company. In our illustrative example, we assume that the minimization of the number of additional worker skills is a major objective. In practice, managers strive to balance the positive performance effects of cross training and the integral costs of addi- tional worker skills.

1.2.2.2   The concept of chaining

In the initial cross-training configuration (Figure 1.2), we see that worker A is occupied for 95% of the time, since he/she is responsible for machines I and II. Workers B, C, and D are, on average, busy 75% of the time. The load on machine V can be used to balance the workload of workers B, C, and D. We assume that the objective of the firm in the example is to minimize flow times of jobs and to optimize the labor situation through further cross train- ing of the workers. A first concern in the initial cross-training configuration is the unequal workload of the workers. Worker A is clearly the bottleneck, so it is likely that most queuing time will arise at machine I or II. The initial cross-training configuration does not permit a shift of work from worker A to B, C, or D. In the terms of Lau and Murnighan (1998), the initial distribu- tion of skills is a potential *fault line* in the team. The graph is not connected, since there is no path that connects every pair of vertices. In order to enable such a path, at least one additional worker skill is needed.

Jordan and Graves (1995) stressed the importance of chaining in the case of limited flexibility. They studied the effect of process flexibility, which they define as the ability of plants to produce different types of products. This type of flexibility is conceptually equivalent to labor flexibility, which refers to the ability of workers to operate different machines. Brusco and Johns

(1998) recognized this and used the term *chaining* to explain the preference of some of their cross-training patterns. They presented a linear programming model that minimizes costs associated with workforce staffing, subject to the satisfaction of minimum labor requirements across a planning horizon. They used their model to evaluate eight cross-training structures across various patterns of labor requirements, reaching the important conclusion that "chaining of employee skill classes across work activity categories" is a basic element of successful cross-training structures.

Hopp et al. (2004) studied two cross-training strategies for serial production systems with flexible servers. They stated that the two primary benefits of workforce agility in this environment are *capacity balancing*, which is needed if lines are unbalanced with respect to the average workload of each station, and *variability buffering*, which provides a solution for worker idleness caused by variability in processing times. Hopp et al. (2004) showed that the two-skill chaining strategy is potentially robust and efficient in obtaining workforce agility in serial production lines.

Figure 1.3 shows, by means of a bold line, that worker B is additionally trained for machine II. This enables an equal workload division among the workers. This step can be regarded as a capacity-balancing step (Hopp et al., 2004), since cross training is used here to remove a structural imbalance with respect to the utilization of workers (i.e., decrease the high utilization of 95% of worker A and increase the utilization of the other workers). Assuming a fair distribution of work among the workers, each worker will be occupied 80% of the time. Figure 1.3 shows that all workers and machines are now chained through the worker skills. In terms of graph theory, the graph becomes connected, since the addition of skill B-II creates a path that directly or indirectly connects worker A with the other workers and machines.

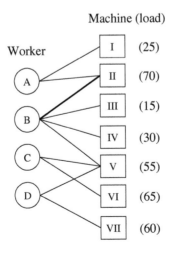

Figure 1.3 Additional training to enable an equal workload division.

Chaining provides the ability to shift work from a worker with a heavy workload to a worker with a lighter workload, leading — directly or indirectly — to a more balanced workload. Chaining, therefore, supports the efficient use of labor capacity and provides sufficient agility to respond to changes in demand, thus enabling fluctuations in the mix of work to be absorbed. Chaining also reduces the likelihood that subgroups may emerge and cause intergroup conflicts, leading to the disintegration of a team (see Wilke and Meertens, 1994). Therefore, chaining is an important aspect to include in developing cross-training policies.

However, training worker B for machine II is only one possibility of realizing a chained cross-training configuration with a minimal number of additional cross trainings. Other possible additional cross trainings, which would create a chained graph, are B-I, C-I, C-II, D-I, and D-II. Several other considerations may play a role in selecting the best additional cross training.

1.2.2.3 *Multifunctionality and redundancy*

Molleman and Slomp (1999) define the flexibility of a labor system in more detail by giving three concepts. Functional flexibility may be defined as the total number of skills in a team. The other two concepts are *multifunctionality* and *redundancy*. The level of multifunctionality is defined as the number of different machines a worker is able to cope with, and redundancy (machine coverage) is defined as the number of operators that can operate a specific machine. In terms of graph theory, multifunctionality and redundancy are represented by the degrees of the vertices. The degree of a vertex is defined as the number of edge ends at that vertex. The degree of an operator vertex represents the multifunctionality of the operator and the degree of a machine vertex represents the redundancy of the machine. With respect to multifunctionality and redundancy, two issues should be addressed. First, setting minimum and/or maximum levels should be considered. Second, consideration should be given to the question of whether the level of multifunctionality/redundancy should be as equal as possible for all workers/machines or if some differentiation should be allowed.

Setting a maximum level of multifunctionality may be appropriate in production environments where learning additional tasks/machines requires extensive (and expensive) training and/or where forgetting aspects play a large role. Minimal levels may be appropriate in DRC systems with low staffing levels (ratio of workers to machines), frequent machine breakdowns, or (large) fluctuations in demand. Boundaries for multifunctionality may also be set individually. Some people are more ambitious than others and like to be able to operate many machines. They may "fly" over the shop floor and stand in wherever they are needed. Others feel most comfortable when they are operating their favorite machine.

As for redundancy, Molleman and Slomp (1999) suggested that, as a general training policy, each task should be mastered by at least two workers in order to reduce the negative impact of absenteeism. Above this minimal level of flexibility, the demand of work should dictate training decisions.

For example, workers should be trained for the task with the highest demand. To what extent should multifunctionality and redundancy be bounded? A certain level of multifunctionality and machine coverage is needed in order to deal with fluctuations. When there is too much multi-functionality and machine coverage, worker skills may remain unused and workers may begin to feel that their contributions to team performance are less unique. In the illustrative example, we assume, for illustration purposes, that the minimal machine coverage is one.

Molleman and Slomp (1999) also concluded that an equal distribution of qualifications among workers creates the best situation to deal with absenteeism of workers. This can be explained by the fact that the absenteeism of highly multifunctional workers deteriorates the performance of a team much more than the absenteeism of less multifunctional workers. An equal distribution of qualifications reduces the negative effect of the absence of the workers with the highest level of multifunctionality. As a result of this consideration, the cross-training B-II is no longer ideal. It is better to cross train, for example, worker C for machine II (Figure 1.4).

This assignment is better with respect to the ability of a team to respond to absenteeism. An equal distribution of qualifications is also better from a social viewpoint, since it enhances feelings of interpersonal justice and equity within a team if workers help each other and share their workloads (e.g., Austin, 1977). The wish to gain an equal distribution of qualifications seems obvious from the viewpoint of the ability to deal with absenteeism and the social viewpoint. Most prior studies on DRC systems focus on single-level labor flexibility, in which workers receive the same degree of cross training and thus are equal in terms of multifunctionality. Little is known about the effects of unequal multifunctionality. An exception is a study by Felan and

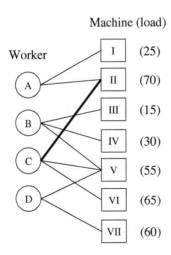

Figure 1.4 Additional skill to enable a more equal level of multifunctionality.

Fry (2001), who focus on multilevel flexibility, where workers are trained to work in a different number of departments. They found that cross-training configurations with unequal levels of cross training lead to better flow times. Because labor learning was included as a factor in the model, their results may be explained by the fact that workers with few skills are able to maximize the task proficiency of those skills, while the few workers with many skills are able to respond to temporary overloads. Felan and Fry (2001) did not consider absenteeism. As a result, the relative benefits of choosing to pursue either equal or unequal multifunctionality remain unclear and likely depend on the specific context. In a situation with absenteeism and without labor learning, for example, we expect equal multifunctionality to be the best option.

In the illustrative example of Figure 1.4, the choice of equalizing or not equalizing redundancy in a manufacturing team does not lead to a different cross-training outcome. Whether the level of redundancy should be as equal as possible for all machines or if some differentiation should be allowed remains an open question. If many workers are able to perform a particular operation, it is likely that some workers will never operate the machine in question. Equal machine coverage, therefore, is likely to minimize the number of unnecessary worker skills. On the other hand, equalizing machine coverage neglects differences in the utilization of machines. A relatively high level of machine coverage for heavily utilized machines may reduce unnecessary idle time due to lack of workers having the necessary skills to operate those machines. Additionally, the unequal division of machine coverage takes the variety of machines in a team into account. Because the required level of learning effort is likely to vary among machines, higher coverage may be more efficient for machines for which workers can be easily trained.

1.2.2.4 *Collective responsibility*

Collective responsibility refers to the distribution of responsibilities within a team. Social comparison theory (as discussed by Jellison and Arkin, 1977, for example) argues that team members prefer complementarity in skill distribution, because they expect this to enhance both their own identity and the performance of the group as a whole. Being a specialist enhances an individual's sense of uniqueness and draws attention to a worker's contribution to group performance (Clark, 1993). Cross training, therefore, may inhibit motivation (Fazakerley, 1976). Furthermore, studies pertaining to diversity show that creativity and motivation are more prevalent in teams whose members have different, but somewhat overlapping, skills (e.g., see Jackson, 1996). Ashkenas et al. (1995) argue that cross training can diminish job boundaries. When more workers are responsible for the same task, the situation may arise in which none of them feels exclusively responsible for that task. This phenomenon is known as social loafing (see Latané, Williams, and Harkins, 1979; Wilke and Meertens, 1994).

In its turn, social loafing may give rise to feelings of inequity and lead to conflicts (Kerr and Bruun, 1983). When cross training workers, therefore, there are reasons to minimize the overlap of responsibilities. Additional cross

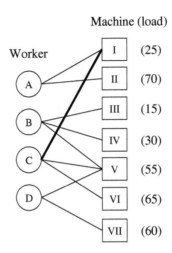

Machine (load)

Figure 1.5 Adding a skill to minimize collective responsibility.

trainings should focus on machines for which the workload is as low as possible. The total workload of the machines to which a worker can be assigned can be regarded as a measure of that worker's responsibility, which may be (partly) shared by other workers who can also be assigned to one or more of these machines. We define collective responsibility as the sum of all worker responsibilities minus the total workload of the machines. In other words, collective responsibility measures the sum of all overlapping responsibilities. Minimizing collective responsibility leads to a situation where workers are most unique and give a specialized contribution to the performance of a team. Figure 1.5 illustrates that this, in addition to the aspects considered before, leads to the situation that worker C needs to be trained for machine I instead of machine II.

On the other hand, policies that minimize the overlap of responsibilities may also minimize the workload that can be assigned to individual workers and thereby the assignment possibilities during working hours. This may lead to situations in which some workers are idle, even as some machines wait for qualified workers. Such a situation is likely to have negative consequences for the flow times of jobs. Moreover, when one or more workers are idle, feelings of inequity may develop among team members. As more responsibilities are shared, more opportunities arise for workers to help each other and to equalize workloads. The foregoing points out that the decision of minimizing or maximizing collective responsibility comprises another nonobvious choice for managers in developing a cross-training policy.

An alternative for cross training worker C for machine I is to cross train worker D for machine I (Figure 1.6). This leads to a better division of worker responsibility (i.e., a more equal amount of (partially) shared workload per worker) and supports equity among workers.

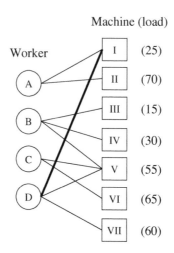

Figure 1.6 Adding a skill to enable a more equal worker responsibility.

1.2.3 *An IGP model and evaluation of specific cross-training policies*

1.2.3.1 *An IGP model to formalize cross-training policies*

The Integer Goal Programming (IGP) model presented in this section (see also Bokhorst et al., 2004b) formalizes various rules for specifying how important aspects should be addressed, and subsequently can be used to support the application of a cross-training policy. It is conceivable that the IGP model forms a useful starting point for developing a decision support tool for cross-training policies in new situations. The IGP model can be solved using a weighted or a lexicographic approach. Here, we applied the lexicographic approach and applied one particular sequence of priorities. Further research is needed to explore the effect of applying different sequences. Additionally, the effect of using the IGP model in different starting situations requires further investigation.

In the IGP model, rules are expressed in terms of goals and constraints. Each cross-training policy requires small alterations in either the goals or constraints (or both) of the IGP model. Table 1.2 summarizes the important aspects to be considered in the development of a cross-training policy and shows which goals and constraints in the IGP model address these aspects.

The objective function (1) minimizes deviation from an optimal cross-training configuration. Constraint (2) demands that all the work be assigned to the various workers. The IGP model is likely to realize a chained graph by means of constraint (3). This constraint demands a cross-training configuration in which all workers can have equal workloads. The basic assumption in our approach is that training should lead to a situation in which all workers can be equally loaded in various circumstances. If that is

Table 1.2 Important Aspects to Consider When Developing a Cross-Training Policy and the Goals and Constraints by Which These Aspects Are Expressed in the Integer Goal Programming (IGP) Model

Aspect	Description (and alternative rules with aspects 3, 4, and 5)	Expression in the IGP model
1	Extent of cross-training Minimize the number of additional cross trainings.	First goal in the objective function; setting *AddCT* to zero in constraint (7)
2	Chaining Enable an equal workload division among the workers to encourage "chaining"	Constraint (3)
3	Multifunctionality Rule 1: Equal multifunctionality per worker Rule 2: Unequal multifunctionality per worker	Second goal in the objective function; constraint (8) supports the realization of an equal distribution; an unequal distribution can probably be realized by neglecting the second goal and constraint (8); an alternative is to give one or more operators more skills than average, before applying the model
4	Machine coverage Rule 1: Equal machine coverage Rule 2: Unequal machine coverage	Third goal in the objective function; constraint (9) supports the realization of an equal distribution; we realize an unequal distribution by neglecting the third goal and constraint (9); an alternative is to cross train a higher-than-average number of workers for particular machines
5	Collective responsibility Rule 1: Minimize collective responsibility Rule 2: Maximize collective responsibility	Fourth goal in the objective function; constraint (10) supports the minimization of collective responsibility; we maximize collective responsibility (or the ease of worker assignment) by giving Φ_4 a negative value
6	Equal worker responsibility Responsibility for an equal amount of (partly) shared workload will support the equity of workers	Fifth goal in the objective function; constraint (11) supports the realization of an equal worker responsibility

Source: Bokhorst J.A.C., Slomp J., and Molleman E., 2004, *IIE Transactions*, 36(10), 969–984. With permission.

the case, then there will be no subgroups under any of these circumstances or, in other words, there is always the possibility of chaining. Constraint (4) forces workers to be or become trained for the machines they must operate. Constraints (5) and (6) concern the minimum levels of multifunctionality and machine coverage, respectively. These two constraints indicate basic choices facing the manager responsible for cross training workers. Additional constraints concerning maximum levels of multifunctionality and machine coverage can be included easily, if necessary.

Constraints (7) to (11) are the goal constraints and indicate other cross-training choices within manufacturing teams. The first goal of the objective function is to minimize the deviation from the desired number of additional cross trainings (AddCTs). A chained graph is easily obtained by fully cross training the team. As mentioned before, however, full cross training is not the ideal situation in many cases. Constraint (7) calculates the deviation from the desired number of AddCT. In reality, however, the training budget may also be an important factor. This is easily expressed by means of constraint (7), using the following procedure: AddCT and Tr_{ij} must be redefined as the training budget and the training costs of cross training worker j for machine i, respectively.

The second goal of the objective function concerns balancing multifunctionality among workers by minimizing the maximal deviation ($d^+_{equalMF}$) from optimal multifunctionality. Constraint (8) calculates this deviation. Optimal multifunctionality is expressed as the configuration in which all workers are skilled for an equal number of machines. The third goal in the objective function minimizes the maximal deviation ($d^+_{equalMC}$) from optimal machine coverage. Constraint (9) calculates this deviation. Optimal machine coverage is expressed as the configuration in which all machines can be operated by the same number of workers. To reduce the overlap of responsibilities, additional cross trainings should concentrate on machines whose workloads are as low as possible.

The fourth goal in the objective function focuses on minimizing deviation (d^+_{CR}) from the optimal situation where each worker has a clear and unique responsibility, or in other words, where collective responsibility is minimized. Constraint (10) is the related goal constraint. The fifth goal of the objective function concerns the equalization of worker responsibility (defined as the sum of the workloads of the machines to which a worker can be assigned) among all workers. This goal supports equity among workers. Constraint (11) calculates the maximum deviation (d^+_{WR}) from the optimal situation in which all workers are responsible for an equal amount of the (shared) workload.

Notation:
Index sets:

$\{i=1,...,I\}$ = Index set of machines.

$\{j=1,...,J\}$ = Index set of workers.

Parameters:

L_i = Workload of machine i, expressed as the percentage of time that the machine will be occupied.

WL = Workload limit of the workers (the workload that needs to be assigned to a worker).

MinMF = Minimal multifunctionality.

MinMC = Minimal machine coverage.

AddCT = Goal with respect to the number of additional cross trainings.

Tr_{ij} = 0, if worker j is already trained for machine i, 1 if not.

M = Constant (large value).

Variables:

X_{ij} = Time assigned to worker j to operate machine i.

Y_{ij} = 1, if worker j needs to be qualified for machine i; 0, if not.

Minimize

$$\Phi_1 d^+_{training} + \Phi_2 d^+_{equalMF} + \Phi_3 d^+_{equalMC} + \Phi_4 d^+_{CR} + \Phi_5 d^+_{WR} \quad (1)$$

subject to:

$$Y_{ij} = 0 \text{ or } 1 \quad \forall i,j \quad (12)$$

1.2.3.2   An evaluation of cross-training policies

As we have seen in Section 1.2.2, research has failed to provide unambiguous rules for addressing multifunctionality, machine coverage, and collective responsibility. In our experience, managers recognize the need for more insight into the effects of alternative cross-training policies in order to create an agile workforce able to respond efficiently and effectively to unplanned changes. We, therefore, have considered eight alternative cross-training policies (Bokhorst et al., 2004b), based on two different choices that can be made with respect to each of the following aspects: multifunctionality, machine coverage, and collective responsibility (Table 1.3). Within each of these cross-training policies, the same rules are included to deal with the other aspects.

Table 1.3 Alternative Cross-Training Policies

Choices	Cross-training policies							
	I	II	III	IV	V	VI	VII	VIII
Equal multifunctionality	NO	NO	YES	YES	NO	NO	YES	YES
Equal machine coverage	YES	YES	NO	NO	NO	NO	YES	YES
Collective responsibility	MIN	MAX	MIN	MAX	MIN	MAX	MIN	MAX

Source: Bokhorst J.A.C., Slomp J., and Molleman E., 2004, *IIE Transactions*, 36(10), 969–984. With permission.

Simple aggregated data from a generic manufacturing team is used for applying cross-training policies to create cross-training configurations. Information concerning the workloads of various machines and the current skill matrix of workers is used as a starting point.

Using the IGP model introduced in Section 1.2.3.1, we formally applied the eight cross-training policies of Table 1.3 to a system with 5 workers and 10 machines, with specific machine workloads (defined as the percentage of the time that the machine is occupied during the presence of the 5 workers). To create an unequal distribution of machine coverage, we neglected the third goal of the IGP model. To create an unequal distribution of worker skills, we fully cross trained two workers before applying the IGP model.

We evaluated the eight resulting cross-training configurations (Table 1.4) by means of a simulation study. We used mean flow time (MFT) from an operations management viewpoint and the standard deviation of the distribution of workload among workers ($SD_{workload}$) from a human resource management viewpoint. Almost all simulation studies include MFT as a major performance measure. $SD_{workload}$ relates to the social dimension of a manufacturing team. The higher the standard deviation, the more variation there will be in the workloads of the various workers. Because of the pressure toward equity, workers in a manufacturing team will attempt to ensure as little variation as possible in the distribution of the workload.

Further, three routing structures were examined as a contextual factor: a parallel routing structure, a serial routing structure, and a job shop routing structure. Within the parallel structure, each part-type visits 1 of the 10 machines randomly. Within the serial structure, all part-types must visit all machines in a fixed order (Machine 1, 2, ... 10). Finally, within the job shop structure, the routing length of part types is uniformly distributed between 1 and 10 machines, while the order of the routing steps is random. As a fixed contextual factor, short, temporary absenteeism (1 to 5 days) was modeled, since the consequences of this type of absence are much more disruptive than are those of medium, long-term, or planned absenteeism.

The results show (see Bokhorst et al., 2004b) that within the parallel structure, it is important for MFT that either multifunctionality or machine coverage be equal. Configurations in which both of these components are equal (as in configurations VII and VIII) perform the best. With respect to $SD_{workload}$, it is best to have equal multifunctionality and maximum collective responsibility (as in configurations IV and VIII). Further, configuration III, representing equal multifunctionality, unequal machine coverage, and minimum collective responsibility, also performs well in this respect. In this configuration, a few machines with low workloads connect all workers, enabling an effective management of workload imbalances. The goals of Operations Management (OM) and of Human Resource Management (HRM) are integrated in configuration VIII, which is the result of applying the cross-training policy of equal multifunctionality, equal machine coverage, and maximal collective responsibility.

Table 1.4 Eight Cross-Training Configurations

Machine (load)	I Worker					II Worker				
	W1	W2	W3	W4	W5	W1	W2	W3	W4	W5
M1 (79.1)	1	1				1	1			1
M2 (72.8)	1	1	1			1	1		1	
M3 (64.7)	1	1			1	1	1			1
M4 (55.4)	1	1			1	1	1		1	
M5 (47.5)	1	1	1			1	1	1		
M6 (40.4)	1	1		1		1	1	1		
M7 (30.3)	1	1		1		1	1	1		
M8 (25.8)	1	1		1		1	1	1		
M9 (17.9)	1	1		1		1	1		1	
M10 (6.1)	1	1		1		1	1			

Machine (load)	III Worker					IV Worker				
	W1	W2	W3	W4	W5	W1	W2	W3	W4	W5
M1 (79.1)	1	1				1	1	1	1	1
M2 (72.8)	1			1		1	1	1	1	1
M3 (64.7)			1		1	1	1	1	1	1
M4 (55.4)				1	1	1	1			
M5 (47.5)		1	1						1	1
M6 (40.4)			1		1	1				1
M7 (30.3)		1		1		1	1			
M8 (25.8)	1	1	1	1	1			1	1	
M9 (17.9)	1	1	1	1	1	1			1	
M10 (6.1)	1	1	1	1	1		1	1		

Machine (load)	V Worker					VI Worker				
	W1	W2	W3	W4	W5	W1	W2	W3	W4	W5
M1 (79.1)	1	1				1	1	1	1	1
M2 (72.8)	1	1				1	1	1	1	1
M3 (64.7)	1	1	1			1	1	1		1
M4 (55.4)	1	1		1		1	1		1	
M5 (47.5)	1	1			1	1	1			
M6 (40.4)	1	1			1	1	1			
M7 (30.3)	1	1		1		1	1			
M8 (25.8)	1	1	1			1	1			
M9 (17.9)	1	1				1	1			
M10 (6.1)	1	1	1	1	1	1	1			

Machine (load)	VII Worker					VIII Worker				
	W1	W2	W3	W4	W5	W1	W2	W3	W4	W5
M1 (79.1)			1		1	1		1		1
M2 (72.8)	1	1		1		1		1	1	
M3 (64.7)	1	1		1			1	1		1

(continued)

Table 1.4 (Continued) Eight Cross-Training Configurations

Machine (load)	VII Worker					VIII Worker				
	W1	W2	W3	W4	W5	W1	W2	W3	W4	W5
M4 (55.4)			1	1	1		1		1	1
M5 (47.5)	1	1		1		1	1	1		
M6 (40.4)	1		1		1		1		1	1
M7 (30.3)		1	1		1	1	1	1		
M8 (25.8)		1	1		1	1	1	1		
M9 (17.9)	1		1		1			1	1	1
M10 (6.1)	1	1		1		1				1

Source: Bokhorst J.A.C., Slomp J., and Molleman E., 2004, *IIE Transactions*, 36(10), 969–984. With permission.

Within the serial structure, reaching an optimal MFT requires a focus on bottleneck machines. Heavily utilized machines should receive the most coverage. This is enabled by the combination of unequal machine coverage and maximum collective responsibility. Within the serial structure, therefore, configurations IV and VI perform well with regard to MFT. With respect to $SD_{workload}$, equal multifunctionality is important and maximum collective responsibility is desirable. The cross-training policy of equal multifunctionality, unequal machine coverage, and maximum collective responsibility, which is applied to create configuration IV, integrates the goals of both OM and HRM within the serial routing structure. Interestingly, although configuration VI, representing unequal multifunctionality, unequal machine coverage, and maximum collective responsibility, performs very well with regard to MFT, its performance with respect to $SD_{workload}$ is among the worst. For this cross-training policy, the goals of OM come into conflict with those of HRM.

Within the job shop structure, equal multifunctionality, equal machine coverage, and minimum collective responsibility appear important for MFT. For $SD_{workload}$, equal multifunctionality and maximum collective responsibility should be the norm. If maximum collective responsibility is not possible, achieving an optimal degree of $SD_{workload}$ requires unequal machine coverage, as this allows for balancing workloads among workers. In this routing structure, therefore, the goals of OM and those of HRM apparently cannot be integrated.

The results of the parallel and job shop structure point in the same direction. Within these routing structures, equal multifunctionality and equal machine coverage are important for achieving an optimal MFT. Within the serial structure, more attention should be paid to bottleneck machines. Here, the combination of unequal machine coverage and maximum collective responsibility results in good MFT performance. Within all routing structures, equal multifunctionality, combined with maximum collective responsibility, seems to enable a fair distribution of workload among workers. A good alternative within the parallel and job shop structure is equal multifunctionality combined with unequal machine coverage and minimum collective responsibility.

1.2.4 A tradeoff between training costs and operational performance

Acquiring additional cross trainings in a team may be very costly. In developing cross-training policies, it then may be wise to explicitly include these costs and find a solution that gives a tradeoff between these costs and improved operational performance. In this section, we will present an integer programming model that deals with this tradeoff. A short literature review follows in order to position our model.

Stewart et al. (1994) presented four integer programming (IP) models for developing a flexible workforce. These models attempted to minimize the total cost of training, to maximize the flexibility of the workforce, to minimize the total time required for training, and to optimize the tradeoff between training costs and workforce flexibility. The models' formulations force an optimal assignment of tasks (hours) to workers. Important constraints in the models are the production hours available, the production requirements, and the budget for training. Although the models of Stewart et al. provide a valuable reference for developing a mathematical formulation, there are some significant pitfalls in them. One of their assumptions is that it is not necessary to balance assignments among workers, which we propose in order to enable chaining. The models of Stewart et al. also do not incorporate the issue of fluctuations in the demand and/or supply of labor.

Brusco and Johns (1998) presented a linear programming model that minimizes workforce staffing costs subject to the satisfaction of minimum labor requirements across a planning horizon. They used the model to evaluate eight cross-training structures across various labor requirement patterns. An important result of their study concerned the conclusion that "chaining of employee skill classes across work activity categories" is a basic element of successful cross-training structures. Our model builds further on this result.

Molleman and Slomp (1999) presented a mathematical model to assign multiskilled workers to the various tasks (or machines) in a team. They studied the effect of labor flexibility on team performance. Team performance is measured as the shortage of labor capacity (i.e., no capable worker is present to perform a particular task), the minimum time needed to perform all tasks (i.e., the load of the bottleneck worker), and the cumulative time needed to perform all tasks. Important conditions that affect the required level of labor flexibility include demand variation and worker absenteeism. Molleman and Slomp showed that a uniform distribution of multifunctionality among the workers provides the best team performance. They also indicated that absenteeism should be regarded as a major reason to invest in labor flexibility. As a general statement, they suggested that each task should be mastered by at least two workers in order to reduce the negative impact of low to moderate levels of absenteeism. Above this minimal level of flexibility, labor flexibility needs to co-vary with the demands on capacity for each task. Although the model of Molleman and Slomp provides some general guidelines, it does not give detailed suggestions with respect to cross training. The model presented below may be regarded as a useful extension of their model.

Norman et al. (2002) presented a mixed integer programming formulation for the assignment of workers to operations in a manufacturing cell. Their formulation permits the ability to change the skill levels of workers by providing them with additional training. Training decisions are taken in order to balance the productivity and output quality of a manufacturing cell and the training costs. Norman et al. did not deal with the need for cross training.

Our approach is explicitly focused on gaining sufficient cross training in a manufacturing cell in order to deal effectively with all kinds of fluctuations, while paying attention to the training costs as well. We have developed an integer programming model (see also Slomp et al., 2005) that can be used to select workers to be cross trained for particular machines in a cellular manufacturing environment. A manufacturing cell may be defined as the grouping of people and processes, or machines, into specific areas dedicated to the production of a family of parts. A manufacturing cell requires a cross-trained workforce in order to deal with variations in the demand mix and/or fluctuations in the supply of labor (see, e.g., Molleman and Slomp, 1999). The basic assumption in our approach is that training should lead to a situation in which all workers can be equally loaded in various circumstances. The model further takes into account the training costs as well as the efficiency levels workers can ultimately achieve on machines. Efficiency is defined in our study as the relative speed by which tasks can be performed by workers. This definition conforms to several other studies (e.g., Brusco and Johns, 1998, and Bobrowski and Park, 1993). The efficiency of a worker at a machine depends on the speed that the worker can perform manual tasks, such as machine setups and quality checks. The variety of efficiency levels of the various workers at a particular machine depends on the labor intensiveness of the tasks involved. Highly automated machines may result in less variety in the efficiency levels of operators. Here we assume that additional training cannot further influence the efficiency level of a worker at a particular machine; the efficiency level is already at the end of the worker's learning curve.

We examined the situation of a manufacturing cell of a Dutch company that supplies components for the electrical energy industry. The cell is responsible for rotation symmetric part types and operates in a three-shifts system, each manned by a different team of operators. Our focus will be on one team of the manufacturing cell, working in the day shift. This team of 7 operators works in a cell consisting of 10 major machines.

Table 1.5 shows which operators of the team are currently able to operate which machines (skills are denoted by an X). The multifunctionality (MF) of operators and the redundancy (RD) of machines within the manufacturing cell are also provided in Table 1.5.

As can be seen, the multifunctionality varies from two machines to four machines per operator and the redundancy from one operator to four

Table 1.5 Initial Worker–Machine Matrix

| | Machines 1-10 with machine numbers | | | | | | | | | | |
| | 80127 | 80133 | 80155 | 80205 | 80225 | 81353 | 81354 | 81365 | 81311 | 80610 | |
Operators	1	2	3	4	5	6	7	8	9	10	MF
A		X						X			2
B									X		1
C				X		X	X			X	4
D				X		X	X			X	4
E						X	X			X	3
F	X				X						2
G			X							X	2
RE	1	1	1	2	1	3	3	1	1	4	18

operators per machine. The variance in multifunctionality can be partly explained by differences in the level of basic education. Operator D received the highest level of basic education within the team and operators B and F received the lowest level of basic education within the team. A higher level of basic education seems to encourage higher levels of multifunctionality. For machines 1, 2, 3, 5, 8, and 9, only one operator is trained to operate the machine. In other words, the redundancy for these machines is only one. Thus, the team is rather vulnerable to absenteeism. However, most machines are operated in two or three shifts. This means that there may be possibilities to fulfill the demand for a machine within other teams, if the skills are available in these teams and the demand can be spread over shifts. On the other hand, products often need to be processed on more than one machine, causing dependencies between machines. If machines needed for a particular routing are operated in different shifts, buffers of working stock have to be created and throughput times for the products will increase. Therefore, we will concentrate on creating an optimal cross-training balance within the day shift team and not across teams of all shifts.

The situation described here forms the starting situation from which the management of the firm has to decide how to develop the skills of the team further. Applying the IP model to this starting situation can be regarded as applying a cross-training policy in order to achieve the objectives of being able to fulfill all demand in all situations of absenteeism with a minimum of training costs and high labor flexibility.

Table 1.6 demonstrates the annual demand for the day shift provided by the company. Further, we have set a maximum and minimum redundancy

Table 1.6 Annual Day Shift Demand and Boundaries for Redundancy

Machines	1	2	3	4	5	6	7	8	9	10
Machine nr.	80127	80133	80155	80205	80225	81353	81354	81365	81311	80610
Day shift demand	965	369	860	701	510	913	1057	1126	996	563
Max. redundancy	3	3	3	4	3	4	4	3	3	5
Min. redundancy	2	2	2	2	2	2	2	2	2	3

Table 1.7 Boundaries for Multifunctionality

Operators	A	B	C	D	E	F	G
Max. multifunctionality	5	4	6	6	6	4	3
Min. multifunctionality	2	2	3	3	2	2	2

for each machine within the day shift. As can be seen in Table 1.6, a minimum redundancy of at least two is set for every machine, in order to be able to deal with absenteeism within the team better. Because the actual redundancy of six machines is only one (Table 1.5), training is needed in order to satisfy the constraint of minimum redundancy. The maximum redundancy is set higher for machines required in important product routings than for machines required in less important ones. Important product routings are routings of products with a high total demand for the machines within the manufacturing cell.

Table 1.7 illustrates the maximum and minimum multifunctionality for all operators of the team. The multifunctionality of people is set at two for a minimum, since the operators are not dependent on one machine only. If the machine breaks down or the demand for the machine is low during a period, the operator can be assigned to a different machine. As can be seen, most of the actual values of multifunctionality (Table 1.5) lie between the minimum and maximum multifunctionality set.

The training costs can be split up into training costs required to learn the general skills of a group of related machines (called a machine group, in this case) and training costs to learn the specifics of a single machine within a machine group. In order to operate machines within a machine group, an operator first needs the basic training for the machine group. Once he or she has passed this training successfully, additional on-the-job training is needed for each specific machine within the machine group. An operator who is already skilled for one machine within a machine group only needs the additional on-the-job training when he or she wants to operate an additional machine within the machine group.

Table 1.8 shows the five machine groups that can be distinguished in the manufacturing cell of this case. Machine group 1 contains machines 1 and 2.

Table 1.8 Machine Group Characteristics

Machine group	Contains machines:	Weeks of training required for each machine	Number of days of external training required to learn the general skills for the machine group
1	1 and 2	52	5, 2
2	3	80	8
3	4 and 5	40	4
4	6, 7, 8, 9	120	12
5	10	40	4

This is a group of automatic lathes. Machine group 2 only contains machine 3, which is a CNC-lathe. Machine group 3 contains machines 4 and 5, which are CNC-milling machines. Machine group 4 is the largest machine group and contains machines 6, 7, 8, and 9. This is a group of CNC-machining centers. Finally, machine group 5 is formed by machine 10, which is a CNC-drilling machine.

For all machines within a machine group, it takes about the same number of weeks of on-the-job training before an operator reaches his or her maximum efficiency. In machine group 1, it takes about 52 weeks; in machine group 2, about 80 weeks; in machine groups 3 and 5, around 40 weeks; and, finally, in machine group 4, about 120 weeks. In order to learn the general skills needed for a machine group, operators receive a training course outside the company for a number of days (see Table 1.5). The training costs for these courses are on average €1000 (US$1261) per day. Once the operator has mastered these general skills, another 10% of the initial costs is needed on average each year to maintain the skills. Table 1.9 shows, for each operator, the group training costs in Euros for the five machine groups in case the operator has not received the general training or the annual maintenance costs for these skills otherwise.

The training costs for the specific machines within a machine group are derived from the on-the-job training time in weeks that it takes to learn how to operate the machine. We assumed that, on average, for about 5% of this time, an additional worker (trainer) is needed to teach the job to the trainee. This worker costs on average €25 (US$31.50) per hour. With regard to machines 6 to 9, for instance, a training time of 120 weeks is required (Table 1.8). According to the above logic, a trainer is needed for 6 weeks, which costs the company €6000 in case of a work week of 40 hours. Furthermore, 5% of this amount is needed annually to maintain the skill after the initial training.

Most of the time within the training weeks (95%) the trainee will operate the machine by himself/herself and gain experience by just doing the job. A learning effect will take place, and at the end of the training weeks, the operator will have reached his/her maximum efficiency. We do not account for the throughput losses due to the lower efficiencies of workers while learning the specifics of a machine.

Table 1.9 Group Training Costs or Maintenance Costs (in Italics) of Group Skills (in Euros)

M-groups	Operators						
	A	B	C	D	E	F	G
1	*520*	5200	5200	5200	5200	*520*	5200
2	8000	8000	8000	8000	8000	8000	*800*
3	4000	4000	*400*	*400*	4000	*400*	4000
4	*1200*	*1200*	*1200*	*1200*	*1200*	12,000	12,000
5	4000	4000	*400*	*400*	*400*	4000	*400*

Table 1.10 Inefficiency Factors (in bold), Specific Machine Training Costs and Specific Skill Maintenance Costs (in italics)

Machines	Operators						
	A	B	C	D	E	F	G
1	**1.08**	**0.97**	**0.91**	**1.04**	**1.08**	**1.01**	**0.98**
	2600	*2600*	*2600*	*2600*	*2600*	*130*	*2600*
2	**1.02**	**1.01**	**0.99**	**1.05**	**1.04**	**0.97**	**0.94**
	130	*2600*	*2600*	*2600*	*2600*	*2600*	*2600*
3	**1.09**	**0.93**	**1.02**	**1.08**	**1.06**	**1.05**	**1.01**
	4000	*4000*	*4000*	*4000*	*4000*	*4000*	*200*
4	**0.98**	**1.02**	**1.06**	**0.92**	**0.95**	**1.01**	**1.08**
	2000	*2000*	*100*	*100*	*2000*	*2000*	*2000*
5	**1.03**	**1.02**	**1.06**	**1.02**	**0.99**	**1.03**	**1.08**
	2000	*2000*	*2000*	*2000*	*2000*	*100*	*2000*
6	**1.04**	**1.09**	**1.02**	**1.01**	**0.95**	**0.99**	**1.02**
	6000	*6000*	*300*	*300*	*300*	*6000*	*6000*
7	**0.98**	**1.06**	**0.94**	**0.98**	**1.02**	**1.03**	**1.07**
	6000	*6000*	*300*	*300*	*300*	*6000*	*6000*
8	**0.99**	**1.10**	**1.10**	**1.06**	**1.07**	**1.05**	**0.99**
	300	*6000*	*6000*	*6000*	*6000*	*6000*	*6000*
9	**1.06**	**0.90**	**0.92**	**1.03**	**1.08**	**1.01**	**1.07**
	6000	*300*	*6000*	*6000*	*6000*	*6000*	*6000*
10	**1.04**	**1.06**	**0.97**	**0.99**	**0.91**	**0.94**	**1.00**
	2000	*2000*	*100*	*100*	*100*	*2000*	*100*

In Table 1.10, the inefficiency factors and the training costs of the operators to learn the specifics of a single machine are given. In case the operator already mastered the skill, the maintenance costs for the skill are given instead. The inefficiency factor (in gray) represents the pace of an operator after training. So, an inefficiency factor of 1.08 means that the worker can operate this machine at 1.08 times the normalized time that stands for operating the machine. We have generated these inefficiency levels arbitrarily for this case, but within the limits of 10% below and 10% above the normalized times, which seems to be quite reasonable in practice. This has been verified by the production manager responsible for the specific manufacturing cell in this instance.

In this practical case, four situations of anticipated absenteeism are defined. In the first situation, operator E is absent for 16% of the time. This absence can be explained by the fact that the operator occasionally needs to work in other departments. Another reason can be that the operator does not work full time. The periodical costs of this situation with regard to the workload of the bottleneck worker ($c_1 \times WB_1$) can be calculated as follows. In 16% of the time, all operators, except operator E, are available to do the work. On average, each remaining operator will have to do more work than when all operators are available. All the workloads of the six available operators are added up (which is the same as multiplying the workload of the bottleneck worker by 6, since all workers become bottleneck workers) and multiplied by an average cost factor per hour (€25) and by the chance that this situation occurs (0.16). For the first scenario, c_k will be 24 (i.e., $0.16 \times 6 \times 25$) and the periodical costs will be $24 \times WB_1$. In the second situation of absenteeism, operator F is absent for 16% of the time and c_2 also equals 24. In the third situation, both operator E and operator F are absent. This situation occurs 4% of the time and, in that case, only five operators remain available to do the work. For this third scenario, c_k then equals to 5 (i.e., $0.04 \times 5 \times 25$). Finally, in the fourth scenario, all operators are available. This situation occurs 64% of the time and c_4 then equals to 112 (i.e., $0.64 \times 7 \times 25$). As can be seen, we assume that the costs of production are linearly related to the workload of the bottleneck worker.

We chose the weights of $\pi 1$ and $\pi 2$ to be 1, which means that the objectives minimizing and balancing workloads in all situations of absenteeism and minimizing the group training costs and specific machine training costs are given equal importance. For this case, the desired payback period for all training costs is set at 1 year; no distinctions are made between different groups of machines, different specific machines, or different employees (i.e., all c_{mj} and c_{ij} are set to 1).

Notation
Index sets:

$\{k=0,\ldots,K\}$ = Index set of labor supply situations.

$\{m=1,\ldots,M\}$ = Index set of groups of related machines.

$\{i=1,\ldots,I\}$ = Index set of machines.

$\{j=1,\ldots,J\}$ = Index set of workers.

Model Parameters:

D_i = Annual demand for machine i.

$\{A_k\}$ = Set of absent workers in labor supply situation k.

TCM_{mj} = Training costs for operator j to acquire the general skills needed in group m of related machines.

TC_{ij} = Training costs for operator j to learn the specific skills for machine i.

e_{ij} = Inefficiency factor for machine i when performed by worker j.

R_i^+ = Maximal redundancy for machine i.

R_i^- = Minimal redundancy for machine i.

M_j^+ = Maximal multifunctionality for worker j.

M_j^- = Minimal multifunctionality for worker j.

Ω = Constant (large value).

Parameters for the objective function:

$\pi 1, \pi 2$ = Weight factors.

c_k = Factor to translate the workload of the bottleneck worker in each labor situation (k) into periodic costs.

c_{mj} = Factor to translate the training costs of worker j for acquiring the general skills in group m of related machines (TCM_{mj}) into periodic costs.

c_{ij} = Factor to translate the training costs of worker j for machine i into periodic costs.

Variables:

WB_k = Load of the bottleneck worker for situation (k).

x_{ijk} = Normalized time assigned to worker j to operate machine i in situation (k).

y_{ij} = 1, if worker j has to be trained for machine i; 0, if not.

z_{mj} = 1, if worker j has to be trained for at least 1 machine in group m of related machines; 0, if not.

Within the manufacturing cell, machines can be clustered in groups of related machines, or *machine groups*, denoted by the index set $\{m=1,...,M\}$. Each machine group contains a number of machines, $i \in \{m\}$. We specify a number of labor supply situations, denoted by the index set $\{k=0,...,K\}$. In situation k, the workers of set $\{A_k\}$ are absent. All situations K cover what management regards as situations that need to be dealt with by the particular team, without considering interventions, such as overwork of some workers and/or subcontracting certain tasks.

WB_k denotes the load of the bottleneck worker for situation k. As mentioned earlier, in a balanced situation, WB_k equals the workload of all workers in situation k. The annual demand for machine i is given by D_i. Each worker can be given an inefficiency factor for each machine, denoted by e_{ij}. If $e_{2,1}$ equals 1.05, for example, then worker 1 will operate machine 2 5% slower than the normalized time for operating machine 2, after training.

The factor e_{ij} enables us to distinguish workers who are not equally efficient in performing the various tasks. This is especially relevant in human-paced task environments.

Training costs can be split between all workers j to learn the general skills needed in machine group m, denoted by TCM_{mj}, and specific training costs for machines within a machine group, denoted by TC_{ij}.

In the model, we have set boundaries on the redundancy per machine and the multifunctionality per worker. Let y_{ij} be a binary variable that equals 1 if worker j has to be trained to perform machine i, and 0 otherwise. The variable z_{mj} indicates if worker j has to be trained for at least 1 machine in group m. The normalized time assigned to worker j to operate machine i in situation k is denoted by x_{ijk}. The objective of the model is to determine which workers should be trained for which machines.

The Integer Programming Model presented below calculates which workers have to be trained for which machines so that absenteeism becomes manageable and the available skills are applied as efficiently as possible. Further, machine group training costs and specific machine training costs can be minimized and constraints concerning a maximum and minimum amount of multifunctionality and redundancy can be formulated. This has resulted in the following integer programming model:

$$\text{Minimize } \pi 1 \sum_{k} c_{k} \, WB_{k} + \pi 2 \left(\sum_{m} \sum_{j} c_{mj} TCM_{mj} \, z_{mj} + \sum_{i} \sum_{j} c_{ij} TC_{ij} \, y_{ij} \right) \quad (1.1)$$

subject to:

$$D_{i} - \sum_{j} x_{ijk} \leq 0 \quad \forall i, k \quad (1.2)$$

$$\sum_{i} e_{ij} x_{ijk} \leq WB_{k} \quad \forall j, k \quad (1.3)$$

$$x_{ijk} \leq \Omega y_{ij} \quad \forall i, j, k \quad (1.4)$$

$$\sum_{i \in \{m\}} y_{ij} \leq \Omega z_{mj} \quad \forall m, j \quad (1.5)$$

$$\sum_{j} y_{ij} \leq R_{i}^{+} \quad \forall i \quad (1.6)$$

$$\sum_{j} y_{ij} \geq R_{i}^{-} \quad \forall i \quad (1.7)$$

$$\sum_i y_{ij} \leq M_j^+ \quad \forall j \tag{1.8}$$

$$\sum_i y_{ij} \geq M_j^- \quad \forall j \tag{1.9}$$

$$x_{ijk} = 0 \quad if \quad [j \in A_k] \quad \forall i,j,k \tag{1.10}$$

$$x_{ijk} \geq 0 \quad \forall i,j,k \tag{1.11}$$

$$y_{ij} = 0 \quad or \quad 1 \quad \forall i,j \tag{1.12}$$

$$z_{mj} = 0 \quad or \quad 1 \quad \forall m,j \tag{1.13}$$

The objective function (1.1) concerns a tradeoff between the operating costs of the manufacturing cell and the costs of cross training. The first part of the objective function concentrates on the minimization of the periodic (e.g., annual) operating costs of the system. We assume that the operating costs incurred in each period are linearly related to the workload of the bottleneck worker. This assumption is based on the idea that the bottleneck worker determines the efficiency of the manufacturing cell. That is, the load of the bottleneck worker may be seen as lower bound on the makespan of all jobs included in the model. If the bottleneck load can be decreased, the manufacturing cell will be able to deliver the same set of jobs within a shorter amount of time. This creates the possibility to attract more work and decrease unit operating costs. Because of job interaction and labor blocking, the minimal makespan for which a feasible schedule can be realized will often be greater than the lower bound (see, e.g., Raaymakers and Fransoo, 2000). A minimal lower bound, however, is positively related to the minimum makespan that can be realized, since both the lower bound and the minimum makespan depend on the flexibility gained by chaining workers and machines and the utilization of workers at their highest efficiency. WB_k is the workload of the bottleneck worker in situation k, assuming that all the demand has to be performed in this situation. The parameter c_k transfers WB_k into the periodic (e.g., annual) production costs due to the short-lived existence of situation k. The value of c_k equals the multiplication of the costs per "bottleneck hour" in situation k times the relative presence of this situation.

The second part of the objective function is focused on minimizing the group training costs and specific training costs for the workers. The parameters c_{mj} and c_{ij} convert the training costs into periodic (annual) costs. The value of

c_{ij} depends on the desired payback period for the training costs spent on teaching machine i to worker j. The desired payback period is obviously short for machines that will change in the near future. It is also conceivable that the desired payback period may vary per worker. The use of the concept of "payback period" for training costs is based on the consideration that cross training is only valuable for a certain period of time because of the risk of workers leaving the company and the need for new skills due to new manufacturing technology.

There are two weight factors in the objective function, $\pi 1$ and $\pi 2$. If these factors have the value 1, the objective function will minimize total costs per period. Other arguments may lead to other settings for the weight parameters $\pi 1$ and $\pi 2$. It is, for instance, likely that the load of the bottleneck worker (WB_k) is related to performance factors, such as due date performance and manufacturing flexibility. We expect that the lower the load of the bottleneck worker, the better the performance will be on non–cost performance indicators. This expectation may lead to higher values for $\pi 1$. Limitations to the training budget may lead to higher settings for the values for $\pi 2$.

The objective function may also be seen as a tradeoff between flexibility advantages and training costs. The value of the first part of the objective function may be seen as a measure of inflexibility. This conforms to the flexibility measure developed by Jordan and Graves (1995). Their (in) flexibility measure may be interpreted as the probability of having workers who are fully loaded while simultaneously having underutilized workers. If the workload is balanced in all situations the inflexibility is zero, according to Jordan and Graves (1995), and the value of the first part of our objective function is minimal. In our model, the value of the first part of the objective function is also determined by the efficiency of the workers. The more that workers are deployed at their most efficient task, the better the flexibility will be, i.e., the lower the value of the first term of the objective function. The inclusion of efficiency considerations may be seen as an extension of the work of Jordan and Graves.

Constraint (2) in the IP model forces the demand for the various machines to be assigned to the workers in all situations k. Constraint (3) forces all operators to be equally or less loaded than the bottleneck worker. The workload of a worker depends on the normalized times assigned to that worker and his/her efficiency at performing the tasks. Constraint (4) forces workers to be or become trained on the machines that they have to operate. Constraint (5) forces workers to be or become trained on the machine groups they are or become involved in. Constraints (6) and (7) concern the desired minimum and maximum redundancy of each task. Constraints (8) and (9) concern the boundaries of the multifunctionality of each worker. Constraint (10) forces no work to be assigned to absent workers. Constraints (11), (12), and (13) show the domains of variables x_{ijk}, y_{ij}, and z_{mj}.

We used LINGO, which is a comprehensive tool designed for building and solving linear, nonlinear, and integer optimization models (www.lindo.com), to solve the IP model applied to the specific case presented earlier. Table 1.11 shows the distribution of skills computed by the model (X and $+$). An X in

Table 1.11 Distribution of Skills Given by the Model, Compared to the Actual Distribution

Operators	Machines										MF
	1	2	3	4	5	6	7	8	9	10	
A		X						X			2
B			+						X		2
C				X		X	X		+	X	5
D				X	+	–	X	+		–	4
E						X	–			X	2
F	X	+			X						3
G	+		X							X	3
RE	2	2	2	2	2	2	2	2	2	3	21

the matrix means that the operator already mastered the skill for that machine (see Table 1.5) and that he or she should maintain this skill. A minus (–) means that the operator is currently skilled, but that it is of no use to maintain this skill any longer. A plus (+) indicates that the operator was not skilled to operate the machine thus far, but that he or she should be trained for it now.

The functional flexibility (Molleman and Slomp, 1999) has increased compared to the actual distribution of skills: 21, according to the model, against 18 currently. It can be seen that all the extra training (+) is needed to comply with the constraint of having a minimum redundancy of two for each machine. Machines that currently have a high redundancy (6, 7, and 10) are cut back by the model.

The absolute levels of multifunctionality of the operators do not change much compared to the current situation. The multifunctionality of operators B, C, F, and G is increased by one, the multifunctionality of operator E is decreased by one, and the multifunctionality for the other operators remains at the current absolute level. However, even though the absolute level of multifunctionality remains the same for operator D, the distribution of skills changes quite drastically. The skills for machines 4 and 7 have to be maintained, but the skills for machines 6 and 10 are not required anymore and the skills for machines 5 and 8 have to be mastered. For the other operators, either new skills have to be mastered on top of the original set of skills (B, C, F, and G), the maintenance of certain skills has to be stopped (E), or nothing changes compared to the initial situation (A).

Additional general group trainings are needed for operator B (machine group 2, containing machine 3) and for operator G (machine group 1, containing machine 1). All other skills that have to be mastered only require the specific machine trainings.

The costs associated with the outcome of the cross-training policy for this case are shown in Table 1.12. The annual operating costs are €193,915 ($244,585). The group training costs are €13,200 ($16,649), since operators B and G have to receive basic training for machine groups 2 and 1, respectively. The total annual maintenance costs for all operators to keep their knowledge

Table 1.12 Costs Associated with the Outcome of the Training Policy and Workloads per Situation of Absenteeism

Annual operating costs	€193,915			
Group training plus maintenance costs	€23,440 (€13,200 + €10,240)			
Machine training plus maintenance costs	€26,060 (€23,200 + €2,860)			
Situation of absenteeism	1	2	3	4
Workload	1309.89	1288.70	1579.63	1104.02

of the general skills of the machine groups they already mastered up to date are €10,240 (12,915). The machine training costs total €23,200 ($29,388) for the six new skills that have to be mastered in the team (by operators B, C, D, F, and G). The total annual maintenance costs to keep the knowledge of the specific machines up to date are €2860 ($3607).

We think that it is sufficient to integrate only the major demand/supply situations that differ significantly from each other in the model. It is likely that the resulting ideal cross-training situation will satisfy the needs in many other scenarios. This assumption is based on the chaining situation gained after running the model. The major chains in the cross-training situation are likely to absorb the variations in minor demand/supply situations. Further, the outcome of the model may be regarded as a starting point for the developer of a cross-training program. The manager may decide to add cross training based on his knowledge of the presence of specific situations.

1.2.5 Summary

A cross-training policy directs the design of a cross-trained workforce in line with the firm's strategy. Developing a cross-training policy first entails deciding which performance measures should be targeted. The firm's strategy and the specific context — with respect to jobs, workers, and demand — then determine which aspects are important to include and how to address these aspects. Table 1.13 summarizes the important facets and the considerations in dealing with these aspects as discussed in Section 1.2.2.

Section 1.2.3 presents an Integer Goal Programming model to formalize various rules for specifying how the important aspects should be addressed. The model may be a useful starting point for developing a decision support tool for cross-training policies in new situations. Further, when using the IGP model, several cross-training policies are applied to a generic manufacturing team and evaluated by means of simulation. Routing structure is included as a contextual variable, and absenteeism and periodic demand fluctuations as given context. We used the mean flow time of jobs from an operations management viewpoint, and the standard deviation of the distribution of workload among workers from a human resource management viewpoint. The results show that within the parallel and job shop structure, equal multifunctionality and equal machine coverage are important for achieving an optimal mean flow time. Within the serial structure, more

Table 1.13 Summary of Important Aspects and Considerations

Aspect	Considerations
Extent of cross training The number of (additional) cross trainings required in the team	Full flexibility is not needed nor is it desirable in practical situations. Most of the positive effects on operational performance can be achieved without going to the extreme of full flexibility. Moreover, extensive cross training can be costly (training costs), lead to losses in productivity (transfer times), impair social identity and cause motivational deficits, lead to social loafing, and lower status differentials within teams.
Chaining A chain connects all workers and machines directly or indirectly.	Chaining provides the ability to shift work from a worker with a heavy workload to a worker with a lighter workload, leading — directly or indirectly — to a more balanced workload. It also reduces the likelihood that subgroups may emerge and cause intergroup conflicts, leading to the disintegration of a team.
Multifunctionality The number of different machines with which a worker is able to cope	Maximum and/or minimum levels may be set, depending on contextual factors and/or individual preferences. Further, equal multifunctionality among workers creates a good situation to deal with absenteeism of workers and enhances feelings of interpersonal justice and equity within a team if workers help each other and share their workloads. Unequal multifunctionality may lead to better flow times in some cases where learning and forgetting play a role.
Redundancy/ Machine coverage The number of workers that can operate a specific machine	A minimum redundancy of two workers per machine reduces the negative impact of absenteeism. Too much redundancy may leave worker skills unused. Equal redundancy is likely to minimize the number of unnecessary worker skills but neglects possible differences in the utilization of machines and the level of training effort required.
Collective responsibility A measure of the sum of all overlapping responsibilities. A worker's responsibility is defined as the total workload of the machines to which a worker can be assigned. Collective responsibility equals the sum of all worker responsibilities minus the total workload of the machines	Minimizing collective responsibility leads to complementarity in skill distribution, which is preferred by team members since they expect this to enhance both their own identity and the performance of the group as a whole. It enhances an individual's sense of uniqueness and irreplaceability. Creativity and motivation are more prevalent in teams whose members have different but somewhat overlapping, skills. When more workers are responsible for the same task, social loafing may occur, which may give rise to feelings on inequity and lead to conflicts. Maximizing collective responsibility enlarges the assignment possibilities, which may have positive effects on the flow times of jobs. Moreover, it may prevent idleness of workers and create more opportunities for workers to help each other and to equalize workloads. An equal amount of (partially) shared workload per worker (equal worker responsibility) supports equity among workers.

attention should be paid to the bottleneck machines by combining unequal machine coverage and maximum collective responsibility. Within all routing structures presented, equal multifunctionality (combined with maximum collective responsibility) seems to enable a fair distribution of workload among workers.

Section 1.2.4 presents a practical case involving a manufacturing cell. Based on this case, a model is presented that pays more attention to training costs and includes static efficiency differences between workers and several absenteeism scenarios as given context. The model is helpful when making a tradeoff between training costs and the workload balance among workers in a manufacturing cell. The workload balance indicates the usefulness of labor flexibility in a particular situation.

1.3 Labor assignment in a cross-trained workforce

An effective workforce is one with cross-trained workers and appropriate assignment rules to (dynamically) assign workers to different tasks/ machines. In other words, cross training by itself does not necessarily result in performance improvements when no attention is being paid to effectively make use of the (additional) skills by means of appropriate labor assignment rules. This section deals with the design of labor assignment rules.

Section 1.3.1 focuses on industrial practice with respect to labor assignment. Section 1.3.2 discusses labor assignment and worker differences as modeled in the literature. Previous studies on labor assignment mostly study the "when-rule" and the "where-rule," which decides when a worker is eligible for transfer and where he/she should be transferred to, respectively. Since relatively little attention has been given to the "who-rule" in literature, this rule is explained in more detail. Section 1.3.3 evaluates labor assignment rules in systems with worker differences by means of simulation. Section 1.3.4 is a summary section that also discusses differences and commonalities between labor assignment in practice and in the literature.

1.3.1 Labor assignment in industrial practice

This section is based on case descriptions of several Dutch firms that were directly or indirectly (through master's degree students) involved in our research on workforce agility. In many practical situations, labor capacity and machine capacity are the major constraints, and workforce agility is a major issue. Workers (of the firms involved) typically have a set of skills that differs from other workers with respect to the number of skills, skill types, etc. We further observed that workers also often differ in task proficiencies, indicating their efficiency on the various machines. Machines — especially in job shop environments — differ from each other with respect to average utilization and utilization per machine. In most firms, the utilization varies in time as well.

We will now take a closer look at one of the firms that manufactures parts and small subassemblies used in the electromechanical industry.

Table 1.14 Basic Data Indicating the Labor Situation in the Machining
and Turning Department

Number of workers per shift	14–17
Number of machines	24
Utilization of machines	20–80%
Variation in utilization of each machine during the observation weeks	About 10%
Percentage of workers who have a main machine to work on	90%
Average percentage of time that a worker spends on his/her main machine	77%
Number of different machines to which workers are being assigned	1–7
Proficiency levels of workers	50–100%

Source: Bokhorst J.A.C., Slomp J., and Gaalman G.J.C., 2004, *International Journal of Production Research,* 42(23), 5049–5074. http://www.tandf.co.uk/journals. With permission.

Over 6 weeks, worker flexibility issues in the machining and turning department of this firm were recorded (Slomp, 2000). The department operates in two shifts. Table 1.14 illustrates some basic data indicating the labor situation.

The machining and turning department is a typical DRC system. The department operates in two shifts with 14 to 17 workers per shift and 24 machines. During a shift, (some) workers need to transfer to other machines. During the observation weeks, all machines differed in load, varying by about 10%. The average utilization of machines ranges from 20 to 80%. The majority of the workers (90%) have a main machine on which they usually work (on average, 77% of their time). The number of machines to which workers are assigned ranges from one to seven. Only few workers do not have a main machine and are assigned to machines according to job priorities. Task proficiencies of workers range from 50 to 100%. The foreman of the machining and turning department, who works in the day shift, is responsible for the assignment of workers to machines. He makes labor assignment decisions during the working day, but particularly after the so-called "morning prayer" in which the foreman of the department, the planning manager, and a process planner discuss the job priorities and problems of the department. Based on this information, the foreman makes basic assignment decisions. Roughly, he applies the following rules: He will assign the workers to their main machines if there is sufficient work for these machines. Workers who do not have a main machine and the workers for whom there is not enough work on their main machines are assigned to machines and jobs in such a way that all jobs are performed in time, as efficiently as possible. Individual differences between workers play an important role in this assignment. Basically, jobs and machines will be assigned to the most proficient worker. During the day, decisions are adapted, depending on the situation. In case of urgent jobs or bottleneck of machines, workers may be reassigned (from their main machine). In some other firms, we also encountered the concept of workers having a main machine to work on. A worker then only seems to transfer to another machine in the case that his/her main machine runs out of work or his/her skills are more urgently required elsewhere.

In another firm — a Dutch printing office — there are about 11 different presses, each requiring different worker qualifications due to differences in the complexity of operating them. The printing department operates in three shifts with about seven to eight workers per shift. Daily, a set of orders is planned by the planning department and published on a board in the printing department. The foreman of the department then determines which of the available workers are assigned to which presses, based on this set of orders and on the skill matrices of the available workers. He therefore determines the complexity of the orders and matches this with the available workers and their qualifications. Some workers prefer working at specific presses, but this hardly influences the decision of the foreman. When a worker finishes an order, he or she may be reassigned to another press, based on current information on the set of orders that still needs to be finished that day and the qualifications of this worker. The firm's goal is to create fully multifunctional workers, which possibly explains the absence of the concept of a "main machine" as encountered at other firms.

At yet another firm producing toughened glass products, the most proficient workers are assigned to the most critical machines. The proficiency levels of workers are known to the members of the production team as well as to the planner involved in the detail planning. At the beginning of a shift, it is determined what the most critical machines are with respect to work content, and these machines are assigned to the most proficient workers first. In the night shift, the staffing level is lower and, thus, appointing the most critical machines becomes more important.

In summation, in the practical cases we have encountered, labor assignment decisions are made at the beginning and possibly during a day or shift. In some firms, workers have a main machine at which they work and then they only transfer to another machine when their main machine runs out of work or the worker is more urgently required elsewhere. In other firms, workers transfer more frequently (after each order a decision is made) and in these firms workers typically do not have a main machine. Worker assignment is further based on the state of the set of orders (at the machines) that needs to be finished in a certain period and worker characteristics, such as task proficiency. In practice, worker differences are of major importance in labor assignment decisions. Therefore, in the next section, we focus on labor assignment decisions in environments where workers differ substantially.

1.3.2   Research findings on labor assignment rules

This section focuses on the variety of labor assignment rules. We first give an overview of findings in literature so far (Section 1.3.2.1). Next, we present a simulation study in order to indicate the impact of various rules in a more real-life situation with worker differences.

1.3.2.1   Labor assignment and worker differences in the literature

Labor assignment rules considered in most DRC studies are the when-rule and the where-rule. The when-rule determines at what moment labor becomes eligible for transfer, while the where-rule determines to which work center or machine a worker needs to be transferred. These labor assignment rules have received ample attention in the literature (see, e.g., the literature reviews of Treleven, 1989; Gargeya and Deane, 1996; Hottenstein and Bowman, 1998). The most commonly used when-rules in the literature are (1) the centralized when-rule, which means that a worker is eligible for transfer after each job which he/she has finished in a work center, and (2) the decentralized when-rule, which means that after finishing a job, a worker is only eligible for transfer if the work center should become idle. Examples of where-rules often used in research are the First in System, First Served (FISFS) rule and the Longest Queue (LNQ) rule. The impact of labor assignment rules on system performance seems to depend on the specific DRC shop modeled and the performance measure considered.

Bokhorst et al. (2004a) focus on the who-rule, which is a labor assignment rule that selects one worker out of several workers to be transferred to a work center. The who-rule is different from the when- and where-rule mentioned in the literature. The latter rules do not include comparisons between workers with individual differences. These rules are focused on individual workers and decide when a worker needs to be transferred to which machine. Based on worker differences, the who-rule determines which worker should be transferred to a work center if more than one skilled worker is available.

If workers differ from each other, it is important to pay attention to the who-rule. For instance, a who-rule focusing on task proficiency differences may be able to redirect work from less proficient workers to more proficient workers, which will reduce mean flow time. Who-rules focusing on other worker differences, such as differences in the number of skills per worker, also impact the division of jobs to workers. An objective of a who-rule may be to try to equalize workloads of workers lest worker differences tend to create a disparity of workloads. This may also lead to better system performance. As with other labor assignment rules, the impact of the who-rule depends on the specific DRC shop modeled. The average labor utilization and the types and extent of worker differences determine the impact of the who-rule on shop performance.

In general, labor assignment rules in theoretic models need to be applied during three specific decision moments: (A) when a job arrives at an empty work center, (B) when a job is finished at a work center, and (C) when a worker becomes available for transfer. At decision moment A, when a job comes in at an empty work center, the when- and where-rules do not apply (Table 1.15). Therefore, the only remaining question could be: "who should be transferred to the work center?" in case more than one skilled worker is available. A who-rule will then decide which worker to transfer to the work center. At decision moment B, when a job is finished at a work

Table 1.15 Three Decision Moments and the Labor Assignment Rules Which (May) Have to be Used at Those Moments

Decision moment	When-rule	Where-rule	Who-rule
A A job arrives at an empty work center	no	no	possible
B A job is finished at a work center	yes	no	no
C A worker becomes available for transfer	no	yes	possible

Source: Bokhorst J.A.C., Slomp J., and Gaalman G.J.C., 2004, *International Journal of Production Research*, 42(23), 5049–5074. http://www.tandf.co.uk/journals. With permission.

center, the when-rule decides whether or not the worker who finished the job will become eligible for transfer. The where- and who-rules do not apply to this decision moment. Finally, at decision moment C, a worker working in a specific work center becomes available for reassignment. This is actually one of the two possible outcomes of decision moment B, meaning that decision moment C is always preceded by decision moment B. Here, the where-rule and, possibly, the who-rule, need to be applied. At this decision moment, a who-rule should generally be applied in case there is still work in the queue in the machine, and if more than one skilled worker is available for assignment to that machine. For this decision moment, Bokhorst et al. (2004a) propose a team-based assignment approach that considers assignment or reassignment possibilities for all idle team members to prevent unnecessary idle time of workers and/or to be able to choose the right worker for assignment to the machine.

What choices are actually made during these decision moments depends on the specific state of the system at that particular decision moment as well as on the characteristics of the DRC system. In DRC systems with full flexibility, or with a decentralized when-rule, for instance, the number of possibilities of applying a who-rule decreases. The ratio of the number of workers who can operate in a work center and the number of identical machines in a work center also influences the effect of a who-rule. If this ratio is equal to or smaller than one, all skilled workers can be assigned to a work center with a queue and there is no need to choose among them. Finally, labor utilization influences the frequency of the occurrence of decision moments A, B, and C.

In prior DRC research, labor is often modeled in a quite simplistic manner. That is, all workers are often able to operate all machines (i.e., full flexibility) with the same proficiency (i.e., homogeneous labor). In other words, labor is just considered as a homogeneous resource, without paying attention to differences between individual workers. We think worker skills should be regarded as an individual human factor. Our focus here is on worker differences with respect to their task proficiencies, the number of skills they possess, and the loads of work centers in which they can operate. We will first give a brief literature overview with respect to these worker differences and then evaluate assignment rules within systems with these worker differences in the next section.

With respect to task proficiency, many prior studies model homogeneous labor. This means that all workers perform their assigned tasks with equal proficiency. Other studies incorporate heterogeneous labor, meaning that task proficiency differences exist within the shop (e.g., Nelson, 1970; Rochette and Sadowski, 1976; Hogg et al., 1977; Bobrowski and Park, 1993; Malhotra et al., 1993; Kher and Malhotra, 1994; Malhotra and Kher, 1994; Fry et al., 1995; Felan and Fry, 2001). Note that studies modeling learning and forgetting effects of workers, by definition, deal with heterogeneous labor. We will next discuss studies that evaluate labor assignment rules and model heterogeneous labor.

Nelson (1970) models a system with two work centers, consisting of two machines each, and two workers. Each worker is fully proficient at performing tasks on his/her own work center, and the task proficiency on the other work center is varied. As a where-rule, a worker is always assigned to the work center where he/she is most proficient, unless there is no work there, and there is work at the other work center. Further, three when-rules are evaluated. The decentralized when-rule only allows a labor transfer if the work center is empty. The centralized when-rule allows a labor transfer after completion of each job. Finally, between these two extremes, a third when-rule is modeled that transfers a worker with a 50% chance after completing a job from a work center that still has jobs in queue. Only the where-rule uses task proficiency information. Results show that if task proficiency is decreased resulting in larger task proficiency differences, the choice of a when-rule is more important and a centralized when-rule is preferred.

Hogg et al. (1977) model three different types of heterogeneous systems: labor differential systems (LD), machine center differential systems (MCD), and labor and machine center differential systems (L&MCD). All of these systems are modeled as systems with full flexibility. In LD systems, the task proficiency differs between workers, but a single worker is equally proficient at all machine centers. In MCD systems, the task proficiency differs between work centers, but workers are equally proficient at a work center. Finally, in L&MCD systems, task proficiency depends on both the machine center and the worker who performs the job. For these three types of heterogeneous systems, they study two where-rules, one based on the time a job has been waiting (FCFS) and another based on the proficiency of workers (maximum laborer efficiency: MLE). The size of the system and labor utilization is also varied. They supposedly use a centralized when-rule and use a who-rule that selects the most efficient worker. The MLE where-rule and the who-rule they use take information on task proficiency differences into account. Their results show that the two where-rules do not perform differently in LD systems. This can easily be explained by the fact that a single worker is equally proficient at every work center, and the MLE where-rule then reverts to the FCFS where-rule. For MCD systems, the FCFS where-rule is superior. The MLE where-rule results in good performance for the work centers at which workers are proficient, but at the expense of the performance of work centers at which workers are not proficient. An interesting result is that in

L&MCD systems, the MLE where-rule is far superior to the FCFS where-rule, at least if no clear MCD pattern can be found within the L&MCD scheme.

Bobrowski and Park (1993) study labor assignment rules in a shop with full flexibility and differences in task proficiency. The heterogeneous system they model can be seen as an L&MCD system. They study five different when-rules and seven where-rules and include task proficiency information in two when-rules and six where-rules. They do not mention a who-rule. Their results show that the "efficiency" where-rule, which moves the worker to the work center where he/she is most efficient, dominates all other where-rules they included in their design. This is consistent with the observation of Hogg et al. (1977) that the MLE where-rule is superior in L&MCD systems. Further, they show that as long as the efficiency where-rule is selected, the selection of the when-rule is not important.

Kher and Malhotra (1994) examine whether it is beneficial to cross train workers in DRC job shops in the presence of worker transfer delays, worker learning effects, and worker attrition in order to improve shop performance. From a total of six departments, workers are trained to process jobs in two or in three departments, representing a system of limited flexibility. Even though they include the when-rule and the where-rule as experimental factors, these rules do not use information on task proficiency differences. The centralized and decentralized when-rules and the first-arrived-in-the-system-first-served (FISFS) and the least slack where-rules are evaluated. No mention is made of a who-rule. They conclude that in the presence of learning losses and transfer delays, a decentralized when-rule is superior and the FISFS where-rule is preferred because of ease of implementation and control.

Malhotra and Kher (1994) studied two when-rules and five where-rules in a shop with full flexibility, heterogeneous resources (LD, MCD, and L&MCD systems), and worker transfer delays. Three of the where-rules use information on task proficiency differences. Their results show that the "most efficient" where-rule, which is equal to the MLE where-rule of Hogg et al. (1977) and the efficiency where-rule of Bobrowski and Park (1993), is robust with respect to the when-rule chosen (i.e., centralized or decentralized) and with respect to the heterogeneous system modeled (i.e., LD, MCD, or L&MCD). The study does not explicate a who-rule.

A second factor causing differences between workers is the number of skills of workers (unequal multifunctionality among workers). Felan and Fry (2001) model heterogeneous labor by including labor attrition and learning rates and model workers who are trained to work in a different number of departments (i.e., one, two, or four departments). The focus of their study is on labor flexibility. They evaluated 13 different flexibility configurations with 9 different levels of average flexibility for systems with a learning rate of 85 and 90%. One of these configurations represents full flexibility, the other configurations represent limited flexibility. Labor assignment rules are not varied in this study. They use a decentralized when-rule and a "longest queue" where-rule, which do not use information on worker differences. Further, they did not explicate a who-rule. Their results showed that an average flexibility level of 1.7 performs just as well as incremental flexibility (i.e., a flexibility level of 2.0). Also, their results suggest that it is better to mix the training between workers than to try to provide equal training. In this way, more workers are kept within one department and the efficiency of these workers can then be maximized. A few resulting workers with high levels of flexibility are able to respond to temporary overloads within departments.

A third factor causing differences between workers is the difference with respect to the loads of work centers in which they are able to operate. In practice, this difference is a reason to prefer one worker above another for assignment to a machine or job. It may occur in shops with limited flexibility and a disparity in work center loads. Or, in other words, in systems where not all workers are able to operate in all work centers and where a noticeable difference between the loads of work centers can be found. In these systems, some workers in a team may receive a larger workload than other workers, due to the fact that they are solely responsible for a number of heavily loaded work centers. Note that a workload difference between workers is also possible in systems where all workers have the same level of multifunctionality. Preventing heavily loaded workers from being assigned an equal share of the workload of all the work centers in which they can operate may be advantageous. In other words, it is probably more beneficial if the workload of work centers, which are shared by heavily loaded and less loaded workers, is allocated to a larger degree to the less-loaded workers.

In conclusion, the literature thus far has only paid limited attention to labor assignment in systems with worker differences, and more specifically, to the who-rule. Based upon industrial practice where we have seen worker differences play a significant role in worker assignment decisions, we believe that more research is needed to address this gap in the literature. The next section overviews two studies that have made a start with this.

1.3.2.2   *Evaluation of labor assignment rules in systems with worker differences*

Bokhorst et al. (2004a) explore the flow time effects of applying different who-rules in several DRC systems where labor flexibility is limited and workers differ with respect to task proficiencies, the number of skills they possess, and the loads of work centers for which they are responsible. The smallest possible DRC shop to which the who-rule applies is modeled and consists of three work centers and two workers who are not fully cross-trained. The dispatching rule used is the first-come-first-served (FCFS) rule. A work center thus starts processing the job that arrived at the work center first. The first-in-system-first-served (FISFS) where-rule is used. This where-rule sends workers who are eligible for transfer to the work center with the job in queue that had the earliest entry into the system. As a when-rule, the centralized when-rule is used, which means that a worker is eligible for transfer after each job he/she has finished at a work center.

Two experiments are conducted in order to study the effects of the who-rule on the average flow time of the system. In the two experiments,

Table 1.16 Experimental Factors and Levels in the Two Experiments

First experiment: Homogeneous labor with respect to task proficiencies

Factor	Levels					
	1	2	3	4	5	6
NS	4	5				
DISTR	0.4/0.5/0.1	0.1/0.4/0.5	0.5/0.1/0.4	0.3/0.5/0.2	0.2/0.3/0.5	0.5/0.2/0.3
WHO	LIT	RND	PRIO			
UTIL	60%	75%	90%			

Second experiment: Heterogeneous labor with respect to task proficiencies

Factor	Levels					
	1	2	3	4	5	6
DISTR	0.4/0.5/0.1	0.1/0.4/0.5	0.5/0.1/0.4	0.3/0.5/0.2	0.2/0.3/0.5	0.5/0.2/0.3
WHO	RND	MEF				

Source: Bokhorst J.A.C., Slomp J., and Gaalman G.J.C., 2004, *International Journal of Production Research*, 42(23), 5049–5074. http://www.tandf.co.uk/journals. With permission.

four experimental factors with different levels are examined (Table 1.16): number of skills (NS), disparity in work center loads (DISTR), the who-rule, and labor utilization (UTIL).

The first experiment models DRC systems with homogeneous labor with respect to task proficiencies, single or multilevel flexibility, and a disparity in work center loads under three levels of average labor utilization. The second experiment models a DRC system with heterogeneous labor with respect to task proficiencies, single-level flexibility, and a disparity in work center loads, with a 60% labor utilization.

In the first experiment, the number of skills is an experimental factor. In this experiment, two matrices (Figure 1.7a-b) are examined. The first matrix (1.7a) represents single-level labor flexibility with a total number of four skills and no task proficiency differences, and the second matrix (1.7b) represents multilevel labor flexibility with a total number of five skills and no task proficiency differences. The third matrix (1.7c) is used in the second

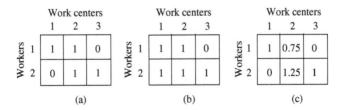

Figure 1.7 Three labor proficiency matrices representing (a) a four-skills configuration with single-level flexibility, (b) a five-skills configurations with multilevel flexibility, and (c) a four-skills configuration with a difference in task proficiencies of workers. (From Bokhorst J.A.C., Slomp J., and Gaalman G.J.C., 2004, *International Journal of Production Research*, 42(23), 5049–5074. http://www.tandf.co.uk/journals. With permission.)

experiment, where the number of skills is a fixed factor and labor is heterogeneous with respect to task proficiencies. In this matrix, the workers have different task proficiencies regarding the second work center. Here, the expected processing time $(1/\mu)$ is multiplied by the proficiency factor p_{nk}. The processing time for work center 2 is adjusted up or down by the proficiency factor p_{nk}, which is 1.25 for worker 2 and 0.75 for worker 1.

In both the first and the second experiments six different divisions of the system arrival rate to the three individual work centers are investigated (see Table 1.16). For instance, at level 1, 40% of the jobs that arrive in the system go to work center 1, 50% to work center 2, and 10% to work center 3. Level 2 and 3 use the same division of the system arrival rate as level 1, but in another order. For levels 4 to 6, the division of the system arrival rate to the individual work centers is less extreme. Even though one work center still gets 50% of the incoming jobs, the work center loads of the other two work centers are more similar to each other than those of levels 1 to 3.

In the first experiment, three who-rules are examined. These who-rules are called (1) the *longest idle time* who-rule (LIT), (2) the *random* who-rule (RND), and (3) the *priority* who-rule (PRIO). The LIT who-rule was already applied in research of Rochette and Sadowski (1976). This who-rule performs well from a social viewpoint in that it seems fair to the team members to assign the worker who has been waiting — or at least not working on a work center — for the longest period of time. The RND who-rule chooses a worker randomly when a choice between workers has to be made. The PRIO who-rule is specifically designed for the shop modeled and it always gives priority to one of the two workers. Which worker is given priority depends on the labor proficiency matrix used (NS), and in case of the four-skills configuration, also on the disparity in work center loads (DISTR). In case of the four-skills configuration, each worker has a unique skill and thus a unique load for one work center, and the other work center is shared with the other worker. The PRIO who-rule compares the unique loads of the two workers and always assigns the worker with the lowest unique load to the shared work center. In case of the five-skills configuration, one worker is able to operate in all work centers and the other in only two work centers. Here, the PRIO who-rule always gives priority to the worker with two skills in case a choice between the workers has to be made. The second experiment compares the RND who-rule (as discussed above) with a who-rule based on task proficiency differences, which is derived from Hogg et al. (1977): the most efficient (MEF) who-rule. The MEF who-rule assigns the worker who is the most efficient at performing the task (i.e., the worker with the lowest proficiency factor).

The first experiment shows that with 60% labor utilization, the effect of the who-rule is relatively larger with multilevel flexibility (i.e., workers possess different numbers of skills), and with distributions of work center loads that create larger worker differences in terms of unique workloads of workers. With 75% labor utilization, only the number of skills of workers affects the flow time effect of the who-rule. On most work center load distributions, the effect of the who-rule is larger in the multilevel flexibility configuration than in the single-level flexibility configuration. Finally, with 90% labor utilization, there are no significant interaction effects of the who-rule and the disparity in work center loads or number of skills, and the main effect of the who-rule is smaller than under lower levels of labor utilization.

The second experiment that models heterogeneous labor with respect to task proficiencies shows larger effects of the who-rule than the first experiment. The higher the load of the shared work center, the more the who-rule is applied and the more often the most efficient worker has to work in the shared work center under the MEF who-rule (this who-rule assigns the worker who is the most efficient at performing the task). This results in better flow time performance.

In Bokhorst (2005), DRC systems are modeled with limited labor flexibility, task proficiency differences, and/or workers who differ in the number of skills they possess. Three simulation experiments are performed to examine the effect of the where-rule, the when-rule, and the who-rule on flow time performance within these systems. Static worker differences in task proficiency are considered, meaning that it is assumed that no worker "learning or forgetting" takes place. While keeping the division of skills equal, three configurations of task proficiency differences are generated by increasing the task proficiency differences between workers. In this way, it can be studied whether the extent of worker differences affects the impact of assignment rules (including rules that make use of those worker differences) on flow time. With respect to differences in the number of skills per worker, a similar approach is followed by including two configurations with increasing differences between the number of skills per worker. Also, a configuration is presented that combines task proficiency differences and worker differences with respect to the number of skills.

With respect to task proficiency, systems that Hogg et al. (1977) call *labor and machine center differential systems* or L&MCD systems are considered. For these systems, task proficiency depends on the specific worker *and* machine performing the task. Nelson (1967, 1970) and Hogg et al. (1977) modeled task proficiency by introducing a factor e_{ji}, representing the relative efficiency of worker j ($j = 1, 2, ..., N$) when using machine i ($i = 1, 2, ..., M$). The expected processing time for a specific job then will be $1/e_{ji}\mu$. The expected processing rate (μ) is adjusted by multiplying it by factor e_{ji}, which is bound to a maximum of 1 in their research. In contrast, we multiply the expected processing time ($1/\mu$) by a proficiency factor p_{ji}. In accordance with work measurement studies, where basic and standard times are assessed, the processing time can be regarded as a distribution of the basic time, which is adjusted up or down by a proficiency factor p_{ji}, ranging from 0.8 to 1.2 within our experiments. For example, a proficiency factor of 0.9 indicates that the task is performed within 90% of the expected basic time. Proficiency factors lower than 1 thus result in processing times that are lower than the basic time, while proficiency factors higher than 1 result in above-average processing times.

Table 1.17 Three Cross-Training Configurations Representing Three Levels of Task Proficiency Differences

	I Worker					II Worker					III Worker				
Machine	W1	W2	W3	W4	W5	W1	W2	W3	W4	W5	W1	W2	W3	W4	W5
M1	1	1	1			0.9	1	1.1			0.8	1	1.2		
M2		1	1	1			0.9	1	1.1			0.8	1	1.2	
M3			1	1	1			0.9	1	1.1			0.8	1	1.2
M4	1			1	1	1.1			0.9	1	1.2			0.8	1
M5	1	1			1	1	1.1			0.9	1	1.2			0.8
M6	1	1	1			0.9	1	1.1			0.8	1	1.2		
M7		1	1	1			0.9	1	1.1			0.8	1	1.2	
M8			1	1	1			0.9	1	1.1			0.8	1	1.2
M9	1			1	1	1.1			0.9	1	1.2			0.8	1
M10	1	1			1	1	1.1			0.9	1	1.2			0.8

Table 1.17 shows three cross-training configurations (I, II, and III) for systems with 10 machines and 5 workers. The three configurations show the same division of skills. Further, each worker is able to operate six machines (i.e., each worker has a multifunctionality of six) and each machine may be operated by three workers (i.e., each machine has a redundancy of three). Since not all workers are able to operate all machines, these configurations represent limited labor flexibility. The basic configuration, which is configuration I, models homogeneous labor; the task proficiency for each skill is the same. Configurations II and III model heterogeneous labor, with task proficiency deviations from the basic time of 10% in the second configuration and of 20% in the third configuration. To make the three configurations comparable, the following holds true for each configuration: $\frac{1}{MF}\sum_i p_{ji} = 1 \forall j$, and $\frac{1}{RE}\sum_j p_{ji} = 1 \forall i$, where MF denotes the multifunctionality of workers j and RE denotes the redundancy of machines i. The average task proficiency of each machine and the average task proficiency of each worker thus equal 1. Further, $p_{ji} \geq 0$, and at least one $p_{ji} = 1 \forall i$. This means that workers cannot be negatively skilled and of the three workers who are able to operate a machine, at least one worker performs conform the basic times (average times). In case of task proficiency differences, one of the other workers performs below average and another performs above average, since the average is set at 1.

With respect to the number of skills, two cross-training configurations (IV and V in Table 1.18) are used in this study to model systems where each worker possesses a different number of skills. Note that even though the multifunctionality of each worker is different, we kept the redundancy of all machines at three. In configuration IV, the first worker possesses four skills, the second worker five, and so on. The number of skills per worker thus ranges from four to eight. The differences in number of skills per worker are somewhat higher in the fifth configuration, where the number of skills ranges from 2 to 10. The fifth worker is fully cross trained in this configuration.

Table 1.18 Two Cross-Training Configurations with Differences in the Number of Skills Per Worker (IV and V) and a Configuration Representing Large Task Proficiency Differences and Large Differences in the Number of Skills Workers Possess (VI)

	IV Worker					V Worker					VI Worker				
Machine	W1	W2	W3	W4	W5	W1	W2	W3	W4	W5	W1	W2	W3	W4	W5
M1	1		1	1		1		1		1	0.8		1.2		1
M2	1		1	1		1		1		1	1.2		1		0.8
M3	1		1		1		1		1	1		0.8		1	1.2
M4	1		1		1		1		1	1		1.2		0.8	1
M5		1	1		1		1		1	1		1		1.2	0.8
M6		1		1	1		1		1	1		1		0.8	1.2
M7		1		1	1			1	1	1			0.8	1.2	1
M8		1		1	1			1	1	1			0.8	1	1.2
M9		1		1	1			1	1	1			1.2	0.8	1
M10			1	1	1			1	1	1			1	1.2	0.8

Configuration VI (Table 1.18) is used to model a combination of task proficiency differences and the differences in the number of skills of workers. Actually, it is an adaptation of configuration V to include task proficiency differences of 20%. In this configuration, task proficiency differences as well as differences in the number of skills of workers are large. For this configuration, it will be interesting to see whether the who-rule should incorporate information on task proficiency differences or information on the difference in number of skills per worker. Further, the extent of the effect of where-rules and who-rules that incorporate information on worker differences in a system with large worker differences can be evaluated.

We used the object-oriented simulation software package eM-Plant v. 5.5 (Stuttgart: Tecnomatix) for building the DRC simulation models. The DRC shop consists of 10 machines and 5 workers. Job arrivals follow a negative exponential distribution with a mean interarrival time of 1.294 time units. This is selected to create an average labor utilization of 85% using task proficiency factors of 1 (i.e., basic times). Jobs need to visit between 1 and 10 machines in a random order or an average number of 5.5 machines. Additionally, a job may visit a specific machine only once. Machine processing times are drawn from a gamma [2, 0.5] distribution, which is equal to a 2-Erlang distribution. This distribution is often used to represent operating times. The FISFS dispatching rule is used in the system.

Three experiments are conducted to study the effects of when-, where-, and who-rules in systems with differences in tasks proficiency (experiment I), differences in the number of skills of workers (experiment II), and a combination of differences in task proficiency and the number of skills (experiment III). Table 1.19 shows these three experiments and the experimental factors within each experiment.

The first experiment focuses on the where-rule and the who-rule in systems with increasing differences in task proficiency of workers.

Table 1.19 Three Experiments with Experimental Factors and Levels

Experiment		Experimental factor	Level 1	Level 2	Level 3
1	TP:	task proficiency differences	0%	10%	20%
	WR:	where-rule	FISFS	PL	
	WHO:	who-rule	RND	PL	
2	NS:	differences in number of skills	6–6	4–8	2–10
	WHO:	who-rule	RND	FNS	
3	WN:	when-rule	CEN	DECEN	
	WR:	where-rule	FISFS	PL	
	WHO:	who-rule	RND	FNS	PL

Note: CEN: central; FISFS: first-in-system-first-served; FNS: fewest number of skills; DECEN: decentral; PL: proficiency level; RND: random.

Cross-training configurations I, II, and III (Table 1.17) are used in this experiment to represent three levels of task proficiency differences. The FISFS where-rule is compared with the proficiency level (PL) where-rule. The PL where-rule assigns workers to the machine where they are most proficient. As we have seen in the previous section, this rule has been shown to provide excellent results in L&MCD systems with full flexibility. The random (RND) who-rule is compared with the PL who-rule. The PL who-rule assigns the worker who is most proficient to the machine that requires labor. Within this experiment, the centralized when-rule is used. To measure system performance, the average flow time of jobs is used.

The full factorial results for the first experiment are shown in Table 1.20. The data were analyzed using a $3 \times 2 \times 2$ analysis of variance (ANOVA) between subjects designed with task proficiency (TP) differences, the where-rule and

Table 1.20 Full Factorial Results for Experiment I

TP	WR	WHO	Average flow time	Std. deviation
0%	FISFS	RND	12.13	0.320
		PL	12.20	0.378
	PL	RND	12.52	0.417
		PL	12.60	0.404
10%	FISFS	RND	12.48	0.376
		PL	11.84	0.417
	PL	RND	10.98	0.219
		PL	10.50	0.217
20%	FISFS	RND	12.79	0.315
		PL	11.84	0.407
	PL	RND	10.22	0.146
		PL	9.33	0.129

the who-rule as independent variables, and flow time as dependent variable. The ANOVA results showed that all main effects and two-way interactions are significant. In our further analysis, we will discuss the significant main and interaction effects and, if necessary, we will revise the higher level effects when studying the lower level effects.

The main effect of TP shows that larger differences in task proficiency result in better flow times. This indicates the advantages of taking worker differences into account in the where-rule and who-rule. This will be made clearer in the discussion of the interaction effects between the proficiency level and the where-rule, and the proficiency level and the who-rule. The main effect of the where-rule indicates that the PL where-rule results in better flow times than the FISFS where-rule. This result is in conformity with that of Hogg et al. (1977), Bobrowski and Park (1993), and Malhotra and Kher (1994), who also included a rule based on efficiency and found that such a rule is superior compared with other where-rules in L&MCD systems. Limited flexibility as modeled here apparently does not alter this conclusion. The main effect of the who-rule indicates that the PL who-rule results in better flow times than the RND who-rule. For the who-rule as well as for the where-rule, it seems advantageous to make choices based on individual task proficiency differences.

The effect of task proficiency differences (TP) on the where-rule (WR) is illustrated in Figure 1.8. The simple main effects showed us that the FISFS

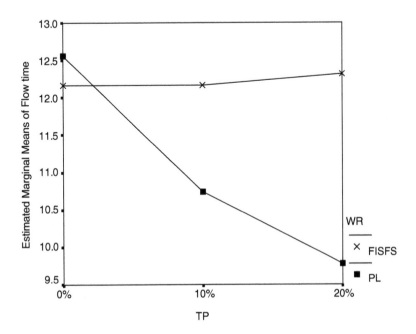

Figure 1.8 The interaction effect of task proficiency (TP) differences and the where-rule.

where-rule and the PL where-rule are significantly different within each level of TP. Without task proficiency differences, the PL where-rule performs worse than the FISFS where-rule. This can be explained by the fact that the PL where-rule chooses a machine randomly if the choice cannot be based on task proficiency differences. In contrast, if task proficiency differences are present, the PL where-rule results in much better flow times than the FISFS where-rule. While the main effect of TP showed a decrease in flow time with larger task proficiency differences, the interaction with the where-rule shows that flow time only decreases if a PL where-rule is used. The PL where-rule, which incorporates task proficiency information, thus can benefit from proficiency differences to reduce system utilization, thereby decreasing average flow time. A further analysis of the significant simple main effects of TP shows that for the PL where-rule, all levels of TP differ significantly, while for the FISFS where-rule, the TP levels of 0 and 10% do not differ significantly. At the TP level of 20%, the average flow time actually increases significantly for the FISFS where-rule, compared with the lower levels of TP. In this experiment, the largest effect of the where-rule is found at a TP level of 20%, resulting in a 20.6% reduction of flow time.

The effect of TP differences on the who-rule is illustrated in Figure 1.9. The simple main effects of the who-rule show that the RND who-rule and the PL who-rule differ significantly at the TP levels of 10 and 20%. As expected, the who-rule is not significant at the TP level of 0%, since the PL who-rule reverts to the RND who-rule if no task proficiency differences exist

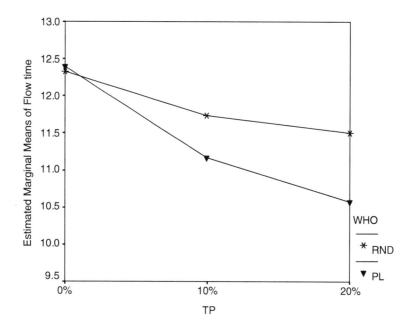

Figure 1.9 The interaction effect of task proficiency (TP) differences and the who-rule.

between workers. At a TP level of 20%, the effect of the who-rule is stronger than at a TP level of 10%. Here, the largest effect of the who-rule results in an 8.0% reduction of flow time. The PL who-rule, which incorporates task proficiency information, can also benefit from task proficiency differences.

Thus, assignment rules that incorporate task proficiency information lead to lower average flow times than rules that do not incorporate that information. For labor assignment rules that incorporate task proficiency information, configurations with larger differences in task proficiency provide more opportunities to improve flow time performance than configurations with lower differences in task proficiency. In contrast, for labor assignment rules that do not incorporate proficiency information, flow time performance may decrease for configurations with larger differences in task proficiency because of the higher variation of processing times of machines and workers.

The second experiment focuses on the who-rule in systems with increasing differences in the number of skills that workers possess. Configurations I, IV, and V (Tables 1.17 and 1.18) are used here to represent three levels of differences in the number of skills per worker. The RND who-rule is compared with the fewest number of skills (FNS) who-rule, which assigns the worker who possesses the fewest number of skills to the machine that requires labor. The FISFS where-rule and the centralized when-rule are used.

The full factorial results for the second experiment are shown in Table 1.21. The data were analyzed using a 3 × 2 ANOVA between subjects designed with differences in number of skills (NS) and the who-rule as independent variables and flow time as dependent variable. The ANOVA results showed that all main effects and the two-way interaction are significant.

All levels of NS are significantly different from each other, while larger differences in the number of skills per worker result in higher flow times. Workers who possess more skills will be assigned more often than workers who possess fewer skills and, therefore, a more unequal division of the number of skills per worker will result in larger differences between the utilization of workers. This result contrasts with that of Felan and Fry (2001), who state that it is better to mix the training between workers than to try to

Table 1.21 Full Factorial Results for Experiment II

NS	WHO	Average flow time	Std. deviation
6–6	RND	12.20	0.323
	FNS	12.12	0.484
4–8	RND	12.49	0.371
	FNS	12.45	0.451
2–10	RND	15.19	0.873
	FNS	14.54	0.665

provide equal training. This contradiction can be explained by the fact that Felan and Fry (2001) also model labor learning. In their experiment, workers with few skills are able to maximize the task proficiency of those skills, while the few workers with many skills are able to respond to temporary overloads.

The significant main effect of the who-rule shows that the FNS who-rule results in better flow times than the RND who-rule. Individual differences with respect to the number of skills thus can be used to base a who-rule on, in order to improve flow time performance. The significant interaction effect of differences in NS on the who-rule shows that the who-rule is only significant at NS 2 to 10, representing large differences in number of skills per worker. Here, the FNS who-rule reduces flow time with 4.3% compared to the RND who-rule. In other words, differences in number of skills per worker must be large in order for the FNS who-rule to have a relatively small (4.3%) but statistically significant effect on flow time.

Finally, the third experiment focuses on the when-rule, the where-rule, and the who-rule in a system with a large difference in task proficiency and a large difference in the number of skills workers possess. Configuration VI in Table 1.18 is used for this. The centralized when-rule is compared to the decentralized when-rule. Further, the FISFS where-rule is compared to the PL where-rule, and the RND, FNS, and PL who-rules are evaluated.

The full factorial results for the third experiment are shown in Table 1.22. The data were analyzed using a $2 \times 2 \times 3$ ANOVA between-subjects design with the when-rule, the where-rule, and the who-rule as independent variables and flow time as dependent variable. The ANOVA results showed that all main effects and two-way interactions are significant.

The significant main effect of the when-rule shows that the centralized when-rule performs considerably better than the decentralized when-rule, which is an improvement of 22.6%. The significant main effect of the where-rule

Table 1.22 Full Factorial Results for Experiment III

WN	WR	WHO	Average flow time	Std. deviation
CEN	FISFS	RND	16.14	0.891
		PL	15.00	0.763
		FNS	15.71	0.817
	PL	RND	10.87	0.162
		PL	10.15	0.158
		FNS	10.73	0.212
DECEN	FISFS	RND	20.47	1.413
		PL	19.74	1.320
		FNS	19.77	1.176
	PL	RND	14.06	0.364
		PL	13.57	0.363
		FNS	13.89	0.420

shows that the PL where-rule performs substantially better than the FISFS where-rule, which is an improvement of 31.4%. All three who-rules differ significantly from each other, where the PL who-rule performs best, followed by the FNS who-rule, and finally the RND who-rule. Here, the who-rule can result in a flow time improvement of 5%.

The significant interaction effect of the when-rule on the where-rule is analyzed using the simple main effects of the where-rule. Even though the where-rule is significant within both levels of the when-rule, for a centralized when-rule the relative flow time difference between a FISFS where-rule and a PL where-rule is slightly larger than for a decentralized when-rule. This result is in conformity with Nelson (1970), who indicates that a centralized when-rule is preferred if task proficiency differs. Also, the simple main effects of the when-rule show that the when-rule is significant within both levels of the where-rule. The flow time difference between the centralized and the decentralized when-rule is 21.9% within a FISFS where-rule, while it is 23.5% within a PL where-rule. This means that the selection of the when-rule is important. This outcome contrasts with that of Bobrowski and Park (1993) and Malhotra and Kher (1994), who claim that as long as the efficiency based where-rule (i.e., "efficiency" or "most efficient," respectively) is selected, the choice of a when-rule does not matter much. The fact that we model limited flexibility may contribute to this contrasting outcome. With limited flexibility, workers have less choice in moving to a work center than with full flexibility. This will reduce the chance that they are allocated to a work center at which they are very proficient. Moreover, with a decentralized when-rule, they can be trapped in work centers at which they are not very proficient until the queue is empty.

The simple main effects of the who-rule are analyzed to explain the significant interaction effect of the when-rule on the who-rule. Since the who-rule is significant for both levels of the when-rule, pair-wise comparisons were made using the Sidak adjustment for multiple comparisons. The results are shown in Table 1.23.

When using a centralized when-rule, the difference between an RND who-rule and an FNS who-rule is not significant; all other differences are significant. This implies that incorporating information on differences in the

Table 1.23 Simple Main Effect of the Who-Rule Within Each Level of the When-Rule and Pair-Wise Comparisons of the Simple Effects Using the Sidak Adjustment for Multiple Comparisons ($\alpha = 0.05$)

WN	WHO	Average flow time
CEN	PL	12.58
	FNS	13.22
	RND	13.50
DECEN	PL	16.66
	FNS	16.83
	RND	17.27

number of skills of workers in the who-rule does not help in reducing flow time. Within an decentralized when-rule, only the difference between a PL who-rule and an FNS who-rule is not significant. The effect of the who-rule on flow time is larger in case of a centralized when-rule (a flow time difference of 6.9%) than in case of a decentralized when-rule (a flow time difference of 3.5%). The explanation may be that the choice between workers has to be made more often, resulting in a better fit of machines and workers who are most proficient for those machines.

In summation, in DRC systems with limited labor flexibility and differences in task proficiency, it is advantageous to base the where-rule and the who-rule on these task proficiency differences. In other words, the where-rule should send a worker to the machine at which he/she is most proficient, which is in conformity with prior research of DRC systems with full flexibility (Hogg et al., 1977; Bobrowski and Park, 1993). Additionally, the who-rule should transfer the worker who is most proficient in case a choice between workers has to be made. Larger differences in task proficiency result in a larger effect of these rules. Further, the flow time effect of the where-rule seems to be larger than the flow time effect of the who-rule.

DRC systems with differences in the number of skills per worker perform worse than systems where skills are divided equally. If differences in the number of skills per worker are present and task proficiencies do not differ, a who-rule based on large differences in the number of skills can slightly improve (4.3%) flow time performance by assigning the worker with the fewest number of skills.

In systems with task proficiency differences and differences in the number of skills per worker, a where-rule based on task proficiency differences performs well. In these systems, it seems more advantageous to base the who-rule on task proficiency differences than on differences in the number of skills per worker. Further, the centralized when-rule is preferred over the decentralized when-rule, even if the where-rule is based on proficiency differences. This result is different from that of prior research (Bobrowski and Park, 1993; Malhotra and Kher, 1994) in a system with full flexibility.

This study suggests that in DRC systems with limited labor flexibility and worker differences, it is advantageous to select where-rules and who-rules that make use of these worker differences.

1.3.3 Summary and discussion

In order to benefit from a cross-trained workforce, it is crucial to think carefully about the design of effective labor assignment rules. In practice, we have noticed that workers differ with respect to the number of skills, skill types, task proficiency, and workload that can be assigned to them. In practical worker assignment decisions, we have seen task proficiency play a role. Worker assignment is done at the beginning of a shift (day) and possibly during a shift (day). In some firms, workers have a main machine at which they normally work in the case there is work for this machine and the worker is not urgently needed elsewhere. In other firms, workers transfer more often and, in these firms, workers typically do not have a main machine. These situations resemble the decentralized and the centralized when-rule as described in the literature, respectively. In most cases, if workers are assigned or reassigned, the state of the orders (at the machines) is considered and worker characteristics, such as task proficiency or the specific qualification of workers, play a role.

In practice, decisions concerning the where-rule and the who-rule seem to be made simultaneously, at least at the beginning of a shift (day). Since the models used in the literature mostly start with an empty system (no work at the machines, no workers assigned), assignment decisions are made more sequentially, at the arrival of jobs. The assignment problem at the beginning of a shift (day), which involves assigning all available workers to machines that already have work in queue, is not specifically addressed in the literature. Further, from a managerial viewpoint, it is important to note that the impact of the who-rule is probably higher in practical situations where future information can be used in assignment decisions. It is evident that the frequency by which the who-rule can be applied in practice is higher in case of practical situations where managers are able to anticipate on future events. That is, if future information is used, workers may, for instance, be kept idle during a short period to wait for a work center to become available for which they are better suited than the worker who is operating the work center at that moment.

In the literature, few studies have investigated labor assignment rules in systems with task proficiency differences, differences in the number of skills of workers, and/or differences with respect to the loads of work centers in which workers are able to operate. In Section 1.3.2.2, several simulation experiments dealing with labor assignment in systems with worker differences are discussed. Two simulation experiments are conducted to study the flow time effects of applying alternative who-rules. Factors likely to affect the extent of the effect of the who-rule are included as experimental factors as well. Next to the who-rule, experimental factors are the number of skills, disparity in work center loads, and labor utilization. The first experiment models DRC systems with homogeneous labor with respect to task proficiencies, single or multilevel flexibility, and a disparity of work center loads, fewer than three levels of average labor utilization. The second experiment models a DRC system with heterogeneous labor with respect to task proficiencies, single-level flexibility, and a disparity of work center loads, with 60% labor utilization. The results show that DRC shop characteristics influence the impact of the who-rule. The impact of the who-rule is larger under lower levels of labor utilization than under higher levels of labor utilization. Furthermore, the first experiment shows that under lower levels of labor utilization, the effect of the who-rule is relatively larger with multilevel flexibility and with distributions of work center loads, which create larger worker differences in terms of unique workloads of workers. The second experiment modeling heterogeneous labor with respect to task proficiencies shows larger effects of the who-rule than the first experiment. The higher the load of the shared work center, the more the who-rule is applied and the more often the most efficient worker is assigned.

Three other simulation experiments examine the flow time effects of the when-rule, the where-rule, and the who-rule in systems with limited labor flexibility with respect to the number of machines workers can operate. We argue that skills of workers should be regarded as an individual human factor, and, therefore, we model workers who differ in their task proficiency and/ or in the number of skills they possess. Different configurations of a DRC system with 5 workers, 10 machines, and 30 skills are simulated in three experiments. The first experiment focuses on the where-rule and the who-rule in three configurations with increasing differences in task proficiency of workers. The results show that where-rules and who-rules that base their choice on task proficiency differences result in better flow times than a simple FISFS where-rule and an RND who-rule in case there are task proficiency differences. The second experiment focuses on the who-rule in three configurations with increasing differences in the number of skills workers possess. The results show that with relatively large differences in the number of skills per worker, a who-rule that assigns the worker with the fewest number of skills results in better flow time performance than an RND who-rule. The third experiment focuses on the when-rule, the where-rule, and the who-rule in a configuration with a large difference in task proficiency and a large difference in the number of skills workers possess. The results show that a centralized when-rule performs considerably better than a decentralized when-rule. Furthermore, the where-rule and who-rule that base their choice on task proficiency perform better than an FISFS where-rule and an FNS and RND who-rule, respectively. Finally, the effect of the where-rule and the who-rule seems to be larger in case of a centralized when-rule.

1.4   Future research issues

1.4.1   Short summary of current work

In dealing with the design and operation of a cross-trained workforce, this chapter focused on the issues of cross training and labor assignment. Section 1.2 dealt with the development and evaluation of effective cross-training policies to support an organizations strategy. A firm's strategy and its specific context — with respect to jobs, workers, and demand — determine which aspects are important to include in a cross-training policy and how to address these aspects. Among the aspects discussed are the extent of cross training, chaining, the level and distribution of multifunctionality and redundancy, and collective responsibility. In the evaluation, we mainly focused on improving flow time and minimizing (training) costs, but we also paid attention to the workload balance of workers from a Human Resource Management point of view.

Creating a right set of skills within a team is one challenge on the road to an effective workforce; applying this set of skills effectively is another challenge.

Section 1.3 dealt with this second challenge and explored labor assignment in practice and in the literature. In practice, workers differ with respect to the number of skills, skill types, task proficiency, and workload that can be assigned to them. Worker assignment is done at the beginning of a shift (day) and possibly during a shift (day). In most cases, if workers are assigned or reassigned, the state of the orders (at the machines) is considered and worker characteristics, such as task proficiency or the specific qualification of workers, play a role. We examined several labor assignment rules — by means of simulation — in DRC systems within different settings. For instance, we considered homogeneous and heterogeneous labor with respect to task proficiencies, and differences with respect to the number of machines workers can operate.

The remainder of this section discusses future research issues, extending the issues dealt with in Section 1.2 and Section 1.3, but also covering a broader range of issues related to cross training and worker assignment that need to be addressed.

1.4.2 Further research

In order to effectively and efficiently fulfill customer demands in Dual Resource Constrained systems, firms need to tune their labor supply and their demand for labor capacity. While customer orders drive the demand for certain skills, the supply is determined by characteristics of the workforce during the time the demand has to be fulfilled (number of workers, their skills, their working times, their efficiency and effectiveness, etc.). Tuning demand and supply may involve setting appropriate labor assignment rules or, in the longer run, creating more assignment options (labor flexibility) by means of cross training. It may also involve scheduling overtime, hiring additional workers (supply), or applying workload control concepts (leveling demand).

Figure 1.10 shows a small model relating supply and demand for skill capacity. On the supply side, we see that workers all have one or more skills. Cross-training decisions impact the exact number and distribution of skills among workers. Worker availability then further determines the potential skill capacity available in the workforce. This is potential capacity, since a worker can only apply one skill at a time. If someone with skills A and B is

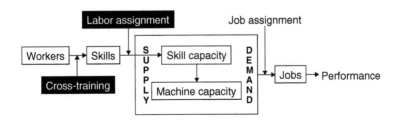

Figure 1.10 A model on demand for and supply of skill capacity.

assigned to a task for which skill A is required, the potential capacity for skill B cannot be used at that moment. This shows that labor assignment actually determines which capacity is allocated (and, therefore, also which potential capacity is not). On the demand side, we see that jobs (customer orders) require machine capacity and skill capacity. Job assignment decisions relate to the possible routing flexibility in the system and alter demand for individual machines (and skills). Most performance measures can be related to attributes of jobs (flow time, quality, dependability, etc.). Finally, the arrow between skill capacity and machine capacity represents the influence of worker capacity on machine capacity in DRC environments. In these systems, actual machine capacity may be limited due to the unavailability of (skilled) workers.

In current research on cross training, the demand for skill capacity is often determined at a high level of aggregation. For instance, the annual demand for a specific machine is estimated and the required capacity for that skill is assumed to be the same. Two assumptions are made here that may go against what is seen in practice and thus requires further research. First, skills are related to machines, meaning that a worker is either able or not able to operate a machine. In practice, we have encountered situations that skills are related to types of orders performed at a machine, or types of tasks performed at a machine (e.g., setting up a machine or keeping it running once it is set up). Second, the demand for machine capacity is set equal to the required demand for skill capacity. In practice, some machines can run unattended for periods of time, enabling workers to attend more than one machine simultaneously. In developing cross-training policies, the demand for skill capacity and its relation with demand for machine capacity should be further investigated.

Even though the supply of skill capacity is determined by characteristics of the workforce during the time the demand has to be fulfilled, researchers have not yet been able to fully deal with these characteristics and the related variations in human performance. For instance, with cross-training decisions, a static team often is considered with respect to its composition, availability of workers, and worker characteristics, such as the efficiency and effectiveness of workers. However, in reality, team composition generally changes over time due to labor attrition, employment of temporary workers, shift schedule changes, etc. In the short term, even absenteeism alters team composition, which may impact team performance. Further, since all humans are unique, there is worker variety implying differences between workers that may impact cross training and labor assignment decisions. Also, qualification levels of workers are often not static. Learning and forgetting takes place, which results in variation in human performance. To a certain extent, the supply of skill capacity thus seems to be volatile. The impact of these issues on cross training and labor assignment decisions requires further research.

Further, in current research, there is a strong focus on minimizing flow time, while other performance measures may be just as important or even more important in practice. Directing attention to due-date performance or performance with respect to quality are the more obvious suggestions, while focusing on labor utilization or skill utilization may indirectly lead to better performance. In general, there is a need for more empirical research and more interdisciplinary research in the fields of cross training and labor assignment. Up to now, a large body of research has investigated hypothetical shops with (too) many assumptions that may not hold in reality, while approaching the topic from an Operations Management point of view and neglecting Human Resources Management issues. For instance, empirical investigations may focus on the rules with respect to important aspects in a cross-training policy that are used in practice. These can be confronted with the rules we found to be attractive. With respect to labor assignment, empirical investigations may provide parameter settings and configurations that bear more resemblance to realistic operating environments. Another topic for empirical research is how to implement and apply labor assignment rules, such as the who-rule, in real industrial situations.

Another interesting direction for further research is to integrally study cross training and labor assignment decisions. Optimal labor assignment rules should be designed for each cross-training configuration developed. If labor assignment rules are fixed in the comparison of cross-training configurations, there may be an interaction effect with the configurations, since it is conceivable that the fixed assignment rules perform differently within each cross-training configuration. In designing optimal labor assignment rules, it may be worthwhile to consider multidimensional and/or dynamic rules. This means that labor assignment rules may be developed that consider more than one dimension (such as the number of skills or the task proficiency of workers) simultaneously. Further, information on the current state or even the "near future" state of the system may be incorporated to get dynamic rules instead of static rules. Dynamic rules frequently provide better operating performance, but they require more information. Again, this raises the question of how to apply the rules in practical situations.

References

Ashkenas, R., Ulrich, D., Jick, T., and Kerr, S. 1995. *The Boundaryless Organization*, Jossey-Bass Publishers, San Francisco.

Austin, W. 1977. Equity theory and social comparison processes. In Suls, J.M. and Miller, R.L. (Eds.), *Social Comparison Processes: Theoretical and Empirical Perspectives*, Hemisphere, Washington, D.C.

Bobrowski P.M. and Park P.S. 1993. An evaluation of labor assignment rules when workers are not perfectly interchangeable. *Journal of Operations Management*, 11, 257–268.

Bokhorst J.A.C., Slomp J., and Gaalman G.J.C. 2004. On the who-rule in Dual Resource Constrained (DRC) manufacturing systems. *International Journal of Production Research*, 42(23), 5049–5074.

Bokhorst J.A.C., Slomp J., and Molleman E. 2004. Development and Evaluation of Cross-Training Policies for Manufacturing Teams. *IIE Transactions*, 36(10), 969–984.

Bokhorst, J.A.C. 2005. *Shop floor design: layout, investments, cross-training and labor allocation*. Ph.D. thesis. Labyrint Publications, Ridderkerk.

Brusco, M.J., and Johns, T.R. 1998. Staffing a multiskilled workforce with varying levels of productivity: an analysis of cross-training policies. *Decision Sciences*, 29, 499–515.

Campbell, G.M. 1999. Cross-utilization of workers whose capabilities differ. *Management Science*, 45(5), 722–732.

Carnall, C.A. 1982. Semi-autonomous work groups and the social structure of the organization. *Journal of Management Studies*, 19, 277–294.

Clark, J. 1993. Full flexibility and self-supervision in an automated factory, in J. Clark (Ed.), *Human Resource Management and Technical Change*, Cornwell Press Ltd, Broughton Gifford, pp. 116–136.

Cordery, J., Sevaston, P., Mueller, W., and Parker, S. 1993. Correlates of employee attitudes toward functional flexibility. *Human Relations*, 46, 705–723.

Fazakerley, G.M. 1976. A research report on the human aspects of group technology and cellular manufacture. *International Journal of Production Research*, 14, 123–134.

Felan, J.T., and Fry, T.D. 2001. Multi-level heterogeneous worker flexibility in a Dual Resource Constrained (DRC) job-shop. *International Journal of Production Research*, 39(14), 3041–3059.

Fry, T.D., Kher, H.V., and Malhotra, M.K. 1995. Managing worker flexibility and attrition in dual resource constrained job shops. *International Journal of Production Research*, 33(8), 2163–2179.

Gargeya, V.B., and Deane, R.H. 1996. Scheduling research in multiple resource constrained job shops: a review and critique. *International Journal of Production Research*, 34(8), 2077–2097.

Hogg, G.L., Maggard, M.J., and Phillips, D.T. 1977. Parallel-channel, Dual Resource Constrained queuing systems with heterogeneous resources. *AIIE Transactions*, 9(4), 352–362.

Hopp, W.J., and Van Oyen, M.P. 2004. Agile workforce evaluation: a framework for cross-training and coordination. *IIE Transactions*, 36, 919940.

Hopp, W.J., Tekin, E., and Van Oyen, M.P. 2004. Benefits of skill chaining in serial production lines with cross-trained workers. Management Science, 50(1), 83–98.

Hottenstein, M.P., and Bowman, S.A. 1998. Cross-training and worker flexibility: a review of DRC system research. *The Journal of High Technology Management Research*, 9(2), 157–174.

Hut, J.A., and Molleman, E. 1998. Empowerment and team development. *Team Performance Management Journal*, 4, 53–66.

Jackson, S.E. 1996. The consequences of diversity in multidisciplinary work teams, in M.A.West (Ed.), *Handbook of Work Group Psychology*, John Wiley & Sons, New York.

Jellison, J., and Arkin, R. 1977. Social comparison of abilities: a self-presentation approach to decision making in groups, in Suls, J.M., and Miller, R.L. (Eds.), *Social Comparison Processes: Theoretical and Empirical Perspectives*, Hemisphere, Washington, D.C., pp. 235–257.

Jordan, W.C., and Graves, S.C. 1995. Principles on the benefits of manufacturing process flexibility. *Management Science*, 41, 577–594.

Kerr, N.L., and Bruun, S.E. 1983. Dispensability of member effort and group motivation losses: free-rider effects. *Journal of Personality and Social Psychology*, 44, 78–94.

Kher, H.V., and Malhotra, M.K. 1994. Acquiring and operationalizing worker flexibility in Dual Resource Constrained job shops with worker transfer delays and learning losses. *Omega*, 22, 521–533.

Latané, B., Williams, K., and Harkins, S. 1979. Many hands make light work: the causes and consequences of social loafing. *Journal of Personality and Social Psychology*, 37, 822–832.

Lau, D.C., and Murnighan, J.K. 1998. Demographic diversity and fault lines: the compositional dynamics of organizational groups. *The Academy of Management Review*, 23, 325–340.

Malhotra, M.K., Fry, T.D., Kher, H.V., and Donohue, J.M. 1993. The impact of learning and labor attrition of worker flexibility in dual resource constrained job shops. *Decision Sciences*, 24(3), 641–663.

Malhotra, M.K., and Kher, H.V. 1994, An evaluation of worker assignment policies in dual resource-constrained job shops with heterogeneous resources and worker transfer delays. *International Journal of Production Research*, 32(5), 1087–1103.

Molleman, E., and Slomp, J. 1999. Functional flexibility and team performance. *International Journal of Production Research*, 37, 1837–1858.

Nelson, R.T. 1967. Labor and machine limited production systems. *Management Science*, 13(9), 648–671.

Nelson, R.T. 1970. A simulation of labor efficiency and centralized assignment in a production model. *Management Science*, 17(2), B-97–B-106.

Nembhard, D.A. 2000. The effects of task complexity and experience on learning and forgetting: a field study, *Human Factors*, 42(2), 272–286.

Norman, B.A., Tharmmaphornphilas, W., Needy, K.L., Bidanda, B., and Warner, R.C. 2002. Worker assignment in cellular manufacturing considering technical and human skills. *International Journal of Production Research*, 40(6), 1479–1492.

Park, P.S., and Bobrowski, P.M. 1989. Job release and labor flexibility in a Dual Resource Constrained job shop. *Journal of Operations Management*, 8, 230–249.

Raaymakers, W.H.M., and Fransoo, J.C. 2000. Identification of aggregate resource and job set characteristics for predicting job set makespan in batch process industries. *International Journal of Production Economics*, 68, 137–149.

Rochette, R., and Sadowski, R.P. 1976. A statistical comparison of the performance of simple dispatching rules for a particular set of job shops. *International Journal of Production Research*, 14(1), 63–75.

Slomp, J. 2000. Multifunctionaliteit van mensen en machines — praktijkcasus, Internal Research Report (in Dutch).

Slomp J., Bokhorst J.A.C., and Molleman E. 2005. Cross-training in a cellular manufacturing environment. *Computers and Industrial Engineering*, 48(3), 609–624.

Stewart, B.D., Webster, D.B., Ahmad, S., and Matson, J.O. 1994. Mathematical models for developing a flexible workforce. *International Journal of Production Economics*, 36, 243–254.

Treleven, M. 1989. A review of dual resource constrained systems research. *IIE Transactions*, 21, 279–287.

Van den Beukel, A.L., and Molleman, E. 1998. Multifunctionality: the driving and constraining forces. *Human Factors and Ergonomics in Manufacturing*, 8, 303–321.

Wilke, H.A.M., and Meertens, R.W. 1994. *Group Performance*. Routledge, London.

chapter 2

Job assignment

Rafael Pastor and Albert Corominas

Contents

2.1 Introduction

As has been pointed out in Chapter 1, workforce multifunctionality is a very important factor for enterprises that wish to be efficient (some discussions of the benefits of a cross-trained workforce can be found, for example, in McCune, (1994). One important step in workforce management consists of assigning a task to each worker, i.e., resolving the job assignment problem.

  Workers are normally classified into different categories with two main characteristics. First, the workers in one category may be totally or partially cross trained, depending on whether they are qualified to carry out the tasks to be assigned or only a subset of those tasks. In addition, in the event of partial cross training, there may be hierarchical multifunctionality (workers can perform the tasks assigned to their category and also the tasks corresponding to all or any of the lower categories) or general multifunctionality (workers can perform the tasks assigned to their category and tasks corresponding to some lower and/or higher categories). Secondly, workers can perform with greater or lesser efficiency tasks for which they are qualified, so we can define the efficiency of the workers belonging to a given category in performing a given type of task with a value between 0 (indicating that they are not qualified to perform that type of task) and 1 (corresponding to the maximum efficiency). One overall means of presenting all of these situations consists of setting out a table with the degrees of efficiency (between 0 and 1) with which a worker in a given category performs a given type of task. The set characteristics provide flexibility for the workforce but involve increasingly complicated management problems (as observed by Kozlowski, 1996), since it must be decided which specific task is to be performed by each worker. This problem is known as the *job assignment problem.*

  The job assignment problem consists of assigning a type of task to each staff member present at the workplace during each period of the planning horizon. Normally, the assignment of tasks is made once a schedule has been assigned to each worker. However, it is also possible that the assignment of tasks to workers will determine their schedule; in this case, there is a set of additional constraints that must be observed, allowing feasible work schedules to be created.

  The simplest example of the job assignment problem is found when the number of tasks to be assigned and the number of available workers are the same, and when assignment to a period t is independent of the assignment to other periods of the planning horizon (whether prior and/or subsequent to t). Nevertheless, there are problems that are much more complex; for example, if we consider that the workers may be partially cross trained and

Table 2.1 Configuration of the 4 × 100-m Medley Relay

	Freestyle	Backstroke	Butterfly	Breaststroke
N1	48.94	55.22	52.02	1:00.21
N2	48.23	53.60	53.18	1:01.19
N3	47.84	54.94	51.22	59.30
N4	49.24	54.06	50.76	1:03.22

Note: Time for each swimmer style.

	Freestyle	Backstroke	Butterfly	Breaststroke
N1				1:00.21
N2		53.60		
N3	47.84			
N4			50.76	

Note: Optimum assignment.

have nonuniform working efficiency, or that the tasks may take up only a part of the period and, consequently, must be assigned along with others. In general, this is a multicriteria decision based on a solution evaluation function that can take into account various factors: financial concerns, satisfaction of workers' preferences for certain tasks over others, and the target of bringing the percentage of workers' dedication to different types of tasks as close as possible to certain ideal values established in advance. There may also be constraints, e.g., the number of consecutive periods dedicated to a single type of task may have upper and/or lower bounds.

We will now look at several examples that will allow us to introduce concepts and set out how some of the best-known and simplest job assignment problems are solved.

Suppose we have a swim team made up of four swimmers that must configure a 4 × 100-m medley relay. For each swimmer-style pair, we know the time for the best personal record; naturally we must take into account that some swimmers are faster in one style than in another (Table 2.1). The aim is to assign to each swimmer the style with which they are to swim in the relay race so that the assignment gives the best possible time (Table 2.1).

Assume we have a large store selling apparel with n workers who must perform n tasks, such as working the cash register, serving customers, restocking articles and placing them correctly on shelves, etc. We need to assign tasks, one for each period of the planning horizon in such a way, for example, that the time dedicated by each worker to the different types of tasks is as close as possible to certain ideal percentages of dedication (in this case, assignment during a period is not independent of assignments made for preceding periods). The ideal percentages of dedication may be the same for all workers, which would be conducive to maintaining workforce cross training, or they may be different for some workers, which would allow those workers to be trained intensively for one or more tasks.

In the two examples set out (ignoring, in the second case, the implications of assignments made in periods previous to the present one), the aim is to solve the best-known and simplest job assignment problem, the *assignment problem*, which consists of assigning n tasks to n workers while optimizing a given objective function. The assignment problem is one of the most famous and widely studied problems in combinatorial optimization, and powerful and specific procedures have been developed for solving it. Given an assignment matrix C, with n rows and n columns, where c_{ij} is the "cost" of assigning row i to column j, the problem consists of assigning each row to a different column in such a way that the sum of selected costs is minimum. The cost c_{ij} can be the time corresponding to the pair "swimmer i – style j," the difference between the percentages of time dedicated by each worker to the different tasks and their ideal percentages of dedication or, during the examination period, the advisability of having proctor i watch over the examination j. Dell'Amico and Toth (2000) present a survey of the techniques for solving the assignment problem and describe the most efficient codes that can be easily obtained.

Now, suppose we have a workplace with a production system comprising a product assembly line made up of n workstations in series, where workers with different efficiencies perform manual tasks. The product moves to the different workstations on a conveyor belt. The belt moves at the maximum speed that will allow those workers taking the longest to perform their tasks to have sufficient time to complete the process. The objective is to assign n workers to n workstations so throughput of the assembly line be as high as possible.

The above example is once again an assignment problem, although with a different objective, namely, to minimize the value of the highest cost c_{ij} resulting for the assignment carried out. This problem is known as the *linear bottleneck assignment problem* and its objective consists of minimizing the maximum inefficiency resulting from the assignment: The person with the maximum inefficiency performing some job would create a "bottleneck," so the aim is to minimize the maximum inefficiency (minimize the bottleneck). In more formal terms, the linear bottleneck problem consists of assigning each row to one column and each column to one row so that the maximal of coefficients c_{ij} selected in the solution is minimal. Carpaneto and Toth (1981) present different algorithms for solving the linear bottleneck assignment problem efficiently. In any event, this problem can also be solved as a sequence of assignment problems. Given the optimal solution of the p-th assignment problem, we can determine the element $c_{ij}^{+,p}$ with the highest value resulting from that assignment; then all the elements with a value greater than or equal to $c_{ij}^{+,p}$ are made equal to ∞ and the $(p+1)$th assignment problem is solved. This procedure is repeated until the sth assignment problem does not have a solution with a finite maximum value, so the optimal assignment is the one obtained for the $(s-1)$th assignment problem and the optimal value of the linear bottleneck assignment problem is $c_{ij}^{+,s-1}$.

Lastly, assume that we have a hospital analysis laboratory in which the following set of tasks must be performed each working day: take blood samples (both in external surgeries and in the hospital) and perform analyses of blood, urine, saliva, fertility, etc. Normally all those tasks, or the majority of them, share the characteristic that their duration is shorter than the duration of the working day, so it is possible to assign more than one task to each worker on each working day. Obviously, we must observe the capacity constraint (in addition to other possible constraints, such as not assigning tasks that overlap in time), i.e., the condition that the total time for the tasks assigned to a worker must be shorter than the length of their working day.

The above problem can be considered a particular case of one of the best-known problems in operations research, namely the *generalized assignment problem*, which has a large number of applications. Given an assignment matrix C, where c_{ij} is the "cost" of assigning worker i to task j, and the capacity of each worker, the problem consists of assignment of one or more tasks to each worker in such a way that the sum of selected costs is minimum and the capacity constraints are respected. A survey of algorithms for the generalized assignment problem is presented by Cattrysse and Van Wassenhove (1992).

2.1.1 Chapter summary

In Section 2.2, we introduce variations on the job assignment problem. Specifically, in Section 2.2.1 we propose a basic classification as a framework for this type of problem; several assignment evaluation functions are discussed in Section 2.2.2; in Section 2.2.3, we set out constraints that might be encountered when assigning tasks; further practical aspects are discussed in Section 2.2.4; and in Section 2.2.5, we set out several cases from specialized literature. Lastly, future research needs and opportunities are discussed in Section 2.3.

2.2 Variations on the job assignment problem

There are a great many variations on the job assignment problem concerning the objective function, the constraints to be taken into account, and the elements that define the specific problem to be solved, so it is impossible to provide a single procedure for solving all the cases that may arise. As a result, different solution tools and procedures have been developed.

Nevertheless, for all cases and for the whole of the planning horizon, it is normal to assume that we know from the outset which workers are available and their characteristics (multifunctionality, degrees of efficiency, preferences for certain tasks over others, etc.). In addition, it is assumed that we have a defined list of tasks to be performed along with their characteristics (capacity required, duration, etc.).

In the following sections we propose a basic classification of the job assignment problem and its variations and discuss how they have been solved in the specialized literature. Having set out the classification, the variations are divided into three types, depending on whether they concern the evaluation function of the solutions, the constraints to be taken into account in the assignment, or other practical aspects that define the problem. Lastly, we set out several cases that have been presented in the specialized literature.

2.2.1   Classification

In spite of the differences between the job assignment problems that may be defined and arise in actual business situations, we can posit a basic structure into which all such problems will fit. The classification that we present takes into account only two dimensions: (1) it considers whether or not the assignment of tasks in a period is related to the assignment of tasks in the other planning horizon periods, and (2) it considers the number of workers and the number of tasks to be allocated.

2.2.1.1   Independent or related periods

In many cases, the job assignment problem concerns the assignment of tasks to workers for just one period. Nevertheless, in order to carry out the assignment, it is sometimes necessary to have access to and consider information relating to several periods preceding and/or subsequent to the planning period.

- **Independent Periods**: In this case, we assume that the assignment of tasks made in one period of the planning horizon does not influence and is not influenced by assignments made in other periods. This is the simplest case and it consists of resolving a separate assignment problem for each period of the horizon.
- **Related Periods**: In this case, the assignment made in a period depends on the assignments made in preceding and/or subsequent periods, whether on the basis of an objective function (e.g., when the aim is for the time dedicated by workers to the different types of tasks to converge with their ideal percentages of dedication) or on the basis of constraints (e.g., when the number of consecutive periods performing the same type of task has an upper or lower bound).

At the time when the problem is solved, it will also be possible, optionally, to make a simultaneous assignment of tasks in several periods. In other words, tasks may be assigned simultaneously for all the periods of the planning horizon, making it possible to consider the impact of the assignments being made. Alternatively, tasks may be assigned period by period, taking into account the assignments made in the preceding periods and, insofar as possible, the impact of the assignment made for the current period on future assignments.

Taking into account the impact of decisions taken in other periods makes assignment in a single period more difficult. In any event, although simultaneous assignment of tasks to several periods is undoubtedly the most comprehensive, it is also the most difficult to solve.

2.2.1.2 *Number of workers and tasks to be allocated*

The simplest situation is found when the number of tasks to be assigned and the number of available workers are equal and a unique task must be assigned to each worker. Of course the number of workers available and the number of tasks to be allocated, in fact, can be different.

- **One Task Per Worker:** In this case, we consider that each worker is to be assigned one task per period and that the number of workers and the number of tasks are the same, meaning that each task must be assigned to one worker. Depending on the goal that we wish to attain, we can consider two possibilities:
  - *Minimizing the total cost of the assignment:* i.e., making the sum of selected costs minimum (or maximizing the total benefit of it). This is the *assignment problem*.
  - *Minimizing the highest cost resulting from the assignment:* i.e., making the element with the highest value as low as possible. This is the *linear bottleneck assignment problem*.

- **Several Tasks to Each Worker:** In this case, we have several tasks that could potentially be assigned to each worker because their duration is less than the planning horizon. Thus, the problem consists of assigning a set of tasks to each worker, while respecting certain constraints. There are three possibilities that we can encounter, depending on the constraints to be taken into account in the assignment:
  - *With restriction of capacity:* For example, we are bound by the condition that the sum of times of the tasks assigned to a worker must be less than the duration of their workday or that the number of tasks assigned to a worker must be lower than an upper limit value. This is the *generalized assignment problem*.
  - *With conflicts between tasks*: In this case, we have a table of conflicts between the types of tasks to be performed. For example, it might be unfeasible to perform two different tasks on the same workday because stopping performance of one task to perform the other might involve substantial preparation or travel time. In addition, tasks that are programmed to be performed at the same time are obviously in conflict with each other. Also, we may find constraints on capacity, in respect to either time or the number of tasks assigned.
  - *With scheduling of activities:* In this case, which includes working with certain tasks that have already been scheduled, when assigning the tasks to be performed to the workers, we must also schedule their performance in the course of the workday. In other words, we must assign the tasks to the workers and specify at what point during the workday the tasks are to be performed. Obviously, the assignment and scheduling of tasks must abide

by a set of constraints, such as an upper and lower limit for the time at which the tasks must be commenced (blood tests cannot be carried out before the samples for testing arrive from another hospital), or an order for performance of the tasks (blood tests cannot be carried out before samples have been taken at the hospital).

- **Several Tasks to Several Workers:** We assume that there is a set of tasks that must be performed and we must establish the composition of a team to perform them. Suppose that we wish to take a research expedition to the Amazon rainforest. We assume that the different research tasks to be performed and the tasks of organization and logistic support required for the expedition are known, along with the group of people who can be hired for the expedition and the tasks that each of them can perform. The objective consists of defining a team of people to take part in the expedition in such a way that all the required tasks can be performed at the lowest total cost. The cost function will include the wages of the individuals hired and a cost depending on the duration of the expedition. The definition of the group of workers making up the team implicitly involves the assignment of tasks to the members of that team.

2.2.2 *Evaluation function*

There may be several ways in which an assignment of tasks to workers can be evaluated and it is important to take this factor into account, since it may make finding a satisfactory solution for the problem more or less difficult. For instance, the cost or benefit of assigning a worker to a task may be independent of the other assignments, but sometimes it is not (e.g., we know that the performance of a player on a football team depends on the players who are playing — or assigned to — the other positions) and, in this case, evaluating the function and solving the problem are generally much more difficult.

In addition, the evaluation function may provide for one or more criteria to be taken into account (e.g., the cost and workers' preferences for certain tasks over others). Two possible ways of handling the job assignment problem when there is a multiobjective evaluation function are: (1) working with an evaluation function that includes all the objectives, which are weighted on the basis of user-defined parameters; or (2) treating objectives according to a hierarchy, either using weighting parameters that are very different for each of the different objectives, or solving problems in series so that each solution ensures the best value found for higher-ranked objectives.

Among the different ways of evaluating an assignment (i.e., a solution), we may consider those set out below. Obviously, there may be other means of evaluation, depending on the particular characteristics of the enterprise in question.

2.2.2.1 Satisfaction of the desired capacity for the different types of tasks

When there is a lack of capacity for the different types of tasks, it is normal to assume that the tasks that have not been covered may be performed by workers not belonging to the staff, and thus we can associate the lack of capacity with a cost (which is variable, depending on the type of task not covered) to be minimized. However, it is increasingly common, particularly in service enterprises, to consider that a lack of capacity results in the level of service provided to clients being lower than desired.

When a lack of capacity can be covered by hiring personnel not belonging to the staff, we can easily accept a cost that is linear with the lack of capacity attained. On the other hand, when the lack of capacity lowers the level of service provided, the result of the lack of capacity should be nonlinear. If we consider that the tasks to be performed all have the same importance, it is not the same to have a lack of capacity of 2 units in a task as having a lack of capacity of 1 unit in two tasks; if both require the same capacity, the level of service in the first instance drops by more than twice as much as in the second instance. Furthermore, it is not the same to have a lack of capacity of 1 unit when 2 units are needed (i.e., a 50% lack of capacity) or to have a lack of 1 unit of capacity when 10 units are needed (i.e., only a 10% lack of capacity). Thus, besides considering a nonlinear result of a lack of capacity, some authors work with the lack of capacity relative to the required capacity.

When there is excess capacity, arguments similar to those set out for lack of capacity can be applied.

In addition, the satisfaction of the desired capacity for different types of tasks can have differing degrees of importance. This characteristic can be incorporated into the evaluation function, weighting the importance of satisfying the desired capacity for different types of tasks on the basis of user-defined parameters.

2.2.2.2 Satisfaction of workers' preferences for tasks or adaptation of the assignment

The Human Resources Departments of enterprises are increasingly paying attention to the concerns and requests of their employees, and the preferences that workers might express for performing one type of task or another are being increasingly taken into account as an evaluation criterion.

It is also possible to consider the appropriateness of an assignment when workers of different ranks can perform different types of tasks. For example, in a major retail outlet, department managers can — in addition to performing the tasks pertaining to their job — perform other tasks, such as serving customers, restocking, or even cleaning or tidying the store. Nevertheless, assigning the task of cleaning to a manager, though possible, would certainly be considered highly inappropriate. In many cases, the greater or lesser appropriateness of an assignment is very closely related to the efficiency of the different categories of workers in performing different types of tasks. However, there are cases in which the appropriateness of an assignment and efficiency in performing the task are independent of each other: A clerk hired on a temporary basis could perhaps take charge of staff payrolls with the same or greater efficiency in comparison with a clerk on staff, but due to the confidentiality of the information, it would be more appropriate for payrolls to be managed by a someone on staff.

2.2.2.3   *Satisfaction of the ideal percentages of dedication of workers to different types of tasks*

Workforce cross training is a highly valued factor in all enterprises. In order to maintain multifunctionality, it is advisable for workers to be assigned periodically to all tasks that they can perform. In addition, there are certain tasks — owing to the amount of physical or mental exertion that they involve or merely due to the fact that they are not particularly pleasant — that must be assigned to workers in the fairest and most balanced way possible.

On the basis of this premise, we can define ideal percentages of dedication for each worker and for the different types of tasks that they can perform. One way of evaluating the quality of the solutions can be based on the greater or lesser degree of proximity between real and ideal rates of dedication (result of past assignments and present assignments). Obviously, in order to calculate the actual rates of dedication, we need to have access to a record of past assignments.

Ideal rates must be established *a priori* on the basis of a medium-term plan that takes into account both the tasks to be performed and the staff available during the planning horizon. These ideal percentages of dedication can be established as the same for all workers, thereby helping to maintain workforce cross training and favoring the fair and balanced assignment of all types of tasks. Nevertheless, differing percentages can also be defined, even if only on a temporary basis, to facilitate the intensive training of a worker in one or more specific tasks. This strategy is particularly useful in special workplaces employing people with mental or physical handicaps, since such workplaces, on the basis of performance of specific types of tasks, train and teach the handicapped employees to improve and overcome their handicaps.

2.2.3   *Constraints on assignment*

The assignment of workers to tasks may require compliance with a series of conditions that may be considered *hard* or *soft*. While a *hard* condition must be satisfied in full for a solution to be considered feasible, a *soft* condition can allow a certain degree of deviation or noncompliance, albeit with heavy penalization in the objective function. In other words, *soft* conditions can allow a certain degree of noncompliance, but in order to ensure that such noncompliance is minimized, it is subject to heavy, nonlinear penalization.

It is normal for a great many of the constraints found in enterprises to be considered *soft*, since in reality industrial and service centers work with the resources that they have at their disposal, whether or not the constraints are effectively observed.

The constraints that are the most difficult to violate are those relating to the multifunctionality of the workers. If a category of workers c is not qualified to perform a type of task k, the assignment cannot be made. If this assignment were possible on an exceptional basis, it might indicate that the category c can perform tasks of type k, although with a greater or lesser penalization. While, in the first case, the constraint is considered to be *hard*, in the second case, the constraint is considered to be *soft*.

Capacity constraints that arise when tasks have a duration that is shorter than the planning period can be considered to be *hard* if the sum of the times of the tasks assigned to a worker must be shorter than their workday, or *soft* if the number of tasks assigned must be lower than a given maximum value.

Traditionally, the total working capacity assigned to each type of task (which results from the number of workers and their efficiencies) is required to be greater than or equal to given values, which depend on the desired service level. However, one must distinguish between the minimum service level and the target service level, and this distinction leads us to consider minimum capacities as lower bound constraints and desired capacities as targets. This minimum capacity, which must be respected, can be associated with a minimum presence, without which the industrial or commercial center cannot open. Therefore, it is a hard constraint. For example, in a shop with two floors, it might be necessary for four workers to be present: one of them at the cash register, two on the second floor (to prevent theft and replenish stocks, or attend to unforeseen circumstances on the first floor), and another on the first floor. If we do not have a minimum of four workers, we can consider that the commercial outlet cannot open.

In some case, assignment of tasks to workers should take into account the immediate history of tasks that they have already performed, so that, for example, we avoid having the same worker perform consecutive tasks requiring intense physical or mental effort, or tasks that are considered unpleasant. Constraints of this type might include, for example:

1. Lower and upper bounds on the number of consecutive periods performing the same type of task. For example, at a chemical company the amount of time dedicated to a hazardous task that involves entering a white room might be limited. Because the workers entering the white room must be decontaminated, they can be asked to perform the task for a given time, but because the task is hazardous, it can only be assigned for a given maximum time.

2. Minimum number of periods without performing tasks of a given type before performing that type of task again. After having performed a

mentally or physically tiring task, the worker should rest for a minimum number of consecutive periods, e.g., operating the cash register in a shop or handling heavy products.

At times the assignment of tasks will determine the sehedule of workers, so the assignment must respect a set of *hard* constraints allowing admissible working schedules to be generated. For example, the start and finish of the working day must fall between certain given values, the total number of hours is fixed (or limited), or allowance must be made for rest periods or meal periods with a given duration and at reasonable times. When the assignment of tasks includes scheduling, there are additional constraints that must be respected, e.g., earliest and latest dates for commencing tasks and/ or a preestablished order of performance. In other cases, the assignment must respect a set of *hard* constraints that prevent incompatible tasks from being assigned to the same worker.

Lastly, we must take into account other *hard* constraints that arise as a result of how the problem is defined. For example, when tasks occupy a period, all workers who are assigned the task must be assigned only one task and all tasks assigned must be assigned to only one worker.

2.2.4   Other practical aspects

The job assignment problems that we can find in industry and service enterprises have a great many variations in the elements and characteristics that define them. We will now look at various other practical aspects that have not been dealt with in detail previously.

2.2.4.1   Deterministic or random data

So far, we have assumed that all the data are deterministic and known from the outset. However, the number of tasks of each type to be performed, the types of tasks to be performed (particularly in service enterprises), and even the duration of tasks can be considered to be either deterministic or random, according to different laws of probability. The same can be said in respect of the workers present at the workplace during a given period, since it is also possible to take absenteeism from work into account. In all these cases, one possible approach is simulation: We solve several job assignment problems on the basis of different scenarios, analyze the results obtained, and take them into account in the final process of assignment.

2.2.4.2   Required capacity for different types of tasks

The required capacity for the different types of tasks may be homogeneous or heterogeneous for the various periods comprising the planning horizon. Homogeneity can allow us to make an assignment in *standard* periods, which can be repeated throughout the planning horizon and, in any event, means that one of the main parameters is stable, thereby facilitating the solution of the problem. Heterogeneity for periods may be found in two aspects: in the total number of types of tasks to be assigned and in the mix of the tasks, i.e., the need for each type of task in the different periods.

2.2.4.3 *Duration of the task in the period*

As stated above, the simplest case is when tasks are performed throughout the period being planned, i.e., when their duration is the same as that of the planning period. This situation might arise when it is decided at the start of a workday which site to send a construction worker to for that day, on the basis of the need for manpower and the skills (multifunctionality) and efficiency of the worker.

However, we can also consider tasks that take up only a part of the period and, therefore, can be assigned along with other tasks. For example, at a small supermarket there might be periods when the cash register is idle, meaning that the worker assigned to that task can perform other tasks, such as stocking shelves in the aisles nearest the cash register. However, there are other tasks that cannot be assigned to this worker because they cannot be abandoned immediately to perform the main task, such as straightening up the storeroom or helping out in another section of the supermarket. In an office, the person assigned to answering phone calls can, if calls are not continuous, supplement their task by writing up reports or letters on a computer; however, they cannot, for example, take notes at a meeting, since they would not be able to leave that task when required to perform their main task (answering the phone).

In addition, there might be tasks that simply do not take up the whole period and when they are finished, another task can be performed. In a supermarket, a worker can clean the windows and, when they have completed that task, they can help to stock shelves.

The tasks that do not take up the whole planning period can be classified as follows:

- *"Main" tasks performed at intervals*, which are to be assigned along with one or more other tasks considered to be compatible, which are secondary and can also be interrupted. For example, answering the phone.
- *"Main" tasks that are performed continuously* and once completed allow the worker to perform one or more other compatible tasks, which can either be continuous or performed at intervals. For example, cleaning the windows at a supermarket.

2.2.4.4 *Duration of the period*

The duration of the period is a factor to be taken into account when the job assignment problem must be solved for all the periods comprising the planning horizon. For example, the assignment of tasks for a week, divided into half-hour periods, at a workplace that operates Monday through Saturday from 8:00 a.m. to 10:00 p.m., requires the solution of 168 job assignment problems (one for each half-hour). However, if the period is 1 day, then only

6 job assignment problems need to be solved, which can require up to 28 times the calculation time required for the previous case.

2.2.5 *Problems from the literature*

We will now look briefly at certain problems from the specialized literature on the job assignment problem, which will point up most of the variations that we have mentioned in this section. Each problem is set out in a section with a title that is meant to identify one of the main characteristics of the problem examined.

2.2.5.1 *Partial multifunctionality with unequal efficiencies*

Campbell and Diaby (2002) present a multidepartment, labor-intensive service environment for allocating partially cross-trained workers, such as that faced by hospital nurses. The cross-trained workers may have different levels of qualification in the different departments and the authors consider a particular case in which the worker efficiencies are 100% of the fully qualified worker.

The paper deals with the assignment of partially cross-trained workers to multiple departments at the beginning of a shift, and it is assumed that each worker will perform the same task during the whole shift. The aim is to maximize a separable and additive objective function made up of the sum of nonlinear (but concave) departmental objective functions that depend on the labor assigned to each department.

An assignment heuristic algorithm is developed and evaluated to solve the problem, which is based on a linear assignment approximation that takes advantage of the special structure of the allocation problem. The computational experiment shows that the assignment heuristic outperforms two classical heuristics across a variety of problem characteristics, that the heuristic's solutions are nearly optimal, and that the heuristic's performance is found to be fairly robust across a variety of experimental factors; moreover, the calculation times are very short.

2.2.5.2 *Satisfying percentages of dedication and constraints throughout a set of interdependent periods*

Corominas et al. (2006) deals with the assignment of tasks to the members of the cross-trained staff of a service company during each period t (1 hour) into which the planning horizon (1 week) is divided. The company is a retail chain that sells clothes and, at present, has nearly 800 shops in 15 countries and more than 30,000 employees.

It is assumed that each shop's staff is completely cross trained and that worker efficiencies are equal for all. Moreover, for a given period t, the set of workers present and the working capacity necessary for each type of task is known and the number of available workers is equal to the number of tasks. There are additional *soft* constraints required by the service company: (1) lower and upper bounds on the number of consecutive periods dedicated by workers to the same type of task, and (2) a minimum number of periods that must pass before a worker can perform the same type of task again. The objective is that the percentage of working time dedicated by each worker i (during a determined time interval) to each type of task k be the same for all workers and as close as possible to reference values (PT_{ik}).

The procedure consists of solving a sequence of assignment problems. One is solved for each period t in the planning horizon, in chronological order. The elements of the assignment matrixes reflect, as a benefit or as a cost, respectively, the fact that the assignment implies an approximation to or a deviation from PT_{ik} values and if the conditions specified by the company are met:

1. Associate a bonus with the assignment when worker i has not been performing tasks of type k for the established minimum number of consecutive periods.
2. Penalize the assignment when worker i has achieved the maximum number of consecutive periods of performing tasks of type k.
3. Penalize the fact that worker i has not achieved the minimum number of consecutive periods without performing tasks of type k.
4. Penalize the assignment of a task of type k to worker i, if the worker has a period of programmed absence that prevents him from reaching the established minimum number of consecutive periods of performing tasks of type k.
5. Associate a bonus with (or penalize) the contribution of the assignment to approximating (or deviating) from the ideal PT_{ik} values.

The assignment problems are solved using the algorithm set out in Jonker and Volgenant (1987). The obtained results are satisfactory and the proposed approach constitutes a potential tool for assigning tasks to a cross-trained workforce in the service industry. The solutions fulfill the constraints and, after a transient phase, the percentages of dedication of the workers to the different types of tasks approaches satisfactorily the ideal values; moreover, the calculation times are very short.

2.2.5.3 *Fair assignment satisfying workers' preferences*

Martí et al. (2000) present an algorithm for assigning teaching assistants to proctor final exams at a university, taking into account proctors' preferences and ensuring that the final assignment is fair.

Several types of constraints must be fulfilled in the assignment. Each exam must be proctored by a specified number of teaching assistants; in other words, it is necessary to ensure a specific presence for each task (exam). A teaching assistant cannot exceed their maximum number of proctor hours; in other words, there is a constraint on capacity for the total time of the assigned tasks. A teaching assistant cannot proctor more than one exam at the same time; in other words, there is a conflict between tasks that are performed during the same time interval. Finally, a teaching assistant cannot proctor exams during periods that conflict with their own exams; in other words, workers are partially multifunctional. Teaching assistants have preference for some exams, which reflect their desire for proctoring on a given day or avoiding certain days. Another criterion that is considered is to assign exams so the workload is evenly distributed among the teaching assistants (unfair workloads are likely to generate conflicts). The objective function consists of maximizing a weighted function of the preferences (the sum for all teaching assistants of preferences for the exams assigned) and of the workload fairness (defined as the minimum teaching assistant utilization: the fraction of the available hours that a teaching assistant is assigned to proctor exams).

The allocation problem is modeled as a mixed-integer program, but the authors also develop a heuristic procedure based on the scatter search methodology. The data used for the computational experiment correspond to instances of the proctor assignment problems at a university in Barcelona (Spain). The results of the heuristic procedure are compared with assignments found by solving the mathematical program with Cplex© (standard optimization software). With short calculation time, the scatter search provides a set of good solutions (although of lower quality than those found by Cplex© in a limited calculation time) and also provides a set of high-quality solutions from which the decision-maker can choose one to implement.

2.2.5.4 *Number of workers and tasks to be allocated and appropriateness of the assignment*

Corominas et al. (2005) deals with the assignment of a type of task to each member of a partially cross-trained staff in a period of the planning horizon. The approach works regardless of whether the number of workers is greater than, equal to, or less than the number of tasks to be performed, since it generates, as required, *units of shortage* and/or *units of surplus* that are assigned to the workers.

It is assumed that the set of workers and the minimum and desired capacities are known; in addition, the worker efficiencies are equal to 100% of the fully qualified workers and the appropriateness of the allocation is considered. The aim is for the resulting number of workers for each type of task to be not less than a given lower bound and as close as possible to another given ideal value. Specifically, the objectives considered are: (1) to minimize the relative shortages of actual capacities vs. the desired ones for each type of task, and to distribute them homogeneously; (2) to minimize the relative surpluses of actual capacities over desired ones and to distribute them homogeneously among the different types of tasks (i.e., to approach the target service level); and (3) to maximize the appropriateness of the allocation of employees of the staff to the different types of tasks. The objective function is defined to avoid, wherever possible, unfeasible solutions in which the lower bound constraints on the capacities are not fulfilled (i.e., to try to guarantee the minimum service level); thus, the minimum capacity constraint is considered to be *soft*.

The allocation problem is modeled as a nonlinear mixed integer program and it is transformed and solved as a minimum cost flow problem on a network with lower and upper bounds of the arc capacities and with appropriate unit transport costs. A computational experiment is performed and the calculation times are very short. These results allow the authors to emphasize the feasibility of the proposed allocation procedure to be used in the resolution of problems of industrial dimensions, even inside a heuristic or a local optimization procedure to assign schedules.

2.2.5.5 Consideration of switching costs

Batta et al. (2005) studies the problem of balancing staffing and switching costs for cross-trained servers in a service center with multiple groups of customers and time-dependent service demand.

The authors consider service centers that have the following features. The service center provides service to multiple service requests and the different types of requests are classed into multiple groups. The demand of each group is time-dependent and has its own distinct pattern, which is known. A special skill is required to serve a specific group of requests. The workers can serve multiple groups of service requests (they are cross trained) at different service rates. It is assumed that the minimum number of fully qualified workers needed to cover service requests is given and, for each server type, the number of workers remains invariable during the whole planning horizon. Finally, the workers can be switched from serving one group of requests to another at the beginning of a period. The switch takes time and incurs a switching cost (for example, the transfer cost); moreover, the staffing (hiring) cost of a specific type of worker is known. The objective is to minimize the total staffing and switching costs subject to service level constraints.

In order to solve the problem, two column generation heuristics are developed and their performances are tested for different data sets and parameter settings (for example, servers' service capability, switching times, and switching costs). From the results, they concluded that the heuristics perform well with nonzero switching cost and, for problems with zero switching cost, they developed a better solution procedure. Furthermore, an extension of the model is presented, which includes the possibility of a service shortage (specifically, the objective is to minimize the sum of staffing cost, switching cost, and penalty cost due to server shortage). The computational experiments show that the heuristics also perform well.

2.2.5.6 Design of an optimum workforce skill mix

When there are different types of tasks to be performed, workers must be cross trained if a suitable staff is to be maintained. However, training workers to make them cross trained in various skill types increases costs and may reduce service efficiency. Managers should balance a reduction in customer waiting time with high service costs and possibly reduced server efficiency due to cross training. In order to determine the workforce skill mix, job assignment problems need to be solved.

Agnihothri et al. (2003) deals with a field service system with two types of jobs and a fixed number of workers. The problem consists of finding the optimal mix of dedicated workers (specialized in only one skill type) and cross-trained workers (with both skill types), with the aim of minimizing the sum of average service costs and customer delay costs per unit time.

The problem is situated in the field service operations of a leading supplier of capital equipment in the electronics industry. At present, the enterprise has installed over 16,000 machines, which are maintained by an organization of more than 200 engineers. Machine failures can be aggregated into two major categories, electrical and mechanical, and within each category the engineers need to be familiar with some 12 specific tasks in order to be able to remedy any failure. Initially, each dedicated engineer is trained in one of these two categories for some set of machine types, although their training is limited to only a subset of these specific tasks. Training time depends on the number of specific tasks and types of machines they are trained for, and varies between 3 months and 1 year. Thus, the cost of cross training is quite significant. In addition, it is considered that the efficiency of the flexible engineer is inferior to that of a dedicated engineer. In addition, for the service systems studied, cost of equipment is very high, resulting in a higher delay cost per unit time. The aim of this company is to establish the appropriate degree of cross training so as to balance the cost of training and the cost of machine downtimes (i.e., to minimize the sum of average service costs and customer delay costs per unit time).

The authors use simulations to model the problem and investigate the impact of various system parameters, such as the total number of workers, worker utilization, the coefficient of variation of service time, and the efficiency of cross-trained workers on the decision of the optimum workforce mix of flexible and dedicated engineers (i.e., on the optimal workforce mix). Final recommendations are presented.

In a later work, Agnihothri and Mishra (2004) studies a field service system with three job types requiring three different skills and a fixed number of workers. Each worker has a primary skill and up to two secondary skills, which is a managerial decision (thus, each worker can be dedicated to one job type, or cross trained to serve two or three job types). Additionally, worker–job mismatch with the implications for travel and varying efficiency in secondary skills are considered. The authors evaluated the tradeoff between the cost of cross-trained workers and the cost of downtime, using a queuing framework and simulation, and they summarized some conclusions on the number of workers cross trained in secondary skills, the number of secondary skills each worker should have, and efficiency in each secondary skill.

Brusco and Johns (1998) develop an integer linear programming model for evaluating the effects of cross-training configurations on workforce staffing costs. The objective is to minimize workforce staffing costs subject to the satisfaction of minimum labor requirements across a planning horizon of a single work shift. The cross-training configurations, which include different levels of associated productivity and labor requirement forecasts, were established based on interviews and data obtained from the maintenance service operations at a large U.S. paper mill. The results indicate that asymmetric cross-training structures are particularly useful.

Finally, Slomp and Molleman (2002) investigates the impact of four types of cross-training policies on team performance, under fluctuating conditions (absenteeism or change in demand). This work deals with the issue of "who among the members of a team should be cross trained next and for which task." They take into account measures indicating the team's effectiveness and efficiency, and psychological, social, and organizational aspects of team functioning.

A hierarchical heuristic procedure for the assignment of tasks to workers is designed, which is used as a tool for investigating the effect of cross-training policies on the performance of a team. In the first step, the load of the bottleneck worker is minimized by solving a linear programming. In the second step, the workload is shared among the workers. The third step tries to minimize the number of newly used qualifications given the assignment after the second step. The heuristic procedure is applied in an experimental study and the four cross-training policies are compared according to their effects on the following performance measures: the average workload of the bottleneck worker, the average utilization level of workers, the standard deviation of the workload among workers, the relative use of qualification, and the average number of newly used qualifications in case of fluctuations. The results show that a worker-oriented cross-training policy (to spread multifunctionality evenly among employees) performs well, but the results also show a diminishing positive effect of expanding the level of labor flexibility: a fully cross-trained workforce, in many situations, is not desirable.

2.3 *Future research needs and opportunities*

The degree of variation of the job assignment problem is enormous, and new possibilities arise continually that are being included in collective agreements and the ways in which the working time of the employees of enterprises is organized. Consequently, future research should focus on solving actual industrial problems as an opportunity for the improvement of enterprises.

Rodríguez (2006) introduces a detailed classification of job assignment problems and shows that a wide variety of them may be modeled and solved by means of mixed integer linear programming. In this work, the author points out the possibility that the assignment of tasks has some impact on the efficiency of workers. In other words, there may be a positive (or negative) correlation in efficiency when two particular workers are assigned to perform the same type of task.

Campbell and Diaby (2002) suggest for future research the possibility of considering the reassignment of tasks to workers within shifts. In this case, one evaluation criterion might be that the new assignment must be as similar as possible to the previous one. It must involve the lowest possible number of changes, or the changes must affect the tasks that are least sensitive to those changes.

Some authors examine the possibility of finding the optimal number of workers with different cross training (i.e., the makeup of the staff) that minimizes the system-related costs.

As a subject of unquestionable interest for future research, there is the idea of designing work systems taking into account lower-level decisions, such as the assignment of tasks to workers, when higher-level decisions are made.

Lastly, one element that is steadily gaining importance in industry is randomness in the value of the data, in respect of both tasks (type and number of tasks, as well as their duration) and workers (absenteeism and time taken to perform tasks).

Acknowledgment

Work supported by the Spanish MCYT Project DPI 2004–05797 co-financed by the FEDER.

References

Agnihothri, S.R., Mishra, A.K., and Simmons, D.E. 2003. Workforce cross-training decisions in field service systems with two job types, *Journal of the Operational Research Society*, 54 (4): 410–418.

Agnihothri, S.R., and Mishra, A.K. 2004. Cross-training decisions in field services with three job types and server-job mismatch, *Decision Sciences*, 35 (2): 239–258.

Batta, R., Berman, O., and Wang, Q. 2005. *Balancing Staffing and Switching Costs in a Service Center with Flexible Servers*. Working paper. Department of Industrial Engineering, University at Buffalo, NY.

Brusco, M.J., and Johns, T.R. 1998. Staffing a multiskilled workforce with varying levels of productivity: an analysis of cross-training policies, *Decision Sciences*, 29: 499–515.

Campbell, G.M., and Diaby, M. 2002. Development and evaluation of an assignment heuristic for allocation cross-trained workers, *European Journal of Operational Research*, 138: 9–20.

Carpaneto, G., and Toth, P. 1981. Algorithm for the solution of the bottleneck assignment problem, *Computing*, 27: 179–187.

Cattrysse, D.G., and Van Wassenhove, L.N. 1992. A survey of algorithms for the generalized assignment problem, *European Journal of Operational Research*, 60: 260–272.

Corominas, A., Ojeda, J., and Pastor, R. 2005. Multi-objective allocation of multi-function workers with lower bounded capacity, *Journal of the Operational Research Society*, 56 (6): 738–743.

Corominas, A., Pastor, R., and Rodríguez, E. 2006. Rotational allocation of tasks to multifunctional workers in a service industry, *International Journal of Production Economics*, 103: 3–9.

Dell'Amico, M., and Toth, P. 2000. Algorithms and codes for dense assignment problems: the state of the art, *Discrete Applied Mathematics*, 100 (1–2): 17–48.

Jonker, R., and Volgenant, A. 1987. A shortest augmenting path algorithm for dense and sparse linear assignment problems, *Computing*, 38: 325–340.

Kozlowski, S. 1996. Cross-training: concepts, considerations, and challenges, *Medical Laboratory Observer*, 28 (2): 50–52/4.

Martí, R., Lourenço, H., and Laguna, M. 2000. Assigning proctors to exams with scatter search, in *Computing Tools for Modeling, Optimization and Simulation*, 215–227. Kluwer, Boston, MA.

McCune, J.C. 1994. On the train gang, *Management Review*, 83 (10): 57–60.

Rodríguez, E.Z. 2006. *Asignación Multicriterio de Tareas a Trabajadores Polivalentes*. PH.D. Dissertation. Technical University of Catalonia, Barcelona.

Slomp, J., and Molleman, E. 2002. Cross-training policies and team performance, *International Journal of Production Research*, 40 (5): 1193–1219.

chapter 3

Factors affecting cross-training performance in serial production systems

David A. Nembhard and Karndee Prichanont

Contents

3.1   Introduction

In order to meet the challenges of worldwide competitiveness, organizations must be able to provide products and services with equal or better value than their competitors. Moreover, there is a need to respond rapidly and effectively to changes in the marketplace. As product life cycles shorten, production systems must be able to effectively operate and switch between different types of products. Many organizations are moving toward greater flexibility in all stages of production in order to provide customers with greater product variety in less time. Creating a multifunctional workforce is one response to form a flexible organization to meet these market challenges.

Introducing worker cross training (also referred to as multifunctionality) may act as a buffer against uncertainties and the variation in workforce supply, such as an increase in the rate of periodic product revision, absenteeism, and job rotation (Shafer et al., 2001; Bokhorst and Slomp, 2000; Slomp and Molleman, 2000; McCreery and Krajewski, 1999; Uzumeri and Nembhard, 1998; van den Beukel and Molleman, 1998). From the managerial perspective, concerns arise from the potential lost output and increased training costs when workers operate on multiple workstations and spend significant time in the learning process. Several researchers have illustrated the importance of empirical investigation of learning and forgetting distributions (e.g., Shafer et al., 2001; Nembhard and Osothsilp, 2001; Lance et al., 1998). However, quantitative studies of multifunctionality and cross training incorporating individual learning and forgetting behaviors are relatively sparse. Several studies have been based on individual steady-state performance, where workers maintain a constant productivity level over time. Another stream of work has been based on assumed experimental conditions of learning and/or forgetting at an organizational level, where all workers on the shop floor have the same performance level (e.g., Kher, 2000; Kher et al., 1999; McCreery and Krajewski, 1999; Malhotra et al., 1993).

In this chapter, we use a heterogeneous workforce, based on an empirically determined distribution of learning and forgetting in manufacturing, to study the effects of worker multifunctionality in conjunction with several process characteristics and operating decisions in flow line systems. This research aims to provide information to managers and researchers to mitigate productivity losses due to frequent task learning and to better understand how multifunctionality affects system performance in a dynamic workplace. More specifically, this will inform managers in setting worker multifunctionality levels and correspondingly, cross-training levels based on several important process characteristics. The following section provides a review of previous related studies beginning with studies of worker multifunctionality based on individual steady-state performance (i.e., no learning–forgetting effect), followed by a discussion of studies of workforce multifunctionality with learning and forgetting.

3.2 Previous research

The subject of creating a multiskilled workforce has been extensively investigated in the traditional manufacturing job shop environment, based on the assumption of steady-state productivity (e.g., Slomp and Molleman, 2002; Slomp and Molleman, 2000; Molleman and Slomp, 1999; Park, 1991; Russell et al., 1991). In a job shop constrained by both workers and machines, Molleman and Slomp (1999) suggested that the distribution of redundancy and the distribution of multifunctionality had significant impacts on the product makespan and the total production time, where *redundancy* refers to the number of workers capable of performing a task, and *multifunctionality* refers to the number of tasks for which each worker is trained. Their finding indicated that introducing worker redundancy helped decrease the negative impact of system variation, such as worker absenteeism. Moreover, their study showed that uniform skill distribution, where all workers are equally trained on the same number of tasks, resulted in shorter product makespan.

The benefit of uniform skill distribution becomes more significant when absenteeism is high. With variation in worker absenteeism and demand, Slomp and Molleman (2000, 2002) examined the effects of cross-training policies on team performance, including the decision of who should be cross trained. Team performance was defined by the workload of the bottleneck worker, the average utilization of workers, and the standard deviation of the workload among workers. The studies suggested that having redundancy at the bottleneck task and balancing the multifunctionality among workers improved team performance. Park (1991) conducted a simulation study to examine the effects of various degrees of worker cross training in a labor-limited job shop. The various cross-training levels, represented by worker-station efficiency matrices, range from no cross training through a completely redundant workforce where all workers are trained on all tasks. The study concluded that the minimum introduction of worker cross training yielded the most significant improvement in system performance. Several job sequence policies were also investigated under various cross-training policies. The author concluded that the preferred job sequence remained the same as the amount of cross training changed. Through a series of simulation experiments, Russell et al. (1991) examined labor assignment strategies based on a hypothetical group technology shop with three cells. Similarly, the results favored a flexible homogeneous workforce such that each worker should be completely and equally cross trained and allowed to be assigned to any machine in any cell.

In a related stream of research, Brusco et al. (1998) developed an integer linear programming model to examine the cross-training configurations in a maintenance operation in a paper production system, concluding that the appropriate level of cross training should be determined interdependently with the demand variation. The improvement of system performance due to an increase in the cross-training level is greater given a high demand variation relative to a low demand variation. The study also suggested that each worker should not necessarily be fully cross trained in all operations to achieve substantial benefits from cross training. This conclusion is consistent with a number of other studies, including, for example, studies by Slomp and Molleman (2002), Cesani and Steudel (2000a), Cesani and Steudel (2000b), Suresh and Gaalman (2000), Slomp and Molleman (2000), Molleman and Slomp (1999), Campbell (1999), and Park (1991). Dunphy and Bryant (1996) also suggested that overskilling, where workers are trained on too many workstations, may cause frustration among workers that may eventually lead to diminished performance.

Wisner and Pearson (1993) indicated the necessity of incorporating worker learning and relearning behavior into worker allocation studies, wherein worker relearning behavior has a significant impact on the labor assignment policy, especially when workers are required to transfer among tasks regularly. The impacts of task learning and forgetting have received increased attention in the study of worker multifunctionality. Several of these studies have incorporated organizational-level learning and forgetting into their models (e.g., Felan and Fry, 2001; Kher, 2000; Kher et al., 1999; McCreery and Krajewski, 1999; and Malhotra et al., 1993).

Malhotra et al. (1993) suggested that in a high forgetting job shop environment, increasing worker flexibility deteriorates the mean product flow time and the level of work-in-process inventory. Kher (2000) investigated training schemes in a job shop under learning, relearning, and attrition conditions. Based on a homogeneous workforce, Kher suggested that under slight forgetting, workers could be transferred freely to any task without affecting mean flow time and mean tardiness, which were used as surrogates for work-in-process inventory level and customer service performance. Kher et al. (1999) remarked that the combinations of the manner in which workers are trained (i.e., training for several departments simultaneously or training on one department for an extended period of time) and the transferring time (i.e., how long a worker remains in a department) has significant impact on average job processing time and worker efficiency. Also, using a homogeneous workforce, McCreery and Krajewski (1999) conducted an assembly line simulation study and found that when learning and forgetting effects exist, the degree of cross training and the worker deployment policy should depend primarily on the degree of task complexity and the degree of product variety.

We suggest that past studies have largely been based on steady-state performance, learning and forgetting at the organizational level, or a set of homogeneous workers. Shafer et al. (2001) illustrated that using a single composite learning curve for all workers could lead to underestimation of overall performance. Also, Buzacott (2002) demonstrated the importance of individual worker differences on system performance where the operating policy plays a critical role in determining how sensitive a system is to individual differences. Further, Nembhard and Norman (2005) investigated heterogeneous cross-trained teams of varying sizes in a learning/forgetting environment and concluded that high turnover can accelerate on-the-job cross-training costs, and that moderate cross-training levels (e.g., two tasks per worker), in general, tend to outperform systems with no cross training and systems with high cross training. The study focused on the interaction of cross-training effects with varying levels of turnover, absenteeism, new product design life, and the magnitude of product change.

The current study extends the Nembhard and Norman (2005) study by investigating multifunctionality levels across a range of process characteristics and operating decisions. Where Nembhard and Norman examine cross training with respect to labor and product uncertainty, the current study investigates the interactions between worker multifunctionality with several process characteristics. The process characteristics are selected to aid in understanding how optimal cross-training levels are affected by manufacturing process variables, which may not be controllable. These include the staffing level, position of the system bottleneck, and the degree of task similarity. The staffing level is somewhat controllable but may be considered exogenous to the operational control of a particular process. That is, at a point in time, a manager may have a fixed set of human resources available. The position of the system bottleneck and the degree of task similarity are likely to be a function of the design of the product and outside of managerial control. We also include the managerially controllable factors of the level of multifunctionality and the rotation interval between cross-trained tasks. While process characteristics may be constrained by the design of the product and are generally known to be important determining factors in productive output, less well understood is how these interact with worker multifunctionality levels. This study addresses the interactions of these factors with varying levels of multifunctionality to better inform managers and researchers on preferred levels of multifunctionality and cross training under a range of operational conditions. We address these questions assuming a set of fixed straightforward worker-station assignment patterns. We conclude that, in general, determining optimal worker-station assignment in the context of learning and forgetting is nontrivial due to nonlinearity and production rates that are dependent on past assignments.

In summary, we will address the following fundamental questions:

1. How do the process characteristics of staffing level, bottleneck position, and task similarity as well as the operating decisions of rotation interval and multifunctionality affect system performance in a learning–forgetting environment?
2. How does the position of the system bottleneck affect preferred levels of multifunctionality?
3. How does the degree of similarity between tasks affect preferred levels of multifunctionality?
4. How does the worker rotation interval affect preferred levels of multifunctionality?

3.3 Research methodology

3.3.1 Simulation design

This study is based on discrete part production systems, where each material unit is processed distinctly. Production is constructed as a flow line system, wherein units are required to visit all stations in the same sequence. We refer to worker activities as tasks, and the locations in the line where these are performed as stations, assuming a one-to-one correspondence between tasks and stations. We employ an *unpaced flow line* (no maximum limit for the time for a worker to perform a task) that is *asynchronous* (job movement to adjacent stations is not coordinated). The worker begins the next task as soon as the corresponding station is available. The completed work is immediately trans-ferred to the buffer of the next station. The model is based on a noncollab-orative environment, where, at most, one worker can perform a station's task at any given time. A noncollaborative setting is commonly found in the shop floor due to, for example, equipment limitation, task characteristics (e.g., workers perform tasks on small materials), and safety issues. We note that flow line systems of this nature are common in a number of manufacturing and service operations, either as complete production systems or as sub-systems within larger processes.

The simulation consists of six functionally different tasks ($J = 6$) with an input buffer preceding each station as in Figure 3.1. The input buffer is assumed to have an infinite supply of raw material so the system is never starved at the input. In addition, the buffers between each station are unlim-ited to prevent confounding effects with the experimental factors. The system is modeled to closely reflect a labor-intensive manufacturing environment where the job processing times are different among workers and vary over time. Individual worker differences in job processing times are based on a model of learning and forgetting from the literature (see Section 3.6, Appen-dix), where the distribution of learning and forgetting characteristics is informed by an empirically based set of workers. The resulting distribution forms the basis for worker heterogeneity in the simulation. Since the tasks in the model are distinct, each worker has unique learning–forgetting param-eters for each task.

The simulation model is designed to represent several common manu-facturing characteristics including the periodic introduction of new products and worker absenteeism. These are included not only to make the simulation

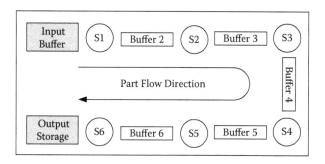

Figure 3.1 Process configuration. Source: Nembhard and Prichanont, IEEE Transac-tions on Engineering Management. 2007.

more representative of practical systems but also to represent two of the motivating rationale for worker multifunctionality. As such, the simulations include a product revision cycle every 3 months, where each revision involves the modification of tasks at the least complex three out of the six stations. The occurrence of product revisions may affect worker learning and forgetting since workers will be exposed to new task requirements. The cumulative experience gained for workers on the obsolete tasks is lost, while experience on the unchanged tasks is retained. We assume any semifinished parts in the buffers are discarded from the system following a product revision.

Worker absenteeism is also modeled since it has been shown to be a motivating influence for implementing cross training and has had an impact on system performance (van den Beukel and Molleman, 1998; Bokhorst and Slomp, 2000). The absenteeism rate is modeled stochastically following a uniform distribution, with a mean level of one absent day per month. This rate reflects 2 to 3 weeks per year of vacation. Each operator has an equal chance of being absent and is absent exactly 1 day per absence period. The model limits the system to one absent worker at a time per worker group. In the case, with no cross training (minimum multifunctionality), the worker in the closest position to the absent worker's station with the least complex task will perform the task of the absent worker.

System performance is measured in terms of productivity and batch time. We define *productivity* as the total amount of finished products during the simulated year, regardless of the product type, and *batch time* as the average time taken to complete 1000 units of work based on customer demand (Molleman and Slomp, 1999; Eckstein and Rohleder, 1998). The quantity of 1000 finished units is selected to represent a few weeks' production, and allows for the examination of system ability to respond rapidly to customer orders. Since the production system is assumed to have a product revision every 3 months, there are four different product types in each simulated year. Thus, batch time is the average time to reach the first 1000 output units across all product types.

3.3.2 Experimental factors

We evaluate the effects of several process characteristics and operating decisions on system performance as well as the effects of such characteristics on the preferred level of worker multifunctionality. The first three factors (staffing level, system bottleneck position, and task similarity) represent process characteristics that are largely determined by the characteristics of the product itself. The operating decisions include worker rotation interval and multifunctionality, since management may have some degree of control over these. The set of experimental factors and levels is summarized in Table 3.1. We also simulate the system with a heterogeneous workforce for a case assuming no learning and forgetting (i.e., workers perform at their own individual steady-state productivity levels) in order to assist in the interpretation of the results.

Table 3.1 Simulation Experimental Factors

Characteristics	Factors	Levels	Level 1	Level 2	Level 3
Process	1. Staffing Level	2	50%	100%	
Characteristics	2. Bottleneck Position	2	Downstream	Upstream	
	3. Task Similarity	2	0%	50%	
Operating	4. Rotation Interval	2	1 hour	8 hours	
Decisions	5. Multifunctionality	3	Minimum	Moderate	Maximum

We simulate a total of 48 unique treatments, with each treatment run for 1 year of simulated time, with 30 randomly selected heterogeneous worker groups at each treatment. The same 30 groups are used at each treatment as a variance reduction technique. A warmup period is not used in this simulation model for two reasons. First, this study focuses on a production line where the introduction of new products occurs regularly four times per year. Second, since this study investigates the effects of learning and forgetting, eliminating the information during the startup would discard crucial transient period dynamics.

3.3.2.1 Staffing level (ST)

Two staffing levels are examined, in order to provide insight into performance in labor-constrained (50% staffing level) and fully staffed (100% staffing level) systems. While broader ranges of staffing exist in practice, the 50 and 100% levels represent a common practical range with a minimum of two stations per operator and one station per operator, respectively. Three operators are assigned to the system in the 50% staffing case ($I = 3$) and six operators in the 100% staffing case ($I = 6$). In labor-constrained systems, a higher staffing level will generally yield proportionally higher system productivity overall. Further, in serial systems, we may expect to observe better batch times since the part delay due to operator unavailability is lower. That is, performance would be expected to be roughly proportional to the staffing level and be the dominant factor if considered alongside other factors. Correspondingly, we consider the staffing levels as two separate cases in order to see whether between these cases there are differences in the way other factors affect performance.

3.3.2.2 System bottleneck position (BN)

In serial production where *homogeneous* workers and steady-state productivity is often assumed (i.e., no learning–forgetting), the position of the bottleneck station may affect work-in-process levels and yet have a lesser effect on system performance because the output cycle time is constrained by the bottleneck station regardless of its position. In the current study with *heterogeneous* worker learning–forgetting, we hypothesize that the bottleneck position may affect the level of work-in-process and the utilization of flow line workers, which potentially leads to significant effects on workers' learning–forgetting. Ultimately, interactions between these factors may be helpful in interpreting the overall results.

Two levels of system bottleneck position are considered: a downstream and an upstream bottleneck. The bottleneck station is indicated by the level of task complexity, where the bottleneck station contains the most complex operations and, thus, requires the longest standard time. In the simulation data and model, we use task complexity factors (c_j) partially to differentiate the tasks at each station (Wood, 1986). Task complexity factors are given in a range from 0.6 to 1.6 in increments of 0.2, where smaller values of c_j represent simpler tasks. With a downstream bottleneck, stations are arranged so that station 1 involves the simplest operations with ($c_1 = 0.6$) and station 6 involves the most complex operations ($c_6 = 1.6$). With an upstream bottleneck, station 1 involves the most complex operations ($c_1 = 1.6$) and station 6 involves the simplest operations ($c_6 = 0.6$).

As a task is more complex (i.e., larger c_j), worker steady-state productivity rates (k_i) decrease, learning rates (r_i) increase, prior experiences (p_i) decrease, and forgetting rates (α_i) increase (Osothsilp, 2002; Nembhard, 2000). In the simulation model, individual learning–forgetting parameters for each station are transformed proportionally as follows: k_i/c_j, $r_i \times c_j$, p_i/c_j, and $\alpha_i \times c_j$.

3.3.2.3 *Task similarity (TS)*

The third factor, task similarity, represents the degree of experience sharing among tasks for workers in the system. Two levels of task similarity are considered: 0% and 50%. The 0% similarity level indicates that the operations required in each station are distinct. The 50% similarity level indicates that half the elements of a station's task are common to both. Thus, at 50% similarity, one unit of experience on task 1 also corresponds to one half unit of experience on task 2. That is, the degree of task similarity has a direct relationship with worker learning behaviors, where a higher similarity among tasks results in faster experience gained by workers. This factor may also affect the choice of multifunctionality level. In a high similarity shop, workers might be assigned to a greater number of stations without losing significant efficiency.

3.3.2.4 *Worker rotation interval (RT)*

Two worker rotation intervals are considered: 1-hour and 8-hour rotations. The rotation interval is modeled deterministically assuming an 8-hour work-day. These levels were chosen to represent a reasonable operational range for cross training workers. A 1-day rotation is a natural break and avoids the potential buildup of large work-in-process (WIP) levels in the system from longer intervals. A shorter natural break occurs at 2 hours, when many work systems have coffee breaks. We have chosen to shorten this further to a 1-hour interval, while maintaining the operational convenience of having even work assignments over time.

We expect that longer intervals will allow workers to increase their learning curves, resulting in high levels of task proficiency and productivity. However, a longer time at a station results in workers being away from other assigned stations for a longer time, which, in turn leads to significant task

Table 3.2 Multifunctionality, Cross Training, and Worker-Station Assignment

Levels	Multifunctionality (stations/worker)	Cross training (workers/station)	Worker-station assignment chain					
(a) 50% Staffing								
1	MIN (2 stations/worker)	1	W1	W2	W3	W3	W2	W1
			S1	S2	S3	S4	S5	S6
2	MOD (4 stations/worker)	2	W2	W3				W2
			W1		W2		W1	
			S1	S2	S3	S4	S5	S6
3	MAX (6 stations/worker)	3	W1					
			W2					
			W3					
			S1	S2	S3	S4	S5	S6
(b) 100% Staffing								
1	MIN (1 station/worker)	1	W1	W2	W3	W4	W5	W6
			S1	S2	S3	S4	S5	S6
2	MOD (2 stations/worker)	2	W6	W2		W4		W6
			W1		W3		W5	
			S1	S2	S3	S4	S5	S6
4	MAX (6 stations/worker)	6	W1					
			W2					
			W3					
			W4					
			W5					
			W6					
			S1	S2	S3	S4	S5	S6

relearning and forgetting effects (Shafer et al., 2001; Uzumeri and Nembhard, 1998). This factor is investigated in order to better understand appropriate rotation intervals and how they interact with levels of multifunctionality.

3.3.2.5 Multifunctionality (MF)

Three levels of multifunctionality are considered (minimum, moderate, and maximum levels), where each also represents a degree of worker cross training, as illustrated in Table 3.2. The first level of minimum (MIN) multifunctionality represents the minimum station coverage to maintain material flow (i.e., two stations and one station in the 50 and 100% staffing cases, respectively). At the MIN level, each station has a single operator and, thus, there is no worker cross training. At the second level of moderate (MOD) multifunctionality, each worker performs on four stations and two stations for 50 and 100% staffing cases, respectively. This level offers greater station coverage and cross training of two operators for each station. The third level of maximum (MAX) multifunctionality offers complete cross training where each operator rotates to all six stations.

Specific worker-station assignment chains for each degree of multifunctionality and staffing level are provided in Table 3.2. This approach is similar to skill chaining described by Daniels et al. (2004). These initial assignment patterns are developed based on the following criteria: (1) an overlapping of workers' skills (van Oyen et al., 2001; Slomp and Molleman, 2000); (2) similar total task complexity level assigned to each worker; and (3) a proximity of assigned stations where workers are assigned such that the distance between stations in a U-shaped line is minimized (Steudel and Desruelle, 1992). For example, in cases of moderate multifunctionality (Table 3.2a, Level 2), an overlapping of workers' skills is modeled such that each station is covered by a different set of workers. The total task complexity per worker is as even as possible within each case while maintaining chaining. However, some chains produce differential workloads. The total task complexity for workers between staffing cases, in general, will not be the same since there are varying levels of multifunctionality involved. That is, stations are not staffed with an effort toward eliminating bottlenecks. Since many production systems have one or more bottlenecks, we allow them in the system in a controlled manner. Determining optimal worker station assignments to balance staffing and eliminate bottlenecks in a learning and forgetting environment is nontrivial due to production rates that are nonlinear and dependent on all past assignments.

3.4 Results and discussion

The analysis of variance (ANOVA) results and least mean square estimates for the main effects and two-way interactions of process characteristics and operating decisions with multifunctionality are shown in Table 3.3 and Table 3.4. The corresponding main effects plots are shown in Figure 3.2 and Figure 3.3. Figure 3.4, by contrast, illustrates the main effects on productivity in a scenario without learning and forgetting. The significant interactions between factors are illustrated in Figure 3.5 to Figure 3.7. Based on the results in Table 3.3 and Figure 3.2, the main effects of the bottleneck (BN) position, task similarity (TS), worker rotation interval (RT), and multifunctionality (MF) are significantly related to the productivity and batch time performance measures at the 50% staffing level. Results for the 100% staffing level show that the BN position, TS, and MF are significantly related to productivity (Table 3.3 and Figure 3.3).

The main effects plots for MF in Figure 3.2 and Figure 3.3 and Table 3.3 show consistently that as multifunctionality is increased to its maximum levels, there is a significant reduction in productivity and increase in batch time. This is in contrast to the "no learning and forgetting" case (see Figure 3.4), in which additional multifunctionality yields greater productivity (Table 3.5, $p < 0.001$). However, we note that this increase comes with a diminishing

Table 3.3 ANOVA Results on Productivity and Batch Time with Learning and Forgetting

| | 50% Staffing | | | | 100% Staffing | | | |
| | Productivity | | Batch time | | Productivity | | Batch time | |
Sources	F	p	F	p	F	p	F	p
Main effects								
Bottleneck position (BN)	171.02	0.000*	116.98	0.000*	7.92	0.005*	0.20	0.657
Task similarity (TS)	274.28	0.000*	206.25	0.000*	71.21	0.000*	33.42	0.000*
Rotation interval (RT)	34.09	0.000*	98.45	0.000*	0.31	0.579	1.21	0.272
Multifunctionality (MF)	27.37	0.000*	16.57	0.000*	9.18	0.000*	1.57	0.209
Two-way interaction effects[1]								
BN × MF	4.94	0.007*	0.27	0.763	1.33	0.266	2.07	0.127
TS × MF	16.04	0.000*	7.03	0.000*	24.64	0.000*	11.96	0.000*
RT × MF	0.18	0.837	4.96	0.009*	0.25	0.782	0.55	0.579

[1]Only interaction effects with multifunctionality are shown.

Asterisk (*) denotes significant at 1%.

marginal return. This suggests that giving workers multiple tasks can help absorb some system variability, which we modeled in the form of worker absenteeism and product change. In this setting, workers eventually lost efficiency at high multifunctionality levels in large part due to significant task relearning and forgetting effects. This result is consistent with several studies showing that workers should not necessarily be fully trained in all operations to achieve significant benefits from cross training (e.g., Cesani and Steudel, 2000a, 2000b; Suresh and Gaalman, 2000; Slomp and Molleman, 2000; and Brusco and Johns, 1998). However, the current results go beyond suggesting diminishing positive returns, to further indicate that when learning and forgetting are considered, there may be negative returns from additional increases in multifunctionality. In the remainder of this section, we address interactions between multifunctionality and the bottleneck position, task similarity, and worker rotation interval process characteristics.

3.4.1 *Effects of bottleneck position on preferred level of multifunctionality*

The main effects for the bottleneck (BN) position indicate significantly higher productivity in the 50% (Table 3.3, $p < 0.001$) and 100% (Table 3.3, $p = 0.005$) staffing cases and shorter batch time in the 50% staffing case (Table 3.3, $p < 0.001$) when the bottleneck is downstream (see also Figure 3.2 and Figure 3.3). However, as illustrated in Figure 3.4, if workers are assumed to have no learning and forgetting, the bottleneck position has similar but smaller, relative effects on productivity (Table 3.5, $p < 0.001$). The significant differences in productivity and batch times between downstream and upstream

Table 3.4 Least Mean Square Estimates for Main Effects and Two-Way Interactions with Multifunctionality

		50% Staffing				100% Staffing			
		Productivity		Batch time		Productivity		Batch time	
		Mean	Standard error	Mean	Standard error	Mean	Standard error	Mean	Standard error
Bottleneck position (BN)									
Downstream		13848.4	101.0	30.9	0.3	25165.7	240.4	17.4	0.2
Upstream		11980.4	101.0	36.4	0.3	24209.1	240.4	17.6	0.2
Task similarity (TS)									
0%		11731.5	101.0	37.3	0.3	23253.1	240.4	18.6	0.2
50%		14097.3	101.0	30.0	0.3	26121.6	240.4	16.3	0.2
Worker rotation interval (RT)									
1 hour		13331.4	101.0	31.2	0.3	24781.6	240.4	17.2	0.2
8 hours		12497.4	101.0	36.2	0.3	24593.1	240.4	17.7	0.2
Multifunctionality (MF)									
Min		13159.1	123.7	32.0	0.4	25015.8	294.4	17.9	0.3
Mod		13403.6	123.7	33.4	0.4	25368.4	294.4	17.0	0.3
Max		12180.5	123.7	35.6	0.4	23677.9	294.4	17.5	0.3
BN × MF									
Downstream	Min	13942.3	175.0	29.4	0.6	25629.2	416.3	18.3	0.4
Downstream	Mod	14655.1	175.0	30.4	0.6	25460.6	416.3	16.8	0.4
Downstream	Max	12947.9	175.0	33.0	0.6	24407.2	416.3	17.0	0.4
Upstream	Min	12375.9	175.0	34.7	0.6	24402.5	416.3	17.4	0.4
Upstream	Mod	12152.1	175.0	36.4	0.6	25276.1	416.3	17.2	0.4
Upstream	Max	11413.1	175.0	38.2	0.6	22948.6	416.3	18.1	0.4
TS × MF									
0%	Min	12548.1	175.0	34.5	0.6	24971.7	416.3	17.9	0.4
0%	Mod	11920.6	175.0	37.1	0.6	24067.1	416.3	18.0	0.4
0%	Max	10725.8	175.0	40.4	0.6	20720.4	416.3	20.0	0.4
50%	Min	13770.1	175.0	29.6	0.6	25059.9	416.3	17.8	0.4
50%	Mod	14886.5	175.0	29.7	0.6	26669.6	416.3	16.0	0.4
50%	Max	13635.2	175.0	30.8	0.6	26635.4	416.3	15.1	0.4
RT × MF									
1 hour	Min	13558.5	175.0	30.2	0.6	25040.2	416.3	17.8	0.4
1 hour	Mod	13779.4	175.0	30.9	0.6	25364.5	416.3	16.9	0.4
1 hour	Max	12656.2	175.0	32.4	0.6	23940.1	416.3	17.0	0.4
8 hours	Min	12759.7	175.0	33.9	0.6	24991.5	416.3	17.9	0.4
8 hours	Mod	13027.8	175.0	35.9	0.6	25372.2	416.3	17.1	0.4
8 hours	Max	11704.8	175.0	38.8	0.6	23415.7	416.3	18.0	0.4

bottlenecks indicate that this process characteristic affects performance but not necessarily directly because of learning and forgetting effects. In the system with a downstream bottleneck, worker idle time is 1.0% on average, while in the system with an upstream bottleneck, worker idle time is 32.8% on average. This suggests that with the downstream bottleneck, work-in-process can enter the system more rapidly, resulting in higher output.

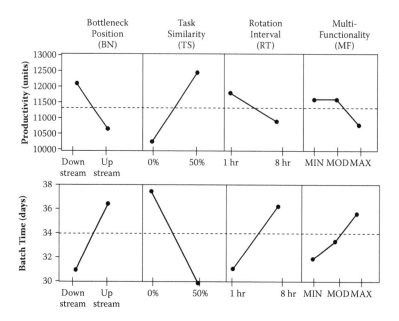

Figure 3.2 Main effects on productivity and batch time in the 50% staffing case. Source: Nembhard and Prichanont, IEEE Transactions on Engineering Management, 2007.

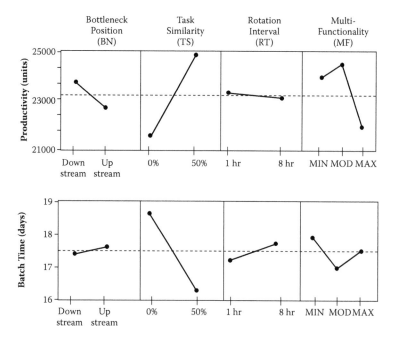

Figure 3.3 Main effects on productivity and batch time in the 100% staffing case. Source: Nembhard and Prichanont, IEEE Transactions on Engineering Management, 2007.

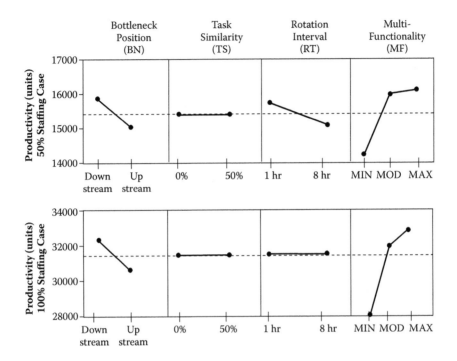

Figure 3.4 Main effects on productivity without learning and forgetting. Source: Nembhard and Prichanont, IEEE Transactions on Engineering Management, 2007.

The interaction between the bottleneck position and worker multifunctionality (BN × MF) for the 50% staffing case is significant for the productivity measure (Figure 3.5, Table 3.3, $p = 0.007$). The corresponding interaction without learning and forgetting is not significant (Table 3.5, $p = 0.585$). That is, the effects of MF on productivity in the 50% staffing case are dependent

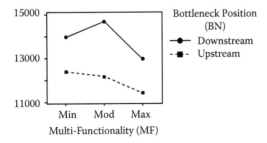

Figure 3.5 Significant interaction effect between bottleneck position and multifunctionality (BN × MF) on productivity in the 50% staffing case. Source: Nembhard and Prichanont, IEEE Transactions on Engineering Management, 2007.

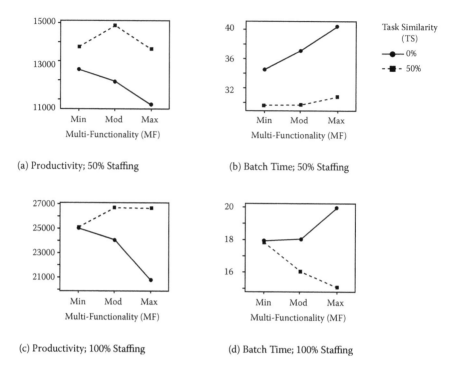

(a) Productivity; 50% Staffing

(b) Batch Time; 50% Staffing

(c) Productivity; 100% Staffing

(d) Batch Time; 100% Staffing

Figure 3.6 Significant task similarity and multifunctionality (TS × MF) interactions on (a) productivity, 50% staffing; (b) batch time, 50% staffing; (c) productivity, 100% staffing; and (d) batch time, 100% staffing. Source: Nembhard and Prichanont, IEEE Transactions on Engineering Management, 2007.

upon the bottleneck position, such that in the 50% staffing case with a downstream bottleneck, introducing worker multifunctionality at a moderate level, improves system productivity, while with an upstream bottleneck, minimal multifunctionality has the highest expected productivity. We saw

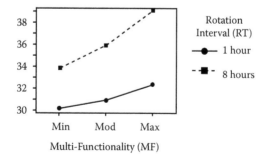

Figure 3.7 Significant interaction effect between rotation interval and multifunctionality (RT × MF) on batch time in the 50% staffing case. Source: Nembhard and Prichanont, IEEE Transactions on Engineering Management, 2007.

Table 3.5 ANOVA Results on Productivity without Learning and Forgetting

Sources	50% Staffing		100% Staffing	
	F	p	F	p
Main Effects				
Bottleneck Position (BN)	8.42	0.004*	16.43	0.000*
Task Similarity (TS)	0.00	1.000	0.00	1.000
Rotation Interval (RT)	15.08	0.000*	0.01	0.936
Multifunctionality (MF)	17.15	0.000*	27.09	0.000*
Two-Way Interaction Effects[1]				
BN × MF	0.54	0.585	1.12	0.342
TS × MF	0.00	1.000	0.00	1.000
RT × MF	0.01	0.990	0.01	0.999

[1]Only interaction effects with multifunctionality are shown.

Asterisk (*) denotes significant at 1%.

that work-in-process levels, in general, remained relatively low (<4% of production), despite having unlimited buffer capacities. These results suggest that, with a downstream bottleneck, workers are less often starved for work and, thus, are able to progress up their learning curves more rapidly with relatively small negative effects of forgetting. That is, in the current setting with periodic product change and absenteeism, workers can be cross trained at a greater number of tasks without losing comparable efficiency. In this case, cross training workers at four of the six tasks is preferred. In contrast, the upstream bottleneck frequently left workers downstream with less work and, consequently, they did not progress up their learning curves as rapidly. In that case, the benefit of cross training did not balance the lost worker efficiency due to multifunctionality, and minimum cross training was preferred. This result indicates that in labor-constrained systems, setting operator multifunctionality levels should be made with knowledge of the current position of the system bottleneck.

In the 100% staffing case, the relationship between the bottleneck position and the preferred multifunctionality level (BN × MF) was not significant (Table 3.3, $p = 0.266$ for productivity, $p = 0.127$ for batch time). The results on average worker idle time do not show a significant difference between the two bottleneck positions (i.e., 0.45 and 0.55% for downstream and upstream bottleneck, respectively). Material can flow through the system without significant interruption, and downstream operators were not starved for work as often as in the 50% staffing case. The minimum and moderate multifunctionality levels yielded higher average productivities and shorter average batch times than the maximum multifunctionality levels (Table 3.4).

3.4.2 Effects of task similarity on preferred multifunctionality

The Task Similarity (TS) process characteristic also had a significant effect on productivity (Table 3.3, $p < 0.001$) and batch time (Table 3.3, $p < 0.001$) at both staffing levels, where greater productivity and shorter batch time corresponded to processes with greater similarity among tasks. In the case with no learning and forgetting, this factor had no effect (Table 3.5, $p \sim 1.0$). With greater task similarity, workers learn and progress toward steady state more rapidly, since learning on one task can also contribute to partial learning on other tasks. These main effects are relatively strong, as illustrated in Figure 3.2 and Figure 3.3. The interactions between task similarity and preferred multifunctionality (TS × MF) were significant in both staffing levels and for both productivity and batch time measures (Table 3.3, $p < 0.001$).

At the 50% staffing level, multifunctionality was not beneficial for a system with dissimilar tasks. With similar tasks, a moderate level of multifunctionality had the highest productivity (Figure 3.6a). However, batch times were comparable under minimum and moderate levels of multifunctionality (Figure 3.6b). Under maximum multifunctionality, productivity decreased and batch times increased in both low and high similarity systems.

At the 100% staffing level, this interaction was mores pronounced for both productivity and batch time. Figure 3.6c/d illustrates that multifunctionality at moderate to maximum levels had the highest productivity and shortest batch times in the high similarity system, with these effects reversed in the low similarity system. It may be that the negative effect of task relearning is not as strong as the negative effect of an individual slow learner in the 100% staffing case with high task similarity, where higher multifunctionality is preferred. This highlights the critical nature of process task similarity levels in developing appropriate cross-training policies. Task similarity allows workers to progress up their learning curves more rapidly as some task experience is shared among similar tasks, faster task learning compensates for the negative effects of forgetting, and task relearning comes with increased multifunctionality. Further, with high task similarity, neither productivity nor batch time was significantly less preferable with maximum multifunctionality. An important implication of this for managerial decision-making is that for high similarity systems, moderate to full cross training has the potential to be an effective policy for both productivity and batch time measures, with the very opposite true for low similarity systems.

3.4.3   Effects of rotation interval on preferred multifunctionality

The rotation interval (RT) had a significant effect on performance in the 50% staffing case where short intervals were related to high productivity (Table 3.3, $p < 0.001$) and short batch times (Table 3.3, $p < 0.001$) (see also Figure 3.2). In the no learning and forgetting case, RT was similarly only significant in the 50% staffing case (Table 3.5, $p < 0.001$), wherein shorter intervals generally improved material flow, yielding greater production and less work in process.

Shorter rotation intervals also provided better batch times in the 50% case (see Figure 3.2). RT had a significant interaction with the preferred level of multifunctionality for batch time in the 50% staffing case (Table 3.3, $p = 0.009$). This interaction is relatively mild, as illustrated in Figure 3.7, and suggests that high multifunctionality, when combined with long rotation intervals, tends to accelerate undesirable increases in batch times. A long rotation interval implies a longer time away from other cross-trained tasks, leading to significant task relearning and forgetting. This suggests a potential benefit of reducing the rotation interval in high multifunctionality work systems in order to mitigate losses due to forgetting.

3.5   Conclusions

This study investigated the benefits of worker multifunctionality under a variety of production process characteristics. In addressing several managerial questions, we conducted a simulation study informed by an empirical distribution of individual worker learning and forgetting. The resulting distribution forms the basis for worker heterogeneity in the simulations. Results suggest that with learning and forgetting effects, managerial decisions on multifunctionality should be made in conjunction with several process characteristics (i.e., staffing level, system bottleneck position, and the degree of task similarity) and operational decisions (i.e., worker rotation rate). In both staffing cases, a small amount of multifunctionality helped to absorb the negative impacts of system variability. However, we noted that workers need not be fully trained at all stations to improve system performance. We conclude that not only are there diminishing returns on increases in multifunctionality, but that in a learning/forgetting environment there may be reduced performance beyond moderate or in some instances minimal cross training.

We examined the effects of process characteristics and operational decisions on system performance and preferred levels of multifunctionality. In a labor-constrained system, a downstream bottleneck (a high similarity among tasks) and a high worker rotation rate tended to provide higher productivity and shorter batch times. These process characteristics and operational decisions may enhance the speed of task learning and diminish the negative effects of task relearning and forgetting.

First, this is most clearly seen for task similarity, where there was no response in the case without learning and forgetting, and a number of significant interactions with multifunctionality when learning and forgetting are considered. When process tasks are more similar, it may be beneficial to increase worker multifunctionality even toward its maximum level under full staffing. The negative impact of lost efficiency due to forgetting was small relative to the benefits gained from increased learning and the ability to respond to the system variability.

Second, the results indicated that the bottleneck position had a significant impact on the preferred level of multifunctionality in one case. While the overall main effect appears not to be directly related to learning and forgetting, the multifunctionality factor results were significantly dependent on the bottleneck position in the labor-constrained case. This may be due to system impacts on learning/forgetting, reduced idle time, or, possibly, a combination of these. Thus, in some labor-constrained cases, decisions on multifunctionality levels should consider knowledge of the bottleneck position and worker idle time. More specifically, when simpler tasks are upstream and parts are able to enter the system more rapidly, introducing multifunctionality is more likely to provide benefits. As interruptions in learning due to part starvation are minimized, workers could have increased multifunctionality without necessarily losing efficiency.

Third, the worker rotation interval had an influence on the preferred multifunctionality level, wherein shorter rotation intervals generally corresponded to better performance in the labor-constrained case. Since there were no significant effects under full staffing, these effects may be due to mitigation of worker starvation by more frequent rotation. However, the interaction with multifunctionality suggests that learning and forgetting may also be involved to a lesser extent. Thus, a manager might be able to obtain shorter batch times from a multifunctional workforce by employing relatively short rotation intervals, keeping in mind that shorter intervals performed better overall and that combining high multifunctionality with long rotation intervals performed particularly poorly.

We suggest that the current study addresses a range of manufacturing conditions, and, to allow for reasonable interpretations of results, we have been selective on the set of factors considered. However, there are numerous other potentially important factors that may affect managerial decisions on setting multifunctionality and cross-training levels. These may include job enrichment, motivation, and employee preferences, among others. In general, the chosen cross-training assignment chains are not necessarily optimal. However, other straightforward patterns examined in the design of this study yielded similar relationships. It may be that optimal cross training affects the current factors in interesting ways. Since determining optimal cross training in a given learning/forgetting environment is a nontrivial matter due to the combination of both nonlinearity and the time dependencies of the production rates, this forms an important topic for future research.

The simulations were based on data from a single workforce, and we note that it will be important for similar examinations to be conducted across a wide variety of task types and environments to gain a fuller understanding of the examined effects. Extensions of this study to other process characteristics, operational design conditions, and performance measures will further aid managers and researchers in understanding and setting multifunctionality and cross-training levels. For example, absenteeism rates, worker turnover rates, frequencies of product revisions, and worker assignment patterns are of general interest. Exploring the effects of various skill distributions along with differential cross training, wherein some workers are specialized and others are cross trained, is another potential extension of the current work.

Appendix

3.6 Learning and forgetting model

The learning and forgetting model used in this study is based on a model described by Nembhard and Uzumeri (2000) that was shown to perform well over a wide range of empirical conditions and has been applied in recent studies (e.g., Shafer et al., 2001). The hyperbolic-recency model with the normally distributed random noise term, $\varepsilon_x \sim N(0, \sigma_x)$, is given by:

$$y_x = k\left(\frac{xR_x^\alpha + p}{xR_x^\alpha + p + r}\right) + \varepsilon_x \qquad (3.1)$$

$$R_x = \frac{\sum_{i=1}^{x}(t_i - t_0)}{x(t_x - t_0)} \qquad (3.2)$$

$$y, x, p, k, \alpha \geq 0, \quad p + r > 0$$

Equation (3.1) represents the hyperbolic-recency model, where y_x is a measure of the productivity rate corresponding to x units of cumulative work, in units per hour.

The hyperbolic-recency model contains four parameters. The three parameters, k, p, and r, reflect a worker's learning behavior, based on the hyperbolic learning model introduced by Mazur and Hastie (1978). Fitted parameter k is the asymptotic steady-state productivity rate, measured in units per hour. Parameter p corresponds to a worker's initial expertise obtained from past similar experience in units of cumulative work. Parameter r represents the learning rate and is defined as the cumulative work and initial expertise required to reach half of k. That is, r is the learning rate relative to the individual's steady-state productivity rate. Hence, lower values of r correspond to more rapid learning, since workers take less cumulative work experience to reach $k/2$.

The component that represents the forgetting effect in this model is termed recency, R_x, shown in Equation (3.2). The recency model describes how recently an individual's practice was obtained. R_x is calculated from the ratio of the average elapsed time to the elapsed time for the most recent unit produced. Additional details may be found in Nembhard and Uzumeri (2000).

References

Bokhorst, J. and J. Slomp. 2000. Long-Term Allocation of Operators to Machines in Manufacturing Cells. *Group Technology/Cellular Manufacturing World Symposium*, San Juan, Puerto Rico, 153–158.

Brusco, M.J., T.R. Johns, and J.H. Reed. 1998. Cross-Utilization of a Two-Skilled Workforce. *International Journal of Operations and Production Management*, 18:6, 555–564.

Brusco, M.J. and T.R. Johns. 1998. Staffing a Multiskilled Workforce with Varying Levels of Perceptivity: An Analysis of Cross-training Policies. *Decision Sciences*, 29:2, 499–515.

Buzacott, J.A. 2002. The Impact of Worker Differences on Production System Output. *International Journal Production Economics*, 78, 37–44.

Campell, G.M. 1999. Cross-Utilization of Workers Whose Capabilities Differ. *Management Science*, 45:5, 722–732.

Cesani, V.I. and H.J. Steudel. 2000a. A Classification Scheme for Labor Assignments in Cellular Manufacturing Systems. *Group Technology/Cellular Manufacturing World Symposium*, San Juan, Puerto Rico, 147–152.

Cesani, V.I. and H.J. Steudel. 2000b. A Model to Quantitatively Describe Labor Assignment Flexibility in Labor Limited Cellular Manufacturing Systems. *Group Technology/Cellular Manufacturing World Symposium*, San Juan, Puerto Rico, 159–164.

Daniels, R.L., J.B. Mazzola, and D. Shi. 2004. Flow Shop Scheduling with Partial Resource Flexibility, *Management Science*, 50(5), 658–669.

Dunphy, D. and B. Bryant. 1996. Teams: Panacea or Prescriptions for Improved Performance? *Human Relations* 49:5, 677–699.

Eckstein, A.L.H. and T.R. Rohleder. 1998. Incorporating Human Resources in Group Technology/Cellular Manufacturing. *International Journal of Production Research*, 36:5, 1199–1222.

Felan, J.T. and T.D. Fry. 2001. Multi-Level Heterogeneous Worker Flexibility in a Dual Resource Constrained (DRC) Job-Shop. *International Journal of Production Research*, 39:14, 3041–3059.

Kher, H.V. 2000. Examination of Flexibility Acquisition Policies in Dual Resource Constrained Job Shops with Simultaneous Worker Learning and Forgetting Effect. *Journal of Operational Research Society*, 51, 592–601.

Kher, H.V., M.K. Malhotra, P.R. Philipoom, and T.D. Fry. 1999. Modeling Simultaneous Worker Leaning and Forgetting in Dual Resource Constrained Systems. *European Journal of Operational Research*, 115, 158–172.

Lance, C.E., W.R. Bennett, M.S. Teachoutt, D.L. Harville, and M.L. Welles. 1998. Moderators of Skill Retention Interval/Performance Decrement Relationships in Eight U.S. Air Force Enlisted Specialties. *Human Performance*, 11:1, 103–123.

Malhotra, M.K., T.D. Fry, H.V. Kher, and J.M. Donohue. 1993. The Impact of Learning and Labor Attrition on Worker Flexibility in DRC Job Shops. *Decision Science*, 24:3, 641–663.

Mazur, J.E. and R. Hastie. 1978. Learning as Accumulation: A Reexamination of the Learning Curve. *Psychological Bulletin*, 85:6, 1256–1274.

McCreery, J.K. and L.J. Krajewski. 1999. Improving Performance Using Workforce Flexibility in an Assembly Environment and Learning and Forgetting Effects. *International Journal of Production Research*, 37:9, 2031–2058.

Molleman, E. and J. Slomp. 1999. Functional Flexibility and Team Performance. *International Journal of Production Research*, 37:8, 1837–1858.

Nembhard, D.A. 2000. The Effects of Task Complexity and Experience on Learning and Forgetting: A Field Study. *Human Factors*, 42:2, 272–286.

Nembhard, D.A. and B.A. Norman. 2005. Worker Efficiency and Responsiveness in Cross Trained Teams, in revision *POMS*.

Nembhard, D.A. and N. Osothsilp. 2001. An Empirical Comparison of Forgetting Model. *IEEE Transaction on Engineering Management*, 48:3, 283–291.

Nembhard, D.A. and M. Uzumeri. 2000. An Individual-Based Description of Learning Within an Organization. *IEEE Transactions on Engineering Management*, 47:2, 370–378.

Osothsilp, N. 2002. Worker-Task Assignment Based on Individual Learning, Forgetting, and Task Complexity. Ph.D. dissertation. University of Wisconsin, Madison.

Park, S.P. 1991. The Examination of Worker Cross-Training in a Dual Resource Constrained Job Shop. *European Journal of Operational Research*, 51, 291–299.

Russell, R.S., P.Y. Huang, and Y. Leu. 1991. A Study of Labor Allocation Strategies in Cellular Manufacturing. *Decision Science*, 22, 594–611.

Shafer, S M., D.A. Nembhard, and M.V. Uzumeri. 2001. An Empirical Investigation of Learning, Forgetting, and Worker Heterogeneity on Assembly Line Productivity. *Management Science*, 47:12, 1639–1653.

Slomp, J. and E. Molleman. 2000. Cross-Training Policies and Performance of Teams. *Group Technology/Cellular Manufacturing World Symposium*, San Juan, Puerto Rico, 107–112.

Slomp, J. and E. Molleman. 2002. Cross-Training Policies and Team Performance. *International Journal of Production Research*, 40:5, 1193–1219.

Steudel, H.J. and P. Desruelle. 1992. *Manufacturing in the Nineties: How to Become a Mean, Lean, World-Class Competitor.* Van Nostrand Reinhold, New York.

Suresh, N.C. and G.J. Gaalman. 2000. Performance Evaluation of Cellular Layouts: Extensions to DRC System Contexts. *International Journal of Production Research*, 38:17, 4393–4402.

Uzumeri, M.V. and D.A. Nembhard. 1998. A Population of Learners: A New Way to Measure Organizational Learning. *Journal of Operations Management*, 16:5, 515–528.

van den Beukel, A.L. and E. Molleman. 1998. Multifunctionality: Driving and Constraining Forces. *Human Factors and Ergonomics in Manufacturing*, 8:4, 303–321.

van Oyen, M.P., E. Senturk-Gel, and W.S. Hopp, 2001. Performance Opportunities for Workforce Agility in Collaborative and Non-Collaborative Work Systems. *IIE Transactions*, 33, 761–777.

Wisner, J.D. and J.N. Pearson. 1993. An Exploratory Study of the Effects of Operator Relearning in a Dual Resource Constrained Job Shop. *Production and Operations Management*, 2:1, 55–68.

Wood, R.E. 1986. Task Complexity: Definition of the Construct. *Organizational Behavior and Human Decision Processes*, 37, 60–82.

chapter 4

Cross training in production systems with human learning and forgetting

David A. Nembhard and Bryan A. Norman

Contents

4.1 Introduction

Organizations have been making use of workforce cross training, also known as job rotation, job flexibility, and worker multifunctionality, for a considerable time. A number of related concepts have been widely recognized by both researchers and practitioners in a variety of theoretical models and organizational designs. In relating to these systems, in which workers are trained on alternate tasks in addition to their primary tasks, much of the workforce flexibility literature has attempted to address the fundamental question of how much cross training should be given to each worker, i.e., how many different tasks each worker should master under certain assumptions (e.g., Bokhorst and Slomp, 2000; Slomp and Molleman, 2000; Cesani and Steudel, 2000a, 2000b; and Russell et al., 1991). A key difficulty in determining the appropriate number of tasks is the fact that there is significant heterogeneity within many organizations with respect to human performance and capabilities (see, for example, Buzzacott, 2002). Specifically, levels of human learning and forgetting play an important role in productivity outcomes (Shafer et al., 2001).

Manufacturing and service industries face increasingly intense competition and must respond to a broad spectrum of market challenges. These forces are moving businesses toward greater agility and flexibility, thereby increasing product and service diversity and accelerating product and service innovation (Pine, 1993). As a result, many organizations are focusing more attention on cross training as an approach to meeting these challenges (McCune, 1994). It has been evident for some time that pure specialization has its limitations; furthermore, there is growing evidence that its modern replacement, job rotation, has its limitations as well. Whereas managers formerly relied on specialization based on the principles of Taylorism, today they hope that the benefits of job rotation will compensate for the loss of long periods of intensive training. However, at present, few guidelines exist to aid managers in decisions on job rotation and job sequencing.

Figure 4.1 illustrates a period of learning, forgetting, and subsequent relearning for one worker. Note that there is a large relative drop in performance after 2000 units of work, corresponding to the worker's 2-month

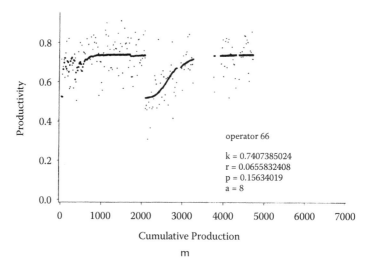

Figure 4.1 Sample learning/forgetting episode with a 2-month break after 2000 units.

absence from the task. Human responses such as these can have both significant and complex effects on productivity as these responses interact with numerous operational characteristics and decisions.

4.2 Overview of the literature

Dar-El (2000) describes a model for optimizing training schedules for a single task. Although the quantitative study of cross training in cellular manufacturing (CM) settings that includes individual learning and forgetting behaviors has been relatively sparse, numerous human factors studies have identified key knowledge and models about human performance. Engineering studies that integrate previously developed models have generally been based on arbitrary experimental conditions of learning, forgetting, and performance. Thus, while many studies have provided some general direction, there continues to be a need for empirically based experiments and simulations, as a number of researchers have pointed out (e.g., Kher, 2000; Kher et al., 1999; McCreery and Krajewski, 1999; and Malhotra et al., 1993).

4.2.1 Dual resource constrained (DRC) problems

The topic of creating a multiskilled workforce has been extensively investigated in a traditional manufacturing job shop system. Several researchers have illustrated the effect of cross training when learning and forgetting effects are present. In a dual resource constrained (DRC) shop setting, Malhotra et al. (1993) investigated the relationship between the learning losses and the benefits of a cross-trained workforce through a series of simulations. McCreery and Krajewski (1999) conducted a simulation study and found that when learning and forgetting effects exist, the degree of cross training and the worker deployment policy should depend primarily on the degree of task complexity and the degree of product variety. Kher (2000) conducted a full factorial designed experiment to investigate the training schemas obtained by cross-trained workers in a DRC shop under learning, relearning, and attrition conditions. Kher (2000) and Kher et al. (1999) concluded that the effectiveness of cross training depended significantly on the existing forgetting rate of the workers. In addition, Kher et al. (1999) remarked on the significant relationship between batch size and worker flexibility. The paper also suggested conditions under which worker flexibility would be suitable or even optimized based on production batch size and learning behaviors.

In a DRC job shop in the presence of operator absenteeism and assumed steady-state productivity, Molleman and Slomp (1999) suggested that the distribution of skills within teams and the degree of workforce cross training had a significant impact on system performance. Their findings indicated that a uniform workforce skill distribution resulted in better system performance. In other words, each worker should, optimally, master the same number of and types of tasks. Slomp and Molleman (2000) compared four cross-training policies based on the workload of the "bottleneck" worker in both static and dynamic circumstances. The results confirmed the intuition that the higher the level of cross training, the better the team performance (i.e., the lower workload for the bottleneck worker). On the other hand, consistent with other cross-training investigations, Slomp and Mollman also indicated that as the degree of cross training increases, the improvement of the system performance decreases, with decreasing marginal returns. Brusco and Johns (1997) developed a mixed integer programming (MIP) model to examine the cross-training configurations in a maintenance operation in a paper production system. The mathematical model was developed based on different levels of worker productivity in order to minimize the training cost. Under the assumption of a heterogeneous workforce in which each worker has a varying productivity level with respect to the task, each worker is assumed to have full capability (i.e., 100% production) in one of four available tasks and less capability in the secondary tasks. They found that, based on the overall cost savings, cross training employees in a secondary task at a 50% production level was adequate when compared to a 100% production level. Brusco et al. (1998) investigated a cross-training policy to minimize workforce requirements based on two-skilled workforce utilization. The result agreed with Brusco and Johns (1997) that the worker should not necessarily be cross trained at the 100% production level. Moreover, they also found that demand variation has a significant impact on the number of workers needed to meet the system requirements.

In the healthcare application domain, Campbell (1999) constructed a nonlinear programming model to explore and quantify the value of cross training 20 workers in four departments. The model was constructed based on departmental requirements and varying workforce capability. In a manner similar to Brusco and Johns (1997), each worker was assigned to be less proficient in secondary departments than in the primary one. A full factorial experimental design was conducted to investigate the effect of numerous factors on system performance. Among those factors were level of cross training, capability level, skill homogeneity, and demand variation. The experimental results concluded that the degree of cross training depends significantly on the demand variation. Under high demand conditions, the improvement of system performance due to an increase in the cross training level was greater than in the low-demand scenario. Moreover, in many circumstances, small amounts of cross training were more beneficial. The result was consistent with previous research (Brusco and Johns, 1997; Brusco et al., 1998) that indicated each worker need not necessarily be fully cross trained in all operations in order to achieve substantial benefits from cross training.

4.2.2   Worker–task assignment

Wisner and Pearson (1993) indicated the necessity of incorporating worker learning and relearning behavior in worker–task assignment studies. Their study showed that worker-relearning behavior has a significant impact on labor assignment policy, especially when a worker is required to transfer between tasks regularly. However, the assignment of *individual* workers to tasks based on learning and forgetting behavioral characteristics has received relatively little attention in the literature. This is partially due to the difficulty in collecting detailed performance data at the individual level. Without a complete understanding of the distribution of individual learning characteristics, worker–task assignment had been based on things that did not necessarily relate to productivity, let alone productivity under conditions of continuous change. As a result, research in the area of worker–task assignment has generally assumed steady-state productivity (i.e., no learning or forgetting). For example, Mazzola and Neebe (1988) modeled the assignment of jobs to agents under agent capacity constraints. More recently, Davis and Moore (1997) constructed an optimization model for assigning workers based on minimizing the number of task reassignments. Brusco and Johns (1997) developed an MIP assignment model in order to minimize total staffing cost. Molleman and Slomp (1999) presented a linear goal-programming method to assign workers to specific tasks in order to minimize shortages, makespan, and production time.

Many optimization-based assignment studies have been based on the classic assignment problem in which people are assigned to perform tasks in order to minimize or maximize a given system performance measure. In addition to the common constraints of worker and task requirements, recent studies have introduced "side constraints" to make the problem more practical (Caron et. al., 1999). For example, Molleman and Slomp (1999) developed a linear goal programming model subject to labor and skill requirements, (e.g., the shortage of capacity and the load of the busiest team member). Human factors, such as learning ability, motivational issues, and worker attitude should also be considered to make the study more applicable. Caron et al. developed an assignment model in order to maximize the number of jobs assigned, assuming two side constraints: seniority and job priority.

Other worker–task assignment studies involved worker heterogeneity, such as varying worker–task efficiency, to represent the productivity level among workers performing a specific task. Several studies defined worker efficiency as the fraction of standard production time (e.g., Nelson, 1970; Park, 1991; Bobrowski and Park, 1993; Kher and Malhotra 1994; Molleman and Slomp, 1999; and Bokhorst and Slomp, 2000). In addition, Campbell (1999) developed a nonlinear model for assigning cross-trained workers in a multidepartment service industry based on worker capability. Nevertheless, these solution methods often assume constant worker–task efficiency.

4.2.3 Cellular task assignment

Both qualitative and quantitative studies have shown that labor-related issues have a significant effect on achieving optimal system performance in a CM shop. Due to the substantial amount of interaction between labor skill and machining technology in a CM system, some researchers have taken into account workforce skill-related issues in the worker–task assignment problem. Warner et al. (1997) proposed a procedure for assigning workers to cells based on their human and technological skills. "Technological skills" were described as mechanical, mathematical, and measurement ability, while "human skills" were referred to as communication skills, leadership, teamwork, and decision-making ability.

There has been increasing interest in the subject of labor–task assignment for manufacturing cells. For example, Askin and Huang (1997) and Tharmmaphornphilas and Norman (2000) proposed solution procedures based on MIP models for assigning workers to cells and selecting a specific training program for each worker. Bhaskar and Srinivasan (1997) presented mathematical models for static and dynamic worker-assignment problems in order to balance workload and to minimize makespan. Through a series of simulation experiments, Russell et al. (1991) examined labor-assignment strategies based on a hypothetical group technology shop of three cells. The results favored a completely flexible, homogeneous workforce. In other words, each worker should be completely and equally cross trained and allowed to be assigned to any machine in any cell. Cesani and Steudel (2000b) conducted an empirical investigation in a two-operator cell and found that when the degree of task sharing (i.e., more than one operator is responsible for a machine) among the operators increases, the system performance improves significantly. Nevertheless, their notable finding, which coincides with the results in the DRC shop, indicates that as the cross-training level increases, the improvement of system performance decreases (Slomp and Molleman, 2000; Cesani and Steudel, 2000b). This raises the question, what is the appropriate level of cross training and worker flexibility in order to achieve optimal cell performance?

There are important limitations in past CM assignment research. For instance, studies have had the objective of balancing workload (an important objective) but not directly improving cell productivity. Also, many have assumed identical productivity levels for each operator over time. The impacts of individual learning and forgetting characteristics have been relatively absent in the investigation of cellular worker–task assignment problems.

A number of studies have suggested that the need for workforce flexibility was not only driven by customer demand fluctuation or low staffing levels but also by the variation in the workforce supply due to operator absenteeism (Bokhorst and Slomp, 2000; Molleman and Slomp, 1999; and van den Beukel and Molleman, 1998). Van den Beukel and Molleman conducted a survey study in 10 different organizations, and results indicated that the need for a flexible, multiskilled workforce in response to the problem of operator absenteeism is more significant when rapid response to customer demand is crucial but relatively less important when external resources, such as temporary workers or other equipment, are available. Though the negative impact of operator absenteeism in work cells has been recognized in the literature, few quantitative studies have been conducted to investigate specifically how cross-training policies could effectively respond to the problem (e.g., Bokhorst and Slomp, 2000; Molleman and Slomp, 1999).

4.2.4 Models of individual learning and forgetting

Although worker-to-task assignment has been considered in the literature, researchers typically assume that workers operate at one productivity level or at a discrete set of productivity levels. However, in reality, workers learn and forget tasks based on how often they perform a task and how much time has elapsed since they last did the task. Thus, it is important to consider when possible the issue of worker learning and forgetting when determining the best worker assignment strategy.

Patterns of learning and forgetting have been studied from a variety of perspectives. For example, psychologists have studied and modeled individual learning processes (Mazur and Hastie, 1978), while engineers have modeled learning as it relates to manufacturing costs (Yelle, 1979), process times (Adler, and Nanda 1974a, 1974b; Axsater and Elmaghraby, 1981; Sule, 1981; Pratsini et al., 1993; Smunt, 1987), setup time learning (Karwan, Mazzola, and Morey, 1988; Pratsini et al., 1994), and line balancing (Dar-El and Rubinovitz, 1991).

At the organizational level, research has centered on the overall implications of learning across large organizational units separated by functional, hierarchical, or geographical boundaries. These organizational learning curves have allowed managers to examine experiential productivity improvements, transfer of knowledge between parallel units (Epple et al., 1991), and the ability of an organization to retain knowledge over time (Argote et al., 1990; Epple et al., 1991). Organizational learning curves are particularly appropriate for measuring improvements in plant-wide productivity over relatively long periods of time. Learning in organizations, characterized by make-to-order operations, has led to the development of more general models of organizational learning (Womer, 1979; Dorroh et al., 1994). A comprehensive view of the organizational learning literature can be found in Argote (1999).

Another stream of research has examined learning at the micro-level in order to investigate the mechanisms by which learning occurs in individuals (e.g., Anderson, 1982; Hancock, 1967; Mazur and Hastie, 1978). Much of this research has examined the determination of a most appropriate mathematical form for individual learning. While the majority of work has centered on the log-linear model (e.g., Badiru, 1992; Buck and Cheng, 1993; Glover, 1966; Hancock, 1967), there have been continuing efforts to identify alternative formulations. Indeed, some alternative forms were found to represent observed behavior better than the log-linear model in their respective industrial settings (e.g., Asher, 1956; Carr, 1946; Ebbinghaus, 1885; Knecht, 1974; Levy, 1965; Pegels, 1969).

Research addressing the retention of learning (forgetting) suggests that the longer a person studies, the longer the retention (Ebbinghaus, 1885). The forgetting process has been shown to be describable by the traditional log-linear performance function (Anderson, 1985). Based on individuals' ability to recall nonsense syllables, Farr (1987) found that as the meaningfulness of the task increased, retention also increased. The retention of learning at an organizational level has also been found to decline rapidly in general (Argote, 1990). Several research efforts have attempted to explain the impact that forgetting has on production scheduling (Smunt, 1987; Fisk and Ballou, 1982; Sule, 1983; Khoshnevis and Wolfe, 1983a, 1983b). Globerson et al. (1989) developed a model of learning and forgetting for a data correction task. The model indicates that forgetting behaves much like the mirror image of the learning process. Bailey (1989) found that the forgetting of a procedural task could be expressed as a linear function of the product of the amount learned and the log of the elapsed time.

Shafer et al. (2001) introduced an approach to measuring organizational learning wherein individual worker heterogeneity is modeled. The resulting distribution of individual learning patterns provides a description of organizational learning as well as the associated detail required to parse the worker population into complementary clusters of workers. Nembhard and Uzumeri (2000a) provide an important extension of this approach by incorporating both learning and forgetting into an individual-based model of productivity. In this model, both the learning and the forgetting components were shown to be preferred models among numerous candidate models (Nembhard and Uzumeri, 2000b; Nembhard and Osothsilp, 2001, respectively). To illustrate, the model shown in Figure 4.2 is designed to measure learning and

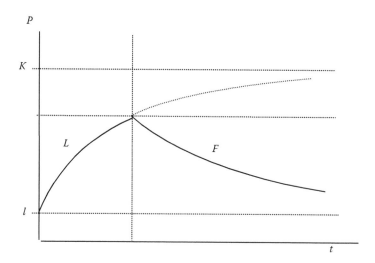

Figure 4.2 Model of individual learning and forgetting characteristics.

forgetting, where P is a measure of the productivity rate corresponding to t time units. In this model, parameter l represents the initial productivity of the worker, K represents the steady-state productivity, L represents the learning rate, and F represents the forgetting rate.

A potentially significant limitation of many learning/forgetting models in the literature is that they commonly are designed as measurement tools and may not lend themselves to use within other analytical or optimization frameworks.

4.3 Models

This section presents several models that may aid in the understanding of productivity and flexibility in systems with significant job rotation. First, we discuss a math programming-based decision model for assigning and scheduling workers to tasks. The model incorporates the effects of individual worker learning and forgetting for worker–task assignment.

4.3.1 A worker–task assignment model

Worker–task assignment models can be represented mathematically and solved in a number of different ways, including math programming, enumerative methods (e.g., dynamic programming), and various meta-heuristics (e.g., tabu search, genetic algorithms, simulated annealing, etc.) The following model utilizes math programming and incorporates individual worker learning and forgetting.

We illustrate the construction of a worker–task assignment model with a manufacturing example, where differences between workers' skills and their task-learning rates are incorporated. The example considers a common type of manufacturing setting in which work is passed from station to station, with intermediate work-in-process (WIP) buffers between stations and the objective of maximizing system output for a given number of workers over a fixed time interval (Figure 4.3). We make the following assumptions in

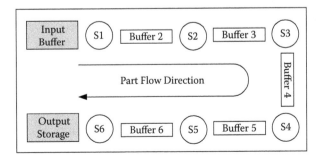

Figure 4.3 Serial process configuration.

order to illustrate the construction of a worker–task assignment model using this approach.

- There are n workers available to perform the tasks.
- There are m sequentially numbered tasks to be completed within the system, and each task is done at a separate workstation. The model can be modified to permit multiple tasks at each workstation.
- Time is separated into p periods, where a worker performs only one task during each time period.
- There is buffer storage between the workstations. However, the size of this buffer storage can be constrained.
- The rate of learning and forgetting is a function of how much time the worker has spent performing a particular task.
- Workers are heterogeneous with respect to initial productivity, learning, and forgetting rates for each task.

Notation I Set of workers $i = 1, 2, \ldots n$.

J Set of tasks $j = 1, 2, \ldots m$.

T Set of time periods $t = 1, 2, \ldots p$.

$X_{i,j,t}$ Binary variable indicating whether worker i does task j during time t.

$P_{i,j,t}$ Productivity if worker i does task j during time t (*based on learning/forgetting model*).

$O_{i,j,t}$ Output from worker i performing task j for period t.

l_{ij} Initial productivity level for worker i on task j (i.e., for the first unit of work).

$K_{i,j}$ Asymptotic learning constant for worker i performing task j (steady-state production rate).

$L_{i,j}, F_{i,j}$ Individual parameters of learning and forgetting, respectively, for worker i on task j.

$B_{j,t}$ Buffer (inventory) level at task j at the end of period t.

BI_j Beginning inventory for task j, (period 0).

Objective

$$\text{Maximize} \sum_{i=1}^{n} \sum_{t=1}^{p} O_{i,m,t} \tag{4.1}$$

Constraints

$$O_{i,j,t} \leq X_{i,j,t} P_{i,j,t} \times (\text{Period Length}) \quad \forall i, \forall j, \forall t \tag{4.2}$$

$$B_{1,1} = BI_1 - \sum_{i=1}^{n} O_{i,1,1} \tag{4.3}$$

$$B_{j,1} = BI_j + \sum_{i=1}^{n} O_{i,j-1,1} - \sum_{i=1}^{n} O_{i,j,1} \quad j = 2,\ldots m \tag{4.4}$$

$$B_{1,t} = B_{1,t-1} - \sum_{i=1}^{n} O_{i,1,t} \quad t = 2,\ldots p \tag{4.5}$$

$$B_{j,t} = B_{j,t-1} + \sum_{i=1}^{n} O_{i,j-1,t} - \sum_{i=1}^{n} O_{i,j,t} \quad j = 2,\ldots m, \ t = 2,\ldots p \tag{4.6}$$

$$\sum_{j=1}^{m} X_{i,j,t} \leq 1 \quad \forall i, \forall t \tag{4.7}$$

$$\sum_{i=1}^{n} X_{i,j,t} \leq 1 \quad \forall j, \forall t \tag{4.8}$$

$$P_{i,j,t} = l_{i,j} + K_{i,j} \left[1 - \exp\left(-\frac{1}{L_{i,j}} \sum_{k=1}^{t} X_{i,j,k} \right) \right] \exp\left[\frac{1}{F_{i,j}} \left(\sum_{k=1}^{t} X_{i,j,k} - t \right) \right] \quad \forall i, \forall j, \forall t \tag{4.9}$$

In this illustrative model of a manufacturing system, the objective (4.1) is to maximize the total output of the last workstation in the system i.e., create as much finished product as possible using n workers over the p periods. Constraint (4.2) provides an approximation of the total output, O, given the period length. The approximation is reasonable for relatively short periods and, in general, will underestimate total productivity. Constraint (4.3) to constraint (4.6) are inventory balance constraints that keep track of the number of units of work available to the adjacent downstream workstation. Constraint (4.7) ensures that each worker is assigned at most one task during each time interval. Constraint (4.8) similarly ensures that each work-station has at most one worker assigned to it. Constraint (4.9) is necessary to determine the production rate for worker i on task j during period t, which depends on the worker's experience performing that task. The current for-mulation is based on log-linear learning and forgetting, and it is assumed that the worker's learning function for a particular task is only related to the worker's initial productivity level for that task and how much time the worker has spent performing that task. Similarly, we assume that the worker's forgetting is only relative to the time spent not performing the task. It is the inclusion of constraint (4.9) that makes this problem nonlinear and this particular formulation a challenge to solve.

Several challenges present themselves in developing these assign-ment models. Learning and forgetting are explicitly characterized for each worker and task in this formulation. This introduces nonlinearity into the model but better represents real work environments and the tradeoffs between flexibility and production. Thus, the form of the model produces a Mixed Integer Nonlinear Program (MINLP), which is a class of problems difficult to solve. Second, the form of the learning and forgetting model in constraint (4.9) significantly affects the level of difficulty in solving the MINLP. As noted above, learning/forgetting models available in the literature were generally not intended for math programming formula-tions. Thus, the development of models that better lend themselves to the MINLP formulation is of considerable interest for future research efforts.

4.3.2   Alternate worker–task assignment models

An assignment model, such as the one described, may be varied to suit more specific scenarios and variants of the stated problem. For instance, if we assume that the $P_{i,j,t}$ values are constant (i.e., equal to K_{ij}), then the resulting problem becomes an MIP problem, which can be solved far more readily than the MINLP problem. Also, since constraint (4.3) to constraint (4.6) models the precedence requirements of the serial production system, a selec-tive removal of some of these constraints would allow some tasks to be performed in parallel with others.

Alternate objective functions can be investigated using this approach, with the idea that there are three main variables involved in the problem: (1) the number of workers available, (2) the time available, and (3) the number of units to be produced. Thus, the model in the preceding section holds the number of workers and the number of time periods constant, and we max-imize the output that is attainable under those constraints. A second objective is to minimize the total number of workers required to produce a predeter-mined number of units over a fixed time horizon. Such an approach might be appropriate in situations where there is a desire to minimize labor require-ments and costs as long as due dates can be met for known demand quan-tities. Thus, objective (4.1) could be replaced with objective (4.10), combined with the addition of constraint (4.11), where M is a large constant and Y_i is a binary variable indicating whether or not worker i is used.

$$\text{Minimize} \sum_{i=1}^{n} Y_i \tag{4.10}$$

$$\sum_{j=1}^{m}\sum_{t=1}^{p} X_{i,j,t} \le Y_i M \quad \forall i = 1,2,\dots n \tag{4.11}$$

With a third type of objective, one can minimize the time required for n workers to meet production requirements. That is, if labor resources are limited and demand is known, management may need to determine effective worker–task schedules to minimize the time needed to produce enough to satisfy the demand level. Using a similar approach to that in the preceding case, in objective (4.12), Y_t is a binary variable indicating that period t is used, and the additional constraint (4.13) must also be added to the model.

$$\text{Minimize} \sum_{t=1}^{p} Y_t \tag{4.12}$$

$$\sum_{i=1}^{n}\sum_{j=1}^{m} X_{i,j,t} \le Y_t M \quad \forall t = 1, 2,\dots p \tag{4.13}$$

Using a quadratic loss approach can also limit the number of time periods required to meet a given demand level. This minimizes the elapsed time to produce a predetermined number of units. Thus, an alternate objective function, such as that in Equation (4.14), will also serve this purpose.

$$\text{Minimize} \sum_{i=1}^{n}\sum_{j=1}^{m}\sum_{t=1}^{p} t^2 X_{i,j,t} \tag{4.14}$$

As a different interpretation on minimizing the time required for production, we may consider objective (4.15), which will minimize the number of periods workers are assigned to tasks. This will minimize the worker resources by limiting the total number of worker periods used. However, this expression will not necessarily minimize the overall elapsed time.

$$\text{Minimize} \sum_{i=1}^{n}\sum_{j=1}^{m}\sum_{t=1}^{p} X_{i,j,t} \tag{4.15}$$

Limits on buffer storage at each workstation can be controlled by adding constraint (4.16), where $a_{j,j+1}$ represents the maximum number of parts that can be placed in the buffer storage between stations j and $j + 1$.

$$B_{j,t} \leq a_{j,j+1} \quad \forall j = 1, 2, \ldots m, \ \forall t = 1, 2, \ldots p \tag{4.16}$$

Pay grades based on an individual's capability might also be incorporated into such a model. In many manufacturing and service settings, worker pay is based on different job grades, which in turn are based on worker skill and task complexity. Workers in higher pay grades can perform tasks that are rated at lower pay grades, but the reverse is generally not acceptable in practice. Therefore, an additional objective is to consider the number of workers of each pay grade required for each assignment and to use this information in the cost evaluation for an assignment.

Task-assignment durations can be set for certain tasks where a minimum number of hours are required for workers on each task. This also has the potential to be used to model the requirement that, for some tasks, the workers should not rotate too often.

Balancing the workload among the workers is a practical requirement in many work settings. That is, it may not be acceptable to have some workers required to perform a greater number of tasks or for a greater length of time. While it may be feasible to achieve a better objective value without such requirements, the need for certain types of equity is a reality in manufacturing environments. For example, a formulation without requirements of this type might assign one worker to four tasks and another to only one task, which would be unacceptable and impractical.

Work layout design is an important consideration in certain industries. For example, in cellular manufacturing, the question of whether to use I-shaped or U-shaped layouts can be modeled, since in practice these may behave differently from one another.

Nonadjacent worker assignments may be limited by incorporating travel distance measures into the assignment model in such a manner that the number and frequency of nonadjacent task assignments will be acceptable. A similar approach was taken by Kher and Malhotra (1994), wherein a worker transfer delay was incorporated into the model.

4.4   An example application

As an illustrative example of the relevance of including learning and forgetting in this decision problem, Table 4.1 summarizes solutions from the MINLP with objective (4.1) and constraint (4.2) through constraint (4.9), using the math programming software (GAMS®, CONOPT®, and SBB®). It illustrates the fact that basing assigned schedules on different assumptions will produce both different worker–task cross-training patterns and different overall

Table 4.1 Optimal Station Assignment/Schedule for Two Workers at Four Stations over 10 Time Periods[1]

Assignment based on	Output[2] (units)	Worker[3]	1	2	3	4	5	6	7	8	9	10
							Period					
Steady-state	18.3	1	2	3	3	1	3	4	4	3	4	4
productivity		2	1	1	2	2	4	2	1	4	2	3
learning	28.3	1	1	1	3	3	1	3	1	3	3	4
only		2	2	2	4	4	2	4	2	4	4	1
learning and	36.1	1	1	1	1	4	4	4	4	1	4	4
forgetting		2	2	2	2	3	3	3	3	4	2	3

[1]To produce each unit of output, workers must pass work through each of the four stations in order. Stations 1 to 4 are ordered from most complex (time-consuming) to least complex. Buffers are assumed to have sufficient availability of work initially.

[2]In all three cases, the assignment output is evaluated assuming that the workers learn and forget. The assignments themselves are based on the criterion shown in the first column.

[3]Worker 1 has a 15% higher steady-state productivity rate. Worker 2 learns 35% faster; a phenomenon not uncommon in the field (Nembhard and Uzumeri, 2000).

productive output. For example, when considering both learning and forgetting in the assignment, we note that worker 1 (the slower learner) cross trains on only two tasks, as opposed to four tasks under the steady-state (no learning or forgetting) assumption. Further, it is notable that when workers rotate between twice as many jobs during the same time, productivity drops substantially, as evidenced by the drop in productivity from 36.1 to 18.3 units. This highlights the importance of considering learning and forgetting during worker–task assignment. The differences in output will tend to be smaller over longer planning periods, as workers progress on the learning curve toward steady state. However, as noted previously, many organizations undergo rapid changes in products, technology, and workers, making shorter planning horizons the norm.

While finding provably optimal solutions to our model would prove to be beneficial, it may not be practical to solve the problems exactly due to the computation time complexity. There are several heuristic approaches developed for similar types of problems (Nembhard, 2001; Tharmmaphornphilas and Norman, 2000; Norman and Bean, 1999; and Carnahan et al. 2000).

4.5   Summary and perspectives

Cross training is both a potential means for creating and transferring technological knowledge and a common method to obtain a partially flexible workforce. Organizations have for some time used cross training as a method for increasing manufacturing flexibility and as a productivity and quality improvement methodology, in order to permit workers to expand their knowledge toward performing multiple tasks. Studies examining the use of cross training to better understand its impacts on individual learning, forgetting, and productivity in the organization may provide clearer technological bases, allowing organizations to implement cross training in an informed manner.

Extensions of existing models to a range of process characteristics, operational design conditions, and production measures will further aid managers and researchers in understanding and setting multifunctionality and cross-training levels. For example, absenteeism rates, worker turnover rates, frequencies of product revisions, worker assignment patterns, and the effects of various skill distributions along with differential cross training, in which some workers are specialized and others are cross trained, are of widespread interest.

Acknowledgments

A portion of this work was supported by grants from the National Science Foundation for the first author under SES9986385 and SES0217666 and for the second author under SES0217189.

References

Adler, G.L. and R. Nanda, The Effects of Learning on Optimal Lot Size Determination – Single Product Case, *AIIE Transactions*, 6(1), (1974a), 14–20.

Adler, G.L. and R. Nanda, The Effects of Learning on Optimal Lot Size Determination – Multiple Product Case, *AIIE Transactions*, 6(1), (1974b), 21–27.

Anderson, J.R., Acquisition of Cognitive Skill, *Psychological Review*, 89 (1982), 369–406.

Anderson, J.R., *Cognitive Psychology and Its Implications* (2nd ed.), W.H. Freeman and Company, New York, (1985).

Argote, L., S.L. Beckman, and E. Epple, The Persistence and Transfer of Learning in Industrial Settings, *Management Science*, 36 (1990), 140–154.

Argote, L., *Organizational Learning: Creating, Retaining and Transferring Knowledge*, Kluwer, Boston, (1999).

Asher, H., Cost-Quantity Relationships in the Airframe Industry, Rep. No. R291, The Rand Corporation, Santa Monica, CA, (1956).

Askin, R.G. and Y. Huang, Employee Training and Assignment for Facility Reconfiguration, *6th Industrial Engineering Research Conference Proceedings, IIE*, (1997), 426–431.

Axsater, S. and S.E. Elmaghraby, A Note on EMQ Under Learning and Forgetting, *AIIE Transactions*, 13 (1981), 86–90.

Badiru, A.B., Computational Survey of Univariate and Multivariate Learning Curve Models, *IEEE Transactions on Engineering Management*, 39 (1992), 176–188.

Bailey, C.D., Forgetting and the Learning Curve: A Laboratory Study, *Management Science*, 35(3), (1989), 340–352.

Bhaskar, K. and G. Srinivasan, Static and Dynamic Operator Allocation Problems in Cellular Manufacturing Systems, *International Journal of Production Research*, 35(12), (1997), 3467–3481.

Bobrowski, P.M. and P.S. Park, An Evaluation of Labor Assignment Rules When Workers Are Not Perfectly Interchangeable, *Journal of Operations Management*, 11 (1993), 257–268.

Bokhorst, J. and J. Slomp, Long-Term Allocation of Operators to Machines in Manufacturing Cells, *Group Technology/Cellular Manufacturing World Symposium*, San Juan, Puerto Rico, (2000), 153–158.

Brusco, M.J. and T.R. Johns, Staffing a Multi-Skilled Workforce with Varying Levels of Perceptivity: An Analysis of Cross-Training Policies,*Decision Sciences*, 29(2), (1997), 499–515.

Brusco, M.J., T.R. Johns, and J.H. Reed, Cross-Utilization of Two-Skilled Workforce, *International Journal of Operations and Production Management*, 18(6), (1998), 555–564.

Buck, J.R. and S.W.J. Cheng, Instructions and Feedback Effects on Speed and Accuracy with Different Learning Curve Models, *IIE Transactions*, 25 (1993), 34–37.

Buzacott, J.A., The Impact of Worker Differences in Production System Output, *International Journal of Production Economics*, 78, (2002), 37–44.

Campbell, G., Cross-Utilization of Workers Whose Capabilities Differ, *Management Science*, 45(5), (1999), 722–732.

Carnahan, B.J., M.S. Redfern, and B.A. Norman, Designing Safe Job Rotation Schedules Using Optimization and Heuristic Search, *Ergonomics*, 43(4), (2000), 543–560.

Caron, G., P. Hansen, and B. Jaumard, The Assignment Problem with Seniority and Job Priority Constraints, *Operation Research*, 47(3), (1999), 449–453.

Carr, G.W., Peacetime Cost Estimating Requires New Learning Curves, *Aviation*, 45 (1946), 76–77.

Cesani, V.I. and H.J. Steudel, A Classification Scheme for Labor Assignments in Cellular Manufacturing Systems, *Group Technology/Cellular Manufacturing World Symposium*, San Juan, Puerto Rico, (2000a), 147–152.

Cesani, V.I. and H.J. Steudel, A Model to Quantitatively Describe Labor Assignment Flexibility in Labor Limited Cellular Manufacturing Systems, *Group Technology/Cellular Manufacturing World Symposium*, San Juan, Puerto Rico, (2000b), 159–164.

Dar-El, E.M. and J. Rubinovitz, Using Learning Theory in Assembly Lines for New Products, *International Journal of Production Economics*, 25 (1991), 103–109.

Dar-El, E.M., *Human Learning: From Learning Curves to Learning Organizations*, Kluwer, Boston, (2000).

Davis, G.S. and J.S. Moore, Manpower Allocation for Engineering in the Product Development Environment, *Proceedings, Annual Meeting of the Decision Sciences Institute*, 2, Orlando, FL, (1997), 918–920.

Dorroh, J., T.R. Gulledge, and N.K. Womer, Investment in Knowledge: A Generalization of Learning by Experience, *Management Science*, 40(8), (1994), 947–958.

Ebbinghaus, H., *Memory: A Contribution to Experimental Psychology* (Translated by H.A. Ruger and C.E. Bussenius, 1964), Dover, New York, (1885).

Epple, D., L. Argote, and R. Devadas, Organizational Learning Curves: A Method for Investigation of Intra-Plant Transfer of Knowledge Acquired through Learning by Doing, *Organization Science*, 2 (1991), 58–70.

Farr, M.J., *The Long-Term Retention of Knowledge and Skills: A Cognitive and Instructional Perspective*, Springer-Verlag, New York, (1987).

Fisk, J.C. and D.P. Ballou, "Production Lot Sizing under a Learning Effect," *IIE Trans.*, 14 (1982), 257.

Globerson, S., N. Levin, and A. Shtub, The Impact of Breaks on Forgetting When Performing a Repetitive Task, *IIE Transactions*, 21(4), (1989), 376–381.

Glover, J.H., Manufacturing Progress Functions I. An Alternative Model and its Comparison with Existing Functions, *International Journal of Production Research*, 4 (1966), 279–300.

Hancock, W.M., The Prediction of Learning Rates for Manual Operations, *Journal of Industrial Engineering*, 18 (1967), 42–47.

Kher, H.V. and M.K. Malhotra, Acquiring and Operationalizing Worker Flexibility in Dual-Resource-Constrained Job Shops with Worker Transfer Delay and Learning Losses, *Omega*, 22(5), (1994), 521–533.

Kher, H.V., Examination of Flexibility Acquisition Policies in Dual-Resource-Constrained Job Shops with Simultaneous Worker Learning and Forgetting Effect, *Journal of Operational Research Society*, 51 (2000), 592–601.

Kher, H.V., M.K Malhotra, P.R. Philipoom, and T.D. Fry, Modeling Simultaneous Worker Learning and Forgetting in Dual-Resource-Constrained Systems, *European Journal of Operational Research*, 115 (1999), 158–172.

Knecht, G.R., Costing Technological Growth and Generalized Learning Curves, *Operations Research Quarterly*, 25 (1974), 487–491.

Knoshnevis, B. and P.M. Wolfe, An Aggregate Production Planning Model Incorporating Dynamic Productivity. Part I Model Development, *IIE Transactions*, 15(2), (1983a), 111–118.

Knoshnevis, B. and P.M. Wolfe, An Aggregate Production Planning Model Incorporating Dynamic Productivity. Part II: Solution Methodology and Analysis, *IIE Transactions*, 15(4), (1983b), 283–291.

Levy, F.K., Adaptation in the Production Process, *Management Science*, 11 (1965), 136–154.

Malhotra, M.K., T.D. Fry, H.V. Kher, and J.M. Donohue, The Impact of Learning and Labor Attrition on Worker Flexibility in DRC Job Shops, *Decision Science*, 24(3), (1993), 641–662.

Mazur, J.E., and R. Hastie, Learning as Accumulation: A Reexamination of the Learning Curve, *Psychological Bulletin*, 85(6), (1978), 1256–1274.

Mazzola, J.B., and A.W. Neebe, Bottleneck Generalized Assignment Problems, *Engineering Costs and Production Economics*, 14(1), (1988), 61–66.

McCreery, J.K., and L.J. Krajewski, Improving Performance Using Workforce Flexibility in an Assembly Environment and Learning and Forgetting Effects, *International Journal of Production Research*, 37(9), (1999), 2031–2058.

McCune, J.C., On the Train Gang, *Management Review*, American Management Association, October, (1994), 57–60.

Molleman, E. and J. Slomp, Functional Flexibility and Team Performance, *International Journal of Production Research*, 37(8), (1999), 1837–1858.

Nelson, R.T., A Simulation of Labor Efficiency and Centralized Assignment in a Production Model, *Management Science*, 17(2), (1970), 97–106.

Nembhard, D.A. and M.V. Uzumeri, Experiential Learning and Forgetting for Manual and Cognitive Tasks, *International Journal of Industrial Ergonomics*, 25(4), (2000a), 315–326.

Nembhard, D.A. and M.V. Uzumeri, An Individual-Based Description of Learning Within an Organization, *IEEE Transactions on Engineering Management*, 47(3), (2000b), 370–378.

Nembhard, D.A., A Heuristic Approach for Assigning Workers to Tasks Based on Individual Learning Rate, *International Journal of Production Research*, 39(9), (2001),1955–1968.

Nembhard, D.A. and Osothsilp, N. An Empirical Comparative Study of Models for Measuring Intermittent Forgetting, *IEEE Transactions on Engineering Management*, 48(3), (2001), 283–291.

Norman, B.A. and J.C. Bean, A Genetic Algorithm Methodology for Complex Scheduling Problems, *Naval Research Logistics*, 46, (1999), 199–211.

Park, P.S., The Examination of Worker Cross-Training in a Dual-Resource-Constrained Job Shop, *European Journal of Operational Research*, 51(2), (1991), 219–299.

Pegels, C.C., On Startup or Learning Curves: An Expanded View, *AIIE Transactions*, 1 (1969), 216–222.

Pine, B.J., Mass Customization: The New Frontier in Business Competition, *Harvard Business School Press*, (1993), Boston.

Pratsini, E., J.D. Camm, and A.S. Raturi, Capacitated Lot Sizing Under Setup Learning, *European Journal of Operational Research*, 72 (1994), 545–557.

Pratsini, E., J.D. Camm, and A.S. Raturi, Effect of Process Learning on Manufacturing Schedules, *Computer Ops. Res.*, 20, (1993), 15–24.

Russell, R.S., P.Y. Huang, and Y. Leu, A Study of Labor Allocation Strategies in Cellular Manufacturing, *Decision Science*, 22, (1991), 594–611.

Shafer, S.M., D.A. Nembhard, and M.V. Uzumeri, Investigation of Learning, Forgetting, and Worker Heterogeneity on Assembly Line Productivity. *Management Science*, 47(12), (2001), 1639–1653.

Slomp, J. and E. Molleman, Cross-Training Policies and Performance of Teams, *Group Technology/Cellular Manufacturing World Symposium*, San Juan, Puerto Rico, (2000), 107–112.

Smunt, T.L., The Impact of Worker Forgetting on Production Scheduling, *International Journal of Production Research*, 25(5), (1987), 689–701.

Sule, D.R., A Note on Production Time Variation in Determining EMQ under Influence of Learning and Forgetting, *AIIE Transactions*, 31 (1981), 91.

Sule, D.R., Effect of Learning and Forgetting on Economic Lot Size Scheduling Problem, *International Journal of Production Research*, 21(5), (1983), 771–786.

Tharmmaphornphilas, W. and B.A. Norman, A Heuristic Search Algorithm for Stochastic Job Rotation Scheduling, *9th Industrial Engineering Research Conference*, May 21–23, Cleveland, OH, CD-Rom format (2000).

Tharmmaphornphilas, W. and B.A. Norman, A Quanitative Method for Determining Proper Job Rotation Intervals, *Annals of Operations Research* 128 (1–4), (2004), 251–266.

van den Beukel and E. Molleman, Multifunctionality: Driving and Constraining Forces, *Human Factors and Ergonomics in Manufacturing*, 8(4), (1998), 303–321.

Warner, R.C., K.L. Needy, and B. Bidanda, Worker Assignment in Implementing Manufacturing Cells, *6th Industrial Engineering Research Conference Proceedings*, IIE, Miami, FL, (1997), 240–245.

Wisner, J.D. and J.N. Pearson, An Exploratory Study of the Effects of Operator Relearning in a Dual-Resource-Constrained Job Shop, *Production and Operations Management*, 2(1), (1993), 55–69.

Womer, N.K., Learning Curves, Production Rate, and Program Costs, *Management Science*, 25(4), (1979), 312–319.

Yelle, L.E., The Learning Curve: Historical Review and Comprehensive Survey, *Decision Sciences*, 10(2), (1979), 302–328.

chapter 5

Valuing workforce cross-training flexibility

Ruwen Qin and David A. Nembhard

Contents

5.1   Introduction

Workforce flexibility is an often sought-after capability of production systems. Manufacturing and service organizations understand that the composition of a workforce (i.e., the number of workers who are competent at specific tasks in a production system and their skill levels on these tasks) may significantly affect productivity. Cross training is one such lever on productivity because it can potentially increase operational productivity by training the workforce in a specific tactical or strategic manner. The capabilities of a workforce given a level of cross training may be considered an important component of workforce flexibility (e.g., see Ebeling and Lee, 1994; Gerwin, 1993; Molleman and Slomp, 1999).

Cross training is often used to ameliorate stochastic system dynamics due to the heterogeneity of workers, variance in worker capacity, changes in demand, technological change, equipment failure, and other related dynamics. It may be possible that by properly diversifying the skills of workers, a system will have a reduced chance of idling. Further, if each task has multiple trained workers available, the possibility that a given task becomes a bottleneck may be substantially reduced. As a result, cross training has the potential to increase a system's average productivity and strengthen management's ability to expand production capacity. From this view, cross training provides an operational flexibility. In a highly uncertain and competitive environment, operational flexibility brought on by cross training has strategic implications. For example, when a product performs better than expected in the marketplace, an expansion of production capacity may allow for a corresponding increase in profits (Russell et al., 1991; van den Beukel and Molleman, 1998).

However, both flexibility, in general, and cross training, more specifically, are difficult to value and are often undertaken on the basis of their own merits. For example, there are numerous motivations for cross training a workforce including increased employee motivation, productivity, job enrichment, reduced physical and mental stress, among them. Many applications of cross training involve relatively static systems or, at least, those for which an assumption of deterministic behavior seems reasonable. It is often assumed that cross training will lead to positive outcomes when applied to such systems. These outcomes may not be realized once market risks are considered. That is, actual demand for a product is generally not precisely predictable *a priori*. As a result, realized profits will generally deviate from these initial expectations. Since one motivation for cross training is to be more flexible and, thus, provide a faster response to these fluctuations, there may be a variety of costs involved with obtaining this additional flexibility. Therefore, from an economic point of view, decisions on setting cross-training levels made at the start of production are not likely to remain robust once market risks come into play. A dynamic cross-training decision with considerations for environmental dynamics is potentially more appropriate. This chapter outlines an approach for valuing cross training as an operational flexibility using a real options approach. The approach is intended to incorporate operational flexibility in an option-pricing framework.

The basic viewpoint of this valuation approach is as follows. From an economic point of view, cross training is an investment in workforce flexibility with quantifiable upfront training costs to obtain a cross-trained workforce. A flexible workforce may provide a benefit to production systems at some unknown times in the future, and these systems may have increased effective worker capacity without additional hiring. Alternately, cross training may lower current production efficiency temporarily as workers become less specialized during cross training, but have later advantages in meeting demand for future products or services. So, the value of cross training dynamically depends on some other underlying assets, and the payoff of cross training may occur at some unknown times in the future. This point of view leads us to model cross training using a real options framework. Thus, cross training is a type of real option and the holder of this option has the right rather than the obligation to conduct cross training. We first briefly describe the real options framework in Section 5.2, followed by several small illustrative examples. Section 5.3 outlines a detailed model for valuing cross training using the real options approach, followed in Section 5.4 by a detailed numerical example.

5.2   What are real options?

An option is defined as the right, but not the obligation, to take an action at a future time. That is, the holder of the option is not obligated to exercise this right. The price paid when the option is exercised is called *exercise price*, and the time when the option is exercised is the *expiration date*. Two common types of options are *call options* and *put options*. Call options give the option holder the right to buy the asset, and put options give the holder the right to sell the asset. Figure 5.1 illustrates call options and put options, respectively, where K is the exercise price of options and T is the expiration date. The bold solid lines represent option values at the expiration date T, whereas dashed lines indicate option values prior to expiration (i.e., $t < T$). Options can also be categorized as European or American. A European option gives the holder the right to exercise the option only on the expiration date. An American option gives the holder the right to exercise the option on or before the expiration date. These basic concepts are discussed in detail by several authors, including Hull (2003) and Mun (2002).

The real options approach is an extension of financial option theory to options on real (i.e., non-financial) assets. Under the real options approach,

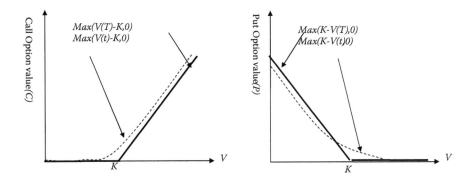

Figure 5.1 Call and put options.

risk-neutral pricing is used as opposed to risk-adjusted discounting to strategically evaluate investment opportunities. The literature on real options is extensive; Dixit and Pindyck (1994) and Schwartz and Trigeorgis (2001) provide useful reviews. Closer to the subject of cross training, several authors have considered using real options valuation methods to assess production systems, flexible production technology, or other machinery that has multiple uses (Aggarwal, 1991; Ernst and Kamrad, 1995; Kulatilaka and Trigeorgis, 1994; Kulatilaka and Perotti, 1998). To address the need of organizations to measure the value of flexibility, Nembhard et al. (2000) propose using a framework for a real options model for managing production system changes. This framework has been developed to use real options to: (1) evaluate the decision to use quality control charts (Nembhard et al., 2002), (2) pursue product outsourcing (Nembhard et al., 2003), (3) evaluate supply chain decisions (Nembhard et al., 2005c), and (4) model strategies aimed at reducing the environmental impact for a manufactured product (Nembhard et al., 2005b). These efforts demonstrate the application of real options in a range of engineering problems.

In traditional discounted cash flow (DCF) techniques, the net present value (NPV) of an investment project is calculated for a stream of expected net cash flows at a "risk-adjusted" discount rate that reflects the risks for those cash flows (Bernhard, 1984; Park and Sharp-Bette, 1990). Fama and French (1997) have emphasized the difficulties of estimating risk-adjusted discount rates. Nevertheless, for this method, the investment decision is made without regard for uncertainty in the production environment and without regard for the idea that projects at different phases represent a sequence of independent investment opportunities. An immediate accept or reject decision is then made by accepting all projects with a positive NPV. This strategy assumes deterministic behavior because at the outset, an irrevocable investment policy is adopted. In the absence of a dynamic investment policy, the probability distribution of the NPV would be reasonably symmetric. However, by cross training, the composition of a knowledge workforce can be adjusted to adapt to future production environments. This aspect introduces an asymmetry in the probability distribution of the NPV that expands the value of investment opportunity by improving its upside potential while limiting downside losses relative to manager's initial expectations under a deterministic investment policy. The strategic value of various projects cannot be properly captured by traditional DCF techniques due to their dependence on future events that are uncertain at the time of the initial decision.

In contrast, a real options framework has the potential to model these scenarios. It is apparent that increasing the flexibility of production systems by cross training is one operational option. That is, managers can decide to cross train when needed. Cross training represents a dynamic investment policy on workforce knowledge, which parallels the concept of real options theory and correspondingly may be associated with option values (Kulatilaka, 1988; Panayi and Trigeorgis, 1998).

5.2.1 Capacity expansion example: traditional DCF valuation

Suppose a manufacturer builds a new production line M_A to augment capacity on a similar production line M_B, where M_B is the sole operation prior to building M_A. At the end of the first year, M_A has an equal chance of yielding values of either $V_u = \$1800k$ or $V_d = \$500k$. Presently, the manufacturer's stock price, S, is equal to $\$10$, and at the end of the year, S will be $S_u = \$18$ or $S_d = \$5$ with equal chance. It is reasonable to assume that the production profit of the new production line M_A is perfectly correlated with the manufacturer's stock price. That is, the production profit of M_A will be $\$1800k$, if the stock price of M_B is $\$18$, and otherwise, $\$500k$. The expected rate of return:

$$r = 0.5 \times \frac{(S_u - S)}{S} + 0.5 \times \frac{(S_d - S)}{S} = 0.5 \times 0.8 + 0.5 \times (-0.5) = 0.15. \quad (5.1)$$

Thus, by traditional DCF analysis, the NPV, V, of M_A can be obtained by discounting possible outcomes with the expected rate of return from Equation (5.1):

$$V = (pV_u + (1-p)V_d)/(1+r) = (0.5 \times 1800 + 0.5 \times 500)/1.15 = \$1000k. \quad (5.2)$$

Traditional DCF does not specify the value of flexibility. In a complex system, one necessity is the ability to react to events as they unfold. In doing so, it becomes clear that flexibility must hold some value. However, the key question remains. How valuable is flexibility? The options framework is designed to address this very question.

5.2.2 Capacity expansions example: real options valuation

We can formulate a framework for option valuation in which the value of cross training, π, is equivalent to the value of a replicated portfolio of the perfectly correlated asset M_B from the previous example. The portfolio is constructed by buying ω shares of stock of the manufacturer and borrowing A at the rate of risk-free rate r_f. At the end of the first year, π has an equal chance of going up to π_u or down to π_d:

$$\pi_u = \omega S_u - (1+r_f)A$$
$$\pi_d = \omega S_d - (1+r_f)A \quad (5.3)$$

By solving the equations in (5.3), ω and A are:

$$\omega = (\pi_u - \pi_d)/(S_u - S_d) \quad (5.4)$$

$$A = (\pi_u S_d - \pi_d S_u)/(1+r_f) \cdot (S_u - S_d) \quad (5.5)$$

and π depends on the movement of S because π is obtained by discounting π_u and π_d back with risk-neutral probability, r_n, which is a function of S_u and S_d as well:

$$r_n = \frac{(1+r_f)S - S_d}{S_u - S_d} = \frac{1.08 \times 1 - 0.5}{1.8 - 0.5} = 0.45, \tag{5.6}$$

$$\pi = \frac{1}{1+r_f}(r_n \pi_u + (1-r_n)\pi_d). \tag{5.7}$$

Without cross training, the NPV, V, of the plant is estimated to be:

$$V = \frac{1}{1+r_f}(r_n V_u + (1-r_n)V_d) = \frac{1}{1.08}(0.45 \times 1,800 + 0.55 \times 500) = \$1,000k, \tag{5.8}$$

which is equal to the result of the traditional DCF method. However, if cross training is applied when needed, the results from the two approaches differ.

Suppose that by cross training, the productivity as well as the value of the plant increases by 20%. However, obtaining this cross training involves a total cost of \$210k. In Figure 5.2, when the additional value created by cross training is greater than the cost (i.e., $0.2 \times 1,800 - 210 = \$150k > 0$), cross training is conducted. If the value is not greater (i.e., $0.2 \times 500 - 210 = -\$110k < 0$), cross training is rejected.

The NPV of the production line, now containing the value of cross training, is called the expanded NPV, V^+:

$$V^+ = \frac{1}{(1+r_f)}(r_n V_u^+ + (1-r_n)V_d^+) = (0.45 \times 1950 + 0.55 \times 500)/1.08 = \$1062k. \tag{5.9}$$

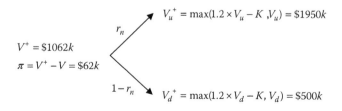

Figure 5.2 The value of cross training using real options.

The corresponding value of cross training is the difference between V^+ and V, the static NPV by following traditional DCF:

$$V^+ - V = 1062k - 1000k = \$62k. \tag{5.10}$$

The value of cross training is the real options value in this example, so it can also be evaluated with Equation (5.7):

$$\pi = \frac{1}{1+r_f}(r_n\pi_u + (1-r_n)\pi_d)$$

$$= \frac{1}{1.08}(0.45 \times \max(150,0) + 0.55 \times \max(-110,0)) = \$62k. \tag{5.11}$$

Note that following the traditional DCF approach, a manager would not conduct cross training because the value of cross training is negative:

$$\pi = 0.2 \times V - K = 0.2 \times 1000 - 210 = -\$10k. \tag{5.12}$$

However, the real options approach suggests cross training with a positive valuation of $62k, rather than –$10k. This example indicates that in the presence of market risks, a static cross-training policy derived from the traditional DCF method is not ideal. Rather, if we follow the dynamic cross-training policy derived from the real options approach, the value of flexibility through cross training will not be ignored. Thus, we suggest that following an option-based cross-training policy may be useful for valuation and decision-making.

5.3 Option-based cross training

5.3.1 How does cross training expand productivity?

Before detailing cross-training valuation, we illustrate how cross training can expand productive potential. Consider a simple sequential production line as shown in Figure 5.3 in which each station involves a unique job. We assume that work in process (WIP) is allowed to accumulate without limit between each of the adjacent stations and is available for immediate processing

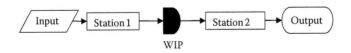

Figure 5.3 Sequential production line.

by the next station. Only after a unit has moved through all workstations is it considered finished.

Each station requires exactly one worker; so two workers are involved in this sequential production system. We assume that prior to cross training, one worker specializes on the base job and the other specializes on the complex job. However, if a worker is absent from the line, the other one will cover both jobs. Thus, during cross training, each worker not only works on their original job but also learns the other less familiar job. They change positions according to a station rotation interval, which we assume to be constant during cross training. After cross training, workers are skilled at each job and do not lose their skills.

We can represent the output of individual workers using a hyperbolic learning model (Mazur and Hastie, 1978; Nembhard et al. 2005a):

$$y(t) = m_k \frac{t + m_p}{t + m_p + m_r} \tag{5.13}$$

where $y(t)$ is the productivity rate corresponding to t time periods of work ($t \geq 0$), m_k is the fitted parameter estimating the steady-state productivity rate, m_p is the fitted parameter representing initial expertise, m_r is the number of time periods of work that initial expertise is required to get half way to m_k, and $m_p + m_r > 0$. Figure 5.4 illustrates the hyperbolic learning model with $m_k = 1$, $m_p = 5$ weeks, and $m_r = 10$ weeks.

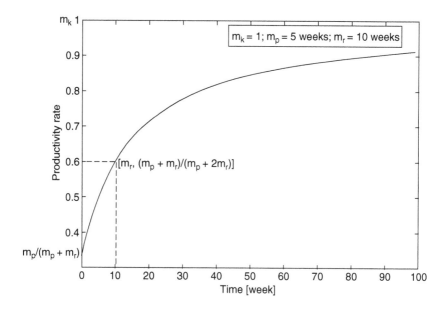

Figure 5.4 Hyperbolic learning model.

There are four factors that may affect the model in Equation (5.13) that are described as follows. The first factor is task heterogeneity. We set the complexity level of the *base task* to be 1. We standardize the task complexity level of the *complex task* correspondingly. To model the task heterogeneity in the production line, we let the job complexity level of the complex task be twice that of the base job.

The second factor is worker heterogeneity, which represents the fact that capabilities can differ substantially among workers in a worker pool (Campbell, 1999). To model worker heterogeneity, we assume that an *average worker* has an average level of capability, based on the set of workers in the pool. A second *random worker* is sampled from the worker pool to complete the team for a sequential production line. We define $m_{\bullet R \bullet}$ to represent randomly selected worker parameters, $m_{\bullet A \bullet}$ as average worker parameters, $m_{\bullet \bullet B}$ as worker parameters on the base job, and $m_{\bullet \bullet C}$ as worker parameters on the complex job. Specifically, this means that m_{kAB} is the steady-state productivity rate of the average worker on the base task (i.e., it is the average value of steady-state productivity rates for the base job of all workers in the worker pool), m_{kRB} is the steady-state productivity rate of the random worker on the base task, m_{kAC} is the steady-state productivity rate of the average worker on the complex task, and m_{kRC} is the steady-state productivity rate of the random worker on the complex task. As a means of comparison, we define Δm_k as the deviation of the steady-state productivity rate for each team and use it as a measure of worker heterogeneity within each team as follows:

$$\Delta m_k = m_{kRB} - m_{kAB}. \tag{5.14}$$

For example, we now have a worker pool with 75 workers, and Figure 5.5 illustrates these workers in terms of their parameter values, wherein the family of dots represents the worker pool's parameter distribution on the basic task and the family of circles represents that on the complex task. The 75 workers' average steady-state productivity rate is 29 on the basic task (i.e., $m_{kAB} = 29$) and 15 (i.e., $m_{kAC} = 15$) on the complex task. Now a worker is randomly selected from the worker pool as illustrated in Figure 5.5. His steady-state productivity rate is 30 on the basic task (i.e., $m_{kRB} = 30$) and 16 on the complex task (i.e., $m_{kRC} = 16$). If the randomly picked worker and the average worker form a team, the deviation of the steady-state productivity of this team is equal to 1 (i.e., $\Delta m_k = m_{kRB} - m_{kAB} = 30 - 29 = 1$).

The third factor is labor dynamics. Worker absenteeism is likely a stochastic event for a dynamic manufacturing system. Suppose each worker in a team has an equal chance of being absent, and absent for exactly 1 day per absence period. During a worker absence, the remaining worker on that team covers the work of the absent worker for that day. We assume the average absentee rate is once per week, modeled stochastically following a uniform distribution. We assume no worker turnover in this system.

The fourth factor is product dynamics. We assume the product is updated in every 3-month period and the degree of product change is 50%.

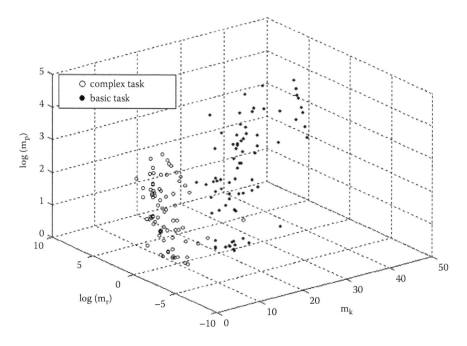

Figure 5.5 Worker pool.

In other words, at the end of each 3-month period, one of the two jobs in a sequential production line has significant change, yet still with the same complexity level. On this point, it is reasonable to equate the changed production system to the original situation with specialized workers. Although each worker is not multiskilled anymore, he/she is still skilled at one job. We also assume the training program begins immediately after the product change and is no longer than the product lifecycle. When production changes occur, skills acquired on jobs that are removed are lost.

Cross training creates a multiskilled workforce. Since they are no longer specialized, each worker's capacity on a specific job is lowered. However, multiskilled workers are good at handling the private risks and heterogeneities mentioned above. Thus, the productivity of the system may potentially increase. In Figure 5.6, suppose the steady-state productivity rate of the production line with specialized workers is 1. Because multiskilled workers can ameliorate some of the effects of system dynamics, the productivity is expanded accordingly. We assume the steady-state productivity rate with well cross-trained workers is e, the expansion rate. Nembhard and Norman (2005) studied the effect of cross training in terms of worker efficiency and responsiveness in the context of several important workplace factors including product, manufacturing process, and workforce dynamics, each of which can cause significant disruptions in productivity. This provides a useful reference for the measurement of e, as well as the cost aspects of cross training.

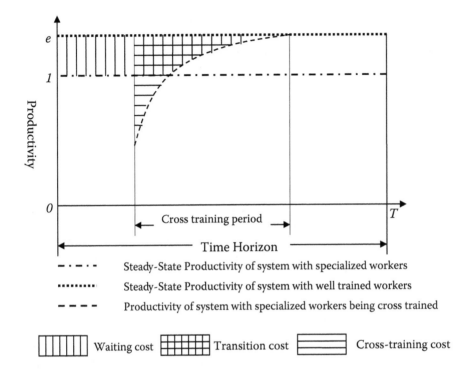

Figure 5.6 Production capability under various workforce scenarios.

5.3.2 The net present value approach: a benchmark

Consider the NPV based on traditional DCF techniques for the production system using a specialized workforce, which can act as a benchmark for comparison as well as an input for the option valuation. In order to use the real options approach to value cross training, we assume that the net value of the production line during a decision time horizon has a probability distribution. We assume the percentage changes in the production value follow an exponential Brownian motion process with a normal distribution, and the uncertain value of the output is perfectly correlated with a portfolio of tradable assets, an approach supported by Arnaiz and Ruiz-Alseda (2003) and Bollen (1999). Let us designate the NPV (based on traditional DCF techniques) of the production line using a specialized workforce, denoted V. This is given by the difference between how much the production line is worth (its present value) and how much they cost during a decision horizon, T

$$V = \lambda_T(P - C)Q_S - n_W L - F \qquad (5.15)$$

where P is the manufacturer's current sales price for unit product, Q_S is the output of a sequential production line during T (at the expected rate of return of the correlated asset r) with a specialized workforce, n_W is the number of workers in the system, L is the present value of workers' salaries, C is current cost per unit product other than the direct labor cost, F is the present value of fixed production cost during T, and λ_T is a discounted factor if the product will be sold at the end of n equal intervals of T:

$$\lambda_T = \frac{1}{n} \sum_{t=1}^{n} \frac{1}{(1+rT/n)^t}. \tag{5.16}$$

For example, if M_A's life is 2 years (i.e., $T = 2$), the previously formed two-worker team operates M_A with the average worker specialized on the basic task and the randomly selected worker specialized on the complex task. By simulating their performance under the system setting as described, their 2-year output is $Q_S = 57,872$. Suppose T is divided into $n = 8$ periods with even length and $r = 0.15$ according to Equation (5.1), and the product is sold at the end of every period, we obtain $\lambda_T = 0.85$ and $V = 49,211(P - C) - (2L + F)$ accordingly.

5.3.3   The expansion rate of productivity

According to Figure 5.6, the evaluation of the NPV of the production line using a well cross-trained workforce is needed to measure the expansion rate e, which is a measure of the NPV by a well cross-trained workforce compared with the NPV by a specialized workforce:

$$e = \frac{\lambda_T Q_C (P - C) - (n_w L + F)}{\lambda_T Q_S (P - C) - (n_w L + F)} \tag{5.17}$$

where Q_C is the output of a sequential production line during T using a previously cross-trained workforce.

For example, assume the two workers still operate M_A for 2 years. However, they are well cross trained. We simulate their performance under the system setting as described and obtain $Q_C = 63,215$. Thus, $e = \frac{49,211(P-C)-(2L+F)}{53,755(P-Q)-(2L+F)}$.

5.3.4   The exercise price

Figure 5.6 further illustrates the capability difference of the production line in terms of a productivity function such as that in Equation (5.13). The production line with cross-trained workers may have different steady-state productivity than that with specialized workers. The production line has opportunities to expand its production capability by cross training. The exercise cost, K, is the cost involved when exercising the option of cross training, which includes three parts:

$$K = K_w + K_c + K_t. \tag{5.18}$$

The first part is the waiting costs K_w, which is the production loss due to delaying cross training. The second part is the cross training cost Kc, which is the production loss from specialized workers while they are being cross trained. That is, while they acquire multiple skills through cross training, they must forgo a portion of production in the short run. The system transition cost, K_t, is the unachievable output during the system transitions. In essence, cross training can potentially expand the future production capability at the expense of some amount of current productivity.

Suppose the system has not conducted cross training until T_w $(0 \leq T_w \leq T)$; the waiting cost (Figure 5.6) is estimated to be:

$$K_w = (P-C)(Q_c - Q_S)T_w / T. \tag{5.19}$$

If a cross-training decision is limited in a finite set of points that are even-interval within the time horizon T,

$$K_{wt} = \frac{t}{n}(P-C)(Q_c - Q_S) \quad t = 0, 1, \cdots, n. \tag{5.20}$$

For example, when $n=8$, at the beginning of period t $(t=0,\cdots,8)$, $K_{wt} = 668(P-C)t$.

Suppose cross training can be finished within T_t $(0 \leq T_t \leq T)$, then the sum of the transition cost and cross training cost can be quantified as:

$$K_t + K_c = e^{-rT_t}(Q_c T_t / T - Q_{St})(P-C) \tag{5.21}$$

where Q_{St} is the output of the production line during a cross-training period T_t using workers being cross trained. For example, cross training is observed to finish in 0.25 years (i.e., $T_t = 0.25$) and $Q_{St} = 4{,}030$. The sum of transition cost and cross training cost $K_t + K_c = 3729(P-C)$. The exercise cost is a function of time. At the end of period i, $K_i = (668i + 3729)(P-C)$.

5.3.5 Expanded NPV

When the production system has the opportunity to expand its productivity using cross training, the NPV includes the value of cross training, referred to as expanded NPV, denoted by V^+.

If $V(t)$ is the net value of M_A at time t based on traditional DCF techniques, then:

$$V = V(0) \tag{5.22}$$

V^+ is obtained via $V(t)$. Suppose the manufacturer successfully can operate production line M_A such that $V(t)$ is nonnegative, and assume $V(t)$ - follows a geometric Brownian motion (GBM) process

$$\frac{dV(t)}{dt} = rdt + \sigma dz(t), \tag{5.23}$$

where r is the drift rate of $V(t)$, σ is the volatility, and $dz(t)$ is a Wiener process (see Mun, 2002). Because M_A is perfectly correlated with the stock price of the manufacturer who only operates M_B before it owns M_A, thus r and σ can be estimated using historical data of M_B. Mun (2002) reviews commonly used methods for estimating σ and discusses their advantages and disadvantages. We follow the corresponding logarithmic method and estimate a value of $\sigma = 0.6$. If the product that M_A produces is a tradable asset, then r can be substituted with r_f the risk-free rate in option valuation.

Equation (5.23) can be approximated by binomial lattices, to model V^+. The lattice approximating $V(t)$ is termed the *underlying asset* lattice, and the other is the *options valuation* lattice. We suppose that the option lasts for time T, and cross-training decisions are made every time interval ΔT. (For example, if an option has a 2-year maturity span and cross-training decisions are made every 0.25 years, then the binomial lattice has 8 steps.) During the life of the option, V can either move up from V to a higher level uV or down from V to a lower level dV, where:

$$u = e^{\sigma\sqrt{\Delta T}} \text{ and } d = 1/u. \tag{5.24}$$

when $\Delta T = T/n = 2/8 = 0.25, u = 1.35$ and $d = 1/u = 0.74$.

The risk-neutral valuation principle is applied in real options valuation, and the risk-neutral probability, r_n, is derived from the up (u) and down (d) factors:

$$r_n = \frac{e^{r_f \Delta T} - d}{u - d}. \tag{5.25}$$

$r_n = 0.46$ when $r_f = 0.08$.

To create the lattice of the underlying asset, we start with V, then multiply it by the up (u) and down (d) factors to create up and down branches $V_t^i = Vu^i d^{t-i}$ ($i = 0,1,\cdots,t$) as illustrated in Figure 5.7. This bifurcation continues at each node to create a binomial lattice for the underlying asset. The intermediate branches all recombine because we are using the risk-neutral rate, r_n. The value of an option is generated by the uncertainties, and risks are captured by the volatility measure σ, where higher values correspond to higher potential option values.

When new information arrives, uncertainty about future cash flows is gradually resolved in the binomial lattice for the underlying variables.

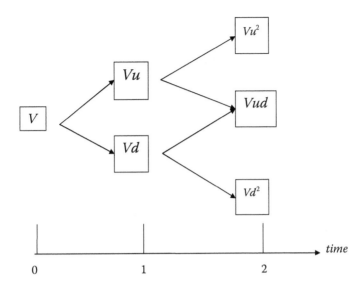

Figure 5.7 Underlying asset lattice (first two steps shown).

Flexibility then can be modeled in the options valuation lattice by allowing varying degrees of flexibility relative to the underlying assets lattice. For the investment problem of cross training, the option is the ability to expand the financial value of production with associated costs at different stages during the time horizon T, where after T the opportunity will disappear. We assume the lattice has n stages, and at the beginning of each stage, according to the state, we determine the lattice values, V_t^{+i}. Each is given by the maximum of the value of expansion (i.e., implementing the cross-training policy) and the value of continuation (i.e., keeping the option open to cross train in the future):

$$V_t^{+i} = \max\left[eV_t^i - K, EV_t^{+i} \right] \quad t = 0,1,\dots,n-1; \quad i = 0,1,\dots,t \qquad (5.26)$$

We calculate the value of keeping the option open at each intermediate node EV_t^{+i} by discounting the one-step-ahead cash flows $\left(V_{t+1}^{+i} \text{ and } V_{t+1}^{+i+1} \right)$ backward with risk-free rate r_n:

$$EV_t^{+i} = e^{-r_f \Delta T}\left[r_n V_{t+1}^{+i+1} + (1 - r_n)V_{t+1}^{+i} \right] \quad t = 0,1,\dots,n-1; i = 0,1,\dots,t$$
$$(5.27)$$
$$EV_n^{+i} = V_n^i \quad i = 0,1,\dots,n$$

With backward induction in the options valuation lattice, we get the NPV of option evaluation lattice at time zero, which is V^+:

$$V^+ = V_0^{+0}. \qquad (5.28)$$

5.3.6   The value of cross training using the real options approach

We may find that the V^+ is different from the NPV without cross training, V. The difference, π, is the financial value of cross training by the real options approach:

$$\pi = V^+ - V \qquad\qquad (5.29)$$

which is the real options value in this framework.

5.4   Numerical example

This section illustrates the procedures described in Section 5.3 through a small but complete example. Table 5.1 lists many of the input production system parameters used.

To illustrate the calculations in relating the cross training to the real options valuation, consider a worker team with slight workforce heterogeneity: the steady-state productivity values are $m_{kAB} = 29$, $m_{kAC} = 15$, $m_{kRB} = 30$, and $m_{kRC} = 16$. We simulate the output of a production line with the two workers under three scenarios. In the first scenario, workers remain specialized in their job, with an expected output over 2 years of $Q_S = 57{,}872$. In the second scenario, workers are initially cross trained, and the expected output over 2 years is $Q_C = 63{,}215$. In the third scenario, specialized workers will be cross trained, and the expected output within a training period ($T_t = 0.25$ year) is $Q_{St} = 4030$.

Let us assume that (1) the initial cost is the sum of the fixed production cost F and labor cost $n_w L$; (2) product is sold at the end of every quarter year; (3) variable cost (e.g., raw material, energy, shipping) of each quarter year is paid with the sale income in that quarter; and (4) cash flows are discounted with the expected rate of return of the correlated asset $r = 0.15$.

Table 5.1 Input Parameters for the Production System

Manufacturer's current sales price for each unit of product (P)	$13.68
Current cost per unit of product other than the direct labor cost (C)	$5.00
Time interval (ΔT)	0.25 year
Time horizon of decision (T)	2 years
Number of workers per unit (n_w)	2
Present value of each worker's salary during the time horizon of decision (L)	$60,000
Present value of fixed production cost during the time horizon of decision (F)	$80,000
Risk-free rate (r_f)	0.08
Working hours per year (H)	2080 h (8 h × 5 d × 52 w)

for $t = 1,2,\cdots,8$

$$c_t = (P - C)(\Delta T / T)Q_S e^{-r(t \times \Delta T)}$$

$$= (13.68 - 5) \times (0.25/2) \times 57872 \times e^{-0.15 \times (0.25t)}$$

$$= e^{-0.15 \times (0.25t)} 62.8k$$

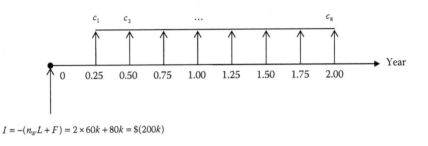

$$I = -(n_W L + F) = 2 \times 60k + 80k = \$(200k)$$

$$r = 0.15$$

Figure 5.8 Net present value (NPV) by traditional discount cash flow (DCF).

So, we calculate the NPV in Equation (5.30) by the traditional DCF illustrated in Figure 5.8

$$V = -(n_W L + F) + \sum_{t=1}^{8} c_t = -(n_W L + F) + (P - C)(\Delta T / T)Q_S \sum_{t=1}^{8} e^{-r(t \times \Delta T)} = \$225.89k.$$

(5.30)

Following the same approach in Figure 5.7, we can get the NPV using initially cross-trained workers. The expansion factor, e, is ratio of these two NPVs:

$$e = \frac{-(n_W L + F) + (P - C)(\Delta T / T)\sum_{t=1}^{8} Q_C e^{-r(t \times \Delta T)}}{-(n_W L + F) + (P - C)(\Delta T / T)\sum_{t=1}^{8} Q_S e^{-r(t \times \Delta T)}} = \frac{\$265.22k}{\$225.89k} = 1.17 \quad (5.31)$$

The exercise price is modeled as a function of time; we will not immediately lose the opportunity if we decide to delay cross training. However, the value of the opportunity will gradually deteriorate. According to Equation (5.19), the waiting cost is estimated to be:

$$K_{wt} = (P - C)(t\Delta T / T)(Q_C - Q_S) = \$5,797t \quad \text{for} \quad t = 0,1,\cdots,7. \quad (5.32)$$

We calculate the cross training cost and transition cost by Equation (5.20):

$$K_t + K_c = e^{-rT_w}(P-C)(\Delta T / T)Q_c e^{-rT_t} = \$32,371. \qquad (5.33)$$

So, the exercise cost is:

$$K_i = K_t + K_c + K_{wi} = 32,371 + 5,797t \quad \text{for } t = 0,1,\cdots,7. \qquad (5.34)$$

Suppose the risk-free rate r_f = 0.08, and the volatility is estimated to be 0.6. Thus, the up and down factors are 1.35 and 0.74, respectively, by Equation (5.24). According to Equation (5.25), the risk-neutral probability is 0.46.

We combine the production value lattice, real option valuation lattice, and decisions as shown in Figure 5.9. That is, there are three elements in each node of the lattice. The element on the top is the NPV of production with specialized workers V_t^i, the middle element is the NPV of production with cross training, V_t^{+i}, and the element at the bottom element represents the corresponding decision. The $V^+ = \$233.89k$, shown in the first node. Thus, the value of cross training in this example is:

$$\pi = V^+ - V = 233.89 - 225.90 = \$7.99k \qquad (5.35)$$

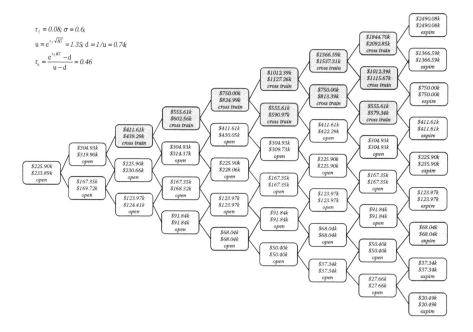

Figure 5.9 The value of cross-training lattice.

That is, the NPV is increased by 3.54%, even though there is a small difference between the two workers. Nembhard et al. (2005a) describe how the value of cross training will generally increase if the workforce heterogeneity and job heterogeneity become larger.

5.5 Summary

Workforce flexibility is not only useful in dealing with private uncertainties, but it also indicates opportunities with regard to market uncertainties. From the economic point of view, cross training is an investment on workforce flexibility under uncertainty. The reason for treating it as an investment is that the costs involved with cross training and the benefits of cross training (i.e., the ability of the generated flexible workforce to deal with private uncertainties and, thus, expand productivity) may occur at some unknown times in the future. Cross-training decisions based on a real options approach allow for the valuation of flexibility that stems from cross training.

Rather than suggest that the traditional DCF method is inappropriate for the problem described, we observe that NPV valuation provides important input to the real options approach. Namely, these methods help to inform the static NPV, expansion rate, and cross-training costs. By realizing the difficulties of purely utilizing real options and the limitations of pure DCF in a given circumstance, these approaches can be considered to overlap in applied settings.

In this chapter, production is assumed to be perfectly correlated with a tradable asset. Thus, the expected rate of return of the production value is substituted via a risk-free rate, and production values are discounted with the risk-free rate. However, if the underlying asset is not tradable, the expected rate of return of the production value is adjusted by subtracting a market risk price and the production values are still discounted back with the risk-free rate. Hull (2003) gives a detailed discussion of proper uses of the risk-free rate of return. Important extensions include using option-based cross training to deal with service congestion and to manage technology supplies with regard to demand within the product lifecycle.

5.6 Acknowledgments

The authors would like to thank Harriet Black Nembhard for her comments and suggestions on preliminary work contained in this chapter. Portions of the authors' research described herein were supported by a grant from the National Science Foundation under SES-0217666 and SES-0435948.

References

Aggarwal, R. 1991. Justifying Investments in Flexible Manufacturing Technology, *Managerial Finance,* 17(2), 77–88.

Arnaiz, O.G., and Ruiz-Alseda, F. 2003. *Entry Patterns of the Product Life Cycle,* unpublished manuscript.

Bernhard, R.H. 1984. Risk-Adjusted Values, Timing and Uncertainty Resolution, and the Measurement of Project Worth, *Journal of Financial and Quantitative Analysis*, 19(1), 83–99.

Bollen, N.P.B. 1999. Real Options and Product Life Cycles, *Management Science*, 45(5), 670–684.

Campbell, G.M. 1999. Cross-Utilization of Workers Whose Capabilities Differ, *Management Science*, 45(5), 722–732.

Dixit, A.K., and Pindyck, R.S. 1994. *Investment Under Uncertainty*. Princeton, NJ: Princeton University Press.

Ebeling, A.C., and Lee, C.Y. 1994. Cross-training Effectiveness and Profitability, *International Journal of Production Research*, 32(12), 2843–2859.

Ernst, R., and Kamrad, B. 1995. Multi-Product Manufacturing with Stochastic Input Process and Output Yield Uncertainty, in *Real Options in Capital Investment: New Contributions*, Trigeorgis, L. (Ed.), New York: Praeger.

Fama, E.F., and French, K.R. 1997. Industry Costs of Equity, *Journal of Financial Economics*, 43(2), 153–193.

Gerwin, D. 1993. Manufacturing Flexibility: A Strategic Perspective, *Management Science*, 39(4), 395–410.

Hull, J. 2003. *Options, Futures, and Other Derivatives*. 5th ed., Upper Saddle River, NJ: Prentice Hall.

Kulatilaka, N. 1988. Valuing the Flexibility of Flexible Manufacturing Systems, *IEEE Transactions in Engineering Management*, 35(4), 250–257.

Kulatilaka, N., and Perotti, E.C. 1998. Strategic Growth Options, *Management Science*, 44(8), 1021–1031.

Kulatilaka, N., and Trigeorgis, L. 1994. The General Flexibility to Switch: Real Options Revisited, *International Journal of Finance*, 6(2), 778–798.

Mazur, J.E., and Hastie, R. 1978. Learning as Accumulation: A Reexamination of the Learning Curve, *Psychological Bulletin*, 85(6), 1256–1274.

Molleman, E., and Slomp, J. 1999 Functional Flexibility and Team Performance, *International Journal of Production Research*, 37(8), 1837–1858.

Mun, J. 2002. *Real Options Analysis: Tools and Techniques for Valuing Strategic Investments and Decisions*, New York: John Wiley & Sons.

Nembhard, D.A., H.B. Nembhard, and R. Qin. 2005a. A Real Options Model for Workforce Cross Training, *The Engineering Economist*, 36(10), 919–940.

Nembhard, H.B., Aktan, M., and Stuart, J.A. 2005b. Real Options for Sustainable Product Design. (Working paper)

Nembhard, H.B., Shi, L., and Aktan, M. 2005c. A Real Options Based Analysis for Supply Chain Decisions, *IIE Transactions (special issue on Financial Engineering)*, 37(10), 945–956.

Nembhard, D.A., and Norman, B.A. 2005. Worker Efficiency and Responsiveness in Cross-Trained Teams. (Working paper)

Nembhard, H.B., Shi, L., and Aktan, M. 2001. A Real Options Design for Quality Control Charts, *The Engineering Economist*, 47(1), 28–59.

Nembhard, H.B., Shi, L., and Aktan, M. 2003. A Real Options Design for Product Outsourcing, *The Engineering Economist*, 48(3), 199–217.

Nembhard, H.B., Shi, L., and Park, C.S. 2000. Real Options Models for Managing Manufacturing System Changes in the New Economy, *The Engineering Economist*, 45(3), 232–258.

Panayi, S., and Trigeorgis, L. 1998. Multi-Stage Real Options: The Cases of Information Technology Infrastructure and International Bank Expansion, *The Quarterly Review of Economics and Finance*, 38(3), 675–692.

Park, C., and Sharp-Bette, G. 1990. *Advanced Engineering Economics*. New York: John Wiley & Sons.

Russell, R.S., Huang, P.Y., and Leu, Y. 1991. A Study of Labor Allocation Strategies in Cellular Manufacturing, *Decision Sciences*, 22(3), 594–611.

Schwartz, E.S., and Trigeorgis, L., Eds. 2001. *Real Options and Investment Under Uncertainty: Classical Readings and Recent Contributions*, Cambridge, MA: MIT Press.

van den Beukel, A.L., and Molleman, E. 1998. Multifunctionality: Driving and Constraining Forces, *Human Factors and Ergonomics in Manufacturing*, 8(4), 303–321.

section II

Cross-trained teams

chapter 6

Best practices for cross-training teams

Kevin C. Stagl, Eduardo Salas, and Stephen M. Fiore

Contents

6.1   Introduction

The milieu in which an increasing number of businesses, alliances, and government-brokered coalitions operate is typified by hyper-competitive global markets, rapidly evolving ambiguous situations, imperfect solutions, information overload, intense time pressure, and where there are severe consequences of error (Orasanu and Salas, 1993). In response to the difficult tasks and complex challenges inherent to today's workplace, organizations often use teams as a performance arrangement to structure work. Teams thrive in this controlled chaos because they can leverage superior resources when identifying emerging opportunities and formulating and executing plans accordingly. In fact, effective team performance can result in improved:

- Financial performance
- Adaptation
- Efficiency
- Productivity
- Safety
- Service quality
- Satisfaction
- Commitment

While teams can yield an impressive array of results, the actual truth is that teams of experts sometimes fail with disastrous consequences for those directly involved and indirectly vested. Tragic team failures abound in the public (e.g., the Federal Emergency Management Agency's leadership team's response to Hurricane Katrina) and private sector (e.g., the 160 fatalities attributed to crew teamwork failures aboard American Airlines flight #965 — December 1995). Why do teams of well-intentioned, highly motivated, seasoned professionals fail on the grandest stages and what can be done about it? Clearly, the answers to these questions are complex, but insight is afforded by both the science of teams and the science of team training (Salas and Cannon-Bowers, 2001). The principles, guidelines, and lessons learned in these domains can be applied to transform a team of experts into an expert team.

6.2 *Theoretical basis and purpose*

One notion that is useful for understanding both exceptional and ineffective team performances is shared mental models. Shared mental models are emergent cognitive states that consist of commonly held organized knowledge structures. These dynamically activated memory structures are representations of reality that allow team members to anticipate the need for, and select actions that are coordinated with, those of their teammates (Cannon-Bowers et al., 1991; Mathieu et al., 2000). Specifically, shared mental models are drawn upon by members before, during, and after they enact the taskwork and teamwork processes that comprise both routine and adaptive team performance (Burke et al., 2006). Thus, shared mental model theory is relevant to training teams to be more effective because it provides a basis for understanding how team members conduct the cognitive components of their jobs (Cannon-Bowers and Salas, 1990). For example, shared mental models can be used to understand

and improve how members recognize, integrate, and apply information to make decisions in naturalistic settings (Klein, 1989).

One human capital management intervention that has successfully been used to foster shared mental models and, thereby, effective team performance is cross training (Salas et al., 1992). Cross training is an instructional strategy that consists of a suite of interrelated techniques (e.g., positional clarification, positional modeling, positional rotation), which are used to reduce interpositional uncertainty by imparting interpositional knowledge (IPK) (Baker, 1991). IPK is a specific type of shared team-interaction mental model that consists of role knowledge of one's teammates' task requirements, functions, information needs, action-outcome contingencies, and coordination demands (Fiore et al., 2005; Kahn et al., 1964). IPK also includes contextually specific knowledge about temporal contingencies and cause-and-effect task associations (Volpe et al., 1996). By fostering IPK, cross training is a means of developing emergent cognitive states, such as shared mental models, which, in turn, facilitate team members' efforts to predict, explain, anticipate, and coordinate within teams. Moreover, because cross training exposes members to a variety of tasks, it can also foster intrinsically motivated, cohesive teams. These cognitive and affective states, in turn, enable the explicit and implicit coordination at the heart of team performance.

While, to date, the majority of scientific and anecdotal evidence suggests cross training is a viable means of fostering effective teams, there is no comprehensive source for best practices in utilizing this instructional strategy to train teams and their members. In an effort to remedy this state of affairs, this chapter briefly reviews the nature of teams and team training in order to set the stage for a more in-depth discussion of the nature of cross training. The theoretical and empirical evidence generated from cross-training interventions conducted in a variety of settings is then reviewed. The findings gleaned from these interventions are summarized in a heuristic framework illustrating the proximal and distal effects of cross training. Moreover, lessons learned are distilled from these initiatives and subsequently framed as 20 best practices for cross-training teams.

6.3   The science of teams

Teams have a long and storied history (Salas et al., 2006a). In fact, the success of teams has been integral to the survival of humankind ever since early tribes used hunting parties to stalk and bag prey. Yet, it was not until the late 20th century that theoretical and empirical work on teams began to flourish. Since that time, research has continued unabated in a number of disciplines and from a variety of perspectives, with the common goal of advancing the science of teams.

Over the course of the past century, much has been learned about the nature of teams. An early description of teams defined them as "interdependent collections of individuals who share responsibility for specific outcomes for their organizations" (Sundstrom et al., 1990, p. 120). More specifically, they were characterized as "two or more individuals who must interact and adapt to achieve specified, shared, and valued objectives" (Salas et al., 1992, p. 4). This definition has matured to include the notion of teams as complex entities, comprising two or more individuals, who interact socially, dynamically, episodically, and adaptively (Stagl, Salas, and Burke, 2007). Team members engage in meaningful dyadic and team-level exchanges before, during, and after the projects or tasks they are charged to complete. Over time, these interactions become entrained to contextual forces like supply and demand such that a pace and tempo of team performance develops (Ancona and Chong, 1999). For example, the members of an Intel flex fab manufacturing team ramp up their efforts to produce components for a client's new product launch. As the half-way point to their deadline approaches, members come together with their leader to reflect upon their progress and adapt to newly arisen issues. The team then reengages, meets its targeted goals, and ramps down until the next set of orders exceed current production levels.

Teams also perform episodically over time. For example, a real estate development team works episodically to create valuable assets from raw land by engaging in taskwork and teamwork processes that convert inputs to outputs. These input–process–output cycles are repeated in action and transition phases that unfold within performance episodes (Marks et al., 2001; Mathieu and Schulze, 2006). The transition phase encapsulates a preliminary meeting, where the team draws upon its shared resources to discuss what it aims to accomplish and what actions are necessary to meet its objectives. During this initial phase, a plan is mapped as the team considers which options are most viable and likely to reach fruition. Once a plan is formulated, the team shifts to an action phase where property is secured, developed, and marketed. After construction is finished and the property has been sold, the team reflects on its shortcomings and accomplishments and the cycle begins anew.

Another characteristic of teams is meaningful levels of task, goal, and feedback interdependencies (e.g., pooled, sequential, reciprocal team) (Saavedra et al., 1993). For example, a medical team composed of surgeons, nurses, rehabilitative therapists, anesthesiologists, and other specialists work toward restoring the infirm back to health. This extended team has sequential interdependencies as each specialist contributes his/her expertise in turn as the patient moves through various levels of need. At the beginning of a serious injury, the individual is traumatized and surgeons work to ensure survival. Afterward, specialty nurses watch over and manage the care. Eventually, physical and occupational therapists teach the injured person how to live and work productively. As each member of this extended team performs his or her respective duties in succession, the patient grows increasingly healthy.

Teams also are often designed with hierarchical structures to accomplish specific tasks over a limited period of time (Salas et al., 1992). These characteristics typify tactical decision-making teams, such as naval gunfire support teams, antisubmarine warfare teams, and guided missile teams (McIntyre and Salas, 1995). For example, guided missile teams operate in a command

and control structure with a formal leader and three to five team members. They interdependently seek out, share, and utilize information from their environment to make decisions about enemy engagements. These teams share the common goal of protecting the interests of the United States at home and abroad. Similarly, hierarchical structures and a limited lifespan also characterize private sector management teams, such as corporate executive teams and regional steering committees (Sundstrom et al., 2000).

The members of teams also typically hold distinct roles. This characteristic can be seen in a football team's offensive squad, which consists of quarterbacks, running backs, receivers, tight ends, centers, guards, and tackles. Each position is typically staffed by members that play unique roles. For example, the 2005 Super Bowl Champion Pittsburgh Steelers of the National Football League had one running back whose role was to play the vast majority of the team's first and second downs, a second running back who specialized in third-down conversions, and a third running back who was often asked to pick up much needed short yardage, particularly at the goal line. The use of each of these three roles was tightly choreographed to provide the team with the optimal mixture of capability given such factors as the down and distance on a play.

Teams also have distributed expertise. Distributed assets are apparent in jazz bands, where each musician leverages his/her experiences, training, and skills to make a unique contribution to a successful performance. Typically, the guitarist of a jazz ensemble does not possess the skills to play a saxophone or drums with professionalism; however, as a team they anticipate each other's actions by knowing enough about the other's capabilities to follow chord changes and instrumental solos. Once a solo is complete, the finished musician steps back and supports the others on stage. Although they possess unique skills, they know enough about each other's capabilities to complete a song seamlessly, with emotion, expertise, and enthusiasm.

While team members often have unique responsibilities and assets, they share common valued goals. Collaborative effort toward common valued goals can be seen in the practices of Solectron Corporation, a customized electronics solutions firm based in Charlotte, North Carolina. Solectron assigns engineers to client company design teams. Solectron embeds its team members in their client's operations because they believe that 80% of a product's price and supply chain performance is determined in the collaborative product design phase. In fact, their customers find that early design involvement with Solectron is crucial to the success and speed of eventual production. By rallying around the common goals of providing innovative, customized, quality electronic solutions via teamwork, Solectron and its clientele are able to truly leverage the benefits of teams.

Teams are also embedded within an organizational/environmental context that influences, and is influenced by, enacted competencies and processes, emergent cognitive and affective states, performance outcomes, and stakeholder judgments of team member and team effectiveness (Salas et al., 2006b). For example, the strategies and decision-making processes enacted by distributed leadership teams at IBM are shaped by their organizational and environmental context (Stagl et al., in press). To capture emerging markets, IBM is moving away from a cost leader strategy emphasizing manufacturing toward an innovator strategy emphasizing the development of information technologies, b2b solutions, and global consulting services. This shift in organizational priorities in response to a changing marketplace affects the types of strategic partners coveted by IBM, as witnessed by the recent $1.75 billion alliance with Lenovo, Asia's largest computer maker.

The handful of team characteristics expounded upon in this preliminary section is just a small sample of the myriad attributes that typify teams. Prior research offers a more comprehensive view of the numerous factors, which contribute to, comprise, flow from, and impinge upon teamwork, team performance, and perceptions of team effectiveness (Figure 6.1) (Tannenbaum et al., 1992). The framework of team effectiveness illustrated in Figure 6.1 depicts multiple inputs, throughputs, and outputs that are shaped and constrained by organizational/situational characteristics.

Tannenbaum et al.'s (1992) framework illustrates some of the core cognitive and behavioral processes that comprise teamwork (e.g., coordination, decision-making, backup behavior, mutual performance monitoring). Team members draw upon their latent resources (e.g., cognitive abilities, tacit knowledge, mental models) and teams draw upon their shared resources (e.g., shared mental models, cohesion) to execute these teamwork and taskwork processes during team performance. These actions transform raw materials

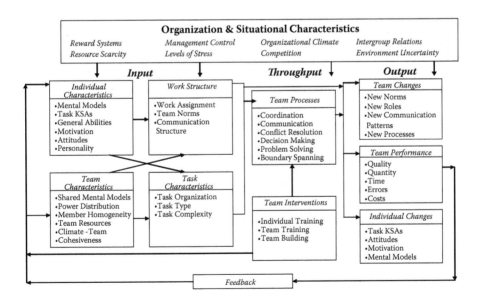

Figure 6.1 Team effectiveness framework. (Adapted from Tannenbaum, S.I., Beard, R.L., and Salas, E. (1992). In K. Kelley (Ed.), *Issue, Theory, and Research in Industrial/Organizational Psychology* (pp. 117–153). Amsterdam: Elsevier. With permission.)

and information resources into products and services. As a team performance episode unfolds, feedback is generated and team members and teams change in terms of the resources they have available for the next phase of performance.

The framework of team effectiveness advanced by Tannenbaum et al. (1992) also depicts a variety of human capital management interventions that can be leveraged by stakeholders charged with developing team members and team performance. These interventions are critical to turning a team of experts into an expert team (see Salas et al., 1997). The developmental solutions highlighted by Tannenbaum and colleagues include individual training, team training, and team building. Of particular note to the current endeavor, a host of team training programs have been advanced for developing teams, one of which is cross training. A brief overview of the nature of team training is provided in the next section and a representative set of best practices are advanced prior to launching into a more targeted treatment of cross training.

6.4   The science of team training

One means of helping to ensure that the benefits of teams are reaped is to prepare them for the operational realities they will encounter via training and development initiatives. Many organizations opt for this approach, as annual domestic expenditures for workforce training and development exceed $200 billion (Carnevale et al., 1990). In fact, survey data suggest that organizations spend approximately $812 to $1189 annually developing each of their employees (American Society of Training and Development, 2004). This booming industry houses a variety of approaches to growing individuals and teams, such as formal training, 360-degree feedback, executive coaching, operational assignments, mentoring programs, guided reflection, action learning, and outdoor challenges (Day and Halpin, 2001).

6.4.1   Best practice #1. Blend scientifically grounded tools, methodologies, competencies, and training objectives to create training programs

Of the above approaches to developing teams and their members, formal training programs are by far the most widely utilized (Salas et al., 2004). Conceptually, formal training programs, or instructional strategies, consist of theoretically grounded tools and methodologies, which are combined with a set of competencies and training objectives to form an instructional strategy (Figure 6.2) (Salas and Cannon-Bowers, 1997). Operationally, training is a planned intervention conducted to enhance the direct determinants of performance, including job-relevant knowledge, skills, and volitional choice behaviors (Campbell and Kuncel, 2001). For example, stress exposure training relies upon decades of stress research to develop the personal resources of team members and, thereby, a team's capacity to buffer the effects of stressors.

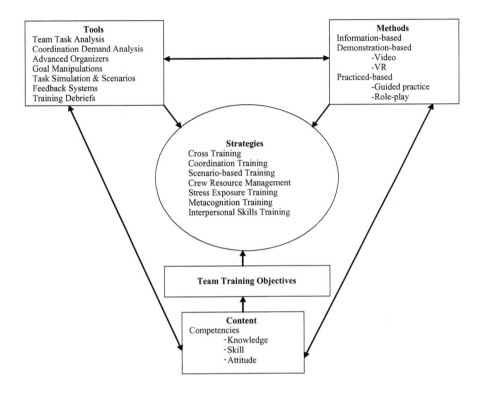

Figure 6.2 Model of structure of team training. (Adapted from Salas, E. and Cannon-Bowers, J.A. (1997). In M.A. Quinones and A. Ehrenstein (Eds.), *Training for a Rapidly Changing Workplace: Applications of Psychological Research* (pp. 249–280). Washington, D.C.: APA. (With permission.)

6.4.2   Best practice #2. Design training to leverage information-based, demonstration-based, and/or practice-based methods

A commonality of all training programs is that they include one of three methods for conveying instructional material (i.e., information-based, demonstration-based, practice-based) (Salas and Cannon-Bowers, 2001). Information-based methods deliver facts, concepts, and tips via lectures, handbooks, advanced organizers, and slide shows. Demonstration-based methods visually depict behaviors, actions, and/or strategies via video or multimedia information technologies. Practice-based methods provide trainees with hands-on rehearsal and real-time feedback via role-play exercises, moderate-fidelity desktop-based computer simulations, or high-fidelity, virtual reality simulations.

6.4.3   Best practice #3. Begin by training team members, then move on to teams

Formal training programs target either individuals and/or teams for development. The emphasis in this chapter is on team training, although team members as individuals are often developed prior to achieving the team-level effects that team-training programs are designed to produce. In fact, it has been suggested that knowledge and skills should be imparted first at the individual level, progress to role dyads, and then extend to the role networks comprising the team (Kozlowski, 1998). In other words, members should first learn to master their own responsibilities before discovering how their activities mesh with those of their teammates.

6.4.4 Best practice #4. Conduct training evaluation to gauge horizontal transfer and vertical transfer

While team training programs vary in content (e.g., crew resource management training, stress exposure training, metacognitive training) and method (e.g., information, demonstration, practice), all of these approaches share the similar aim of leveraging principles of learning to create meaningful growth experiences that benefit both individuals and their employers. This speaks to the purposeful nature of training, it is more than a feel-good intervention; it is a substantial investment that must be evaluated to make decisions concerning the continued use of the program and about the trainees exposed to the learning opportunity. In terms of evaluation, training must impart knowledge, affect, and skills and produce results that horizontally and vertically transfer to the workplace (Kozlowski et al., 2000). This perspective extends the traditional consideration of the mechanisms that facilitate horizontal transfer from the learning environment to the performance environment, to include a specification of the higher level (e.g., team, organizational) characteristics that enhance the expression of newly developed competencies in context (Kozlowski and Salas, 1997).

6.4.5 Best practice #5. Estimate the return on investment from training programs

In regard to evaluating training, organizational stakeholders should also estimate the return on investment from training; utility analysis can be utilized for this purpose (Schmidt et al., 1982; Boudreau, 1983). Mathieu and Leonard (1987) advanced a time-based utility model that can be applied to estimate the dollar value of training. This model is noteworthy because it incorporates additional economic parameters that previous formulas have not taken into account, such as variable costs, marginal tax rates, and discounting. In contrast to a deterministic utility approach that uses discounted cash flows to arrive at the net present value of an investment, training can also be valued using a real options framework (Nembhard et al., 2005). Modeling the value of training with this latter approach acknowledges the volatility of production dynamics, labor dynamics, task heterogeneity, and workforce heterogeneity.

6.4.6   *Best practice #6. Think twice before utilizing traditional team training programs to impart or change knowledge structures*

To this point, we have noted that team training is an effective means of imparting knowledge, skills, and attitudes and therefore is a useful means of shaping workplace behavior. Increasingly, however, the designers and end-users of team training programs are concerned with developing trainee knowledge structures, such as shared team-interaction mental models (Kraiger et al,, 1993; Mathieu et al., 2000). This is in contrast to the traditional emphasis on imparting knowledge of general principles and developing finite skills. Unfortunately, research suggests traditional training programs are often insufficient for changing existing mental models or fostering new ones once(Brigham and Laios, 1974; Morris and Rouse, 1985).

6.4.7   *Best practice #7. Use cross training to impart interpositional knowledge, a specific type of shared team-interaction mental model*

Instructional strategies that are designed to impart mental models must go beyond merely presenting task principles in order to emphasize the components that comprise a system as well as the interdependencies between those components (Kieras and Bovair, 1984). One instructional strategy that is useful for imparting both team member knowledge structures and shared team knowledge structures is cross training (Marks et al., 2002). Cross training has been successfully utilized to train both private sector and military teams (Marks et al., 2002; McCann et al., 2000; Volpe et al., 1996). The results, lessons learned, and best practices that can be distilled from these interventions are discussed in greater detail next.

6.5   *The science of cross training*

Cross training is a team training strategy that is used to train each of several team members in the tasks, roles, and responsibilities of their teammates (Volpe et al., 1996). By providing team members with insight into, and sometimes practice with, their teammates' duties, cross training serves to develop team interpositional knowledge, a specific type of shared team-interaction mental model. Mental models are "mechanisms whereby humans are able to generate descriptions of system purpose and form, explanations of system functioning and observed system states, and predictions of future system states" (Rouse and Morris, 1986, p. 351). These instantiated schemas can be held throughout a range of areas of team functioning, such as a team's task, equipment, team members, and team member interactions (Cannon-Bowers et al., 1993).

In keeping with the definition advanced by Marks et al. (2002, p. 5) in their recent research on cross training, we define a team-interaction mental model as the "content and organization of interrole knowledge held by team members in a performance setting." These cognitive repositories consist of procedural knowledge about how members work together to address problems within a given domain, including information concerning who does what and when those actions are most appropriate. Thus, team-interaction models are cognitive states that members draw upon to represent, explain, and predict the content and sequence of interdependent activities during goal accomplishment.

Shared mental model theory posits that members of effective teams hold *complementary* and, sometimes, *overlapping* representations of their tasks, equipment, members, and member interactions. Whether the procedural knowledge, which comprises shared team-interaction models, must be identical (Cannon-Bowers et al., 1993; Klimoski and Mohammed, 1994) or compatible (Kozlowski et al., 1996) is an issue beyond the scope of this chapter (see Kozlowski and Klein, 2000). For now, we simply assert that team members who understand the demands faced by their colleagues will be better positioned to coordinate with their teammates via the provision of assistance or leadership during routine or adaptive performance.

As evidence mounts for the assertion that effective teams have members with shared knowledge (e.g., Entin and Serfaty, 1994; Marks et al., 2000), shared mental models are increasingly viewed as a powerful theoretical explanation for effective team performance. For example, Hemphill and Rush's (1952) research with B-29 aircrews found a detailed understanding of team member positions (i.e., IPK) was correlated with both individual and crew effectiveness. Specifically, IPK was moderately correlated with crew coordination, crew leadership, and crew initiative. Moreover, in a subset of B-29 crews who completed an extended measure of IPK, interpositional knowledge was found to be correlated with a host of performance indices as rated by supervisors. In this subsample, IPK was positively correlated with competence on arrival in the unit, effectiveness in working with others, performance under stress, conformity with standard operating procedures, attitude and motivation, and overall effectiveness.

With respect to unique theoretical developments in cross training, Fiore et al. (2005) argued that presenting performance information during after-action review via differing *perspectives* represents an important way to potentially build IPK. They devised a technique based upon narrative theory (e.g., Bal, 1997) that would support the development of IPK through a unique form of positional modeling. This form of training may augment views of learning via positional rotation that has been argued to support "greater development of interpositional knowledge, which helps to enhance team functioning in difficult environments that require implicit coordination strategies" (Hollenbeck et al., 2004, p. 360).

An additional theoretical development has arisen out of the study of groups. This is the notion of transactive memory systems within groups, a concept that may similarly contribute to an improved understanding of cross training. Team familiarity in the form of "transactive memory systems" is related to successful task execution (e.g., Liang et al., 1995) in that these systems are described as long-term memory stores. These are essentially a form of shared mental model whereby a team does not store all aspects of knowledge relevant to the team or to their task; instead, members retain which particular person is aware of particular bits of information (Moreland and Myaskovsky, 2000). Fiore and colleagues have argued that this form of shared mental model directly relies on shared episodic memories, that is, they pertain to "interactions with teammates and the scenarios or situations engaged by the team" (Fiore et al., 2003, p. 353). Thus, as teams interact, communication and coordination often proceeds with the development of a shared or "common language for describing tasks" (Liang et al., 1995 p. 386). Cross-training techniques that rely on understanding and developing the forms of communication patterns that are indicative of the development of shared mental models and to effective team coordination represent important developments in team training.

To this point, our discussion has treated cross training as a unified team training approach when, in fact, it consists of three interrelated training techniques (i.e., positional clarification, positional modeling, positional rotation). These three techniques differ in regards to the method used to impart knowledge about teammates' roles and duties (Blickensderfer et al., 1998). For example, *positional clarification* is used to transmit a general understanding to a team member about his or her teammate's position via lecture or discussion. *Positional modeling* extends the technique of positional clarification by adding an observational component to the discussion of teammates' roles and responsibilities. *Positional rotation* is the most complex form of this instructional strategy, as it combines presentation and modeling with actual practice handling of teammates' responsibilities and tasks.

6.5.1 Best practice #8. Consider the pros and cons associated with each of the three cross-training techniques in light of growth objectives and contextual constraints

The three techniques that comprise cross training can be conceptualized as varying along a continuum of learning outcomes (Marks et al., 2002). At one end of the spectrum is positional clarification, which is used to raise awareness of the type and timing of activities that occur within a team. In contrast, positional rotation is at the other extreme because it is used to build knowledge and skill redundancy. While each of the three types of cross training is used with the common objective of increasing IPK, coordination, and thereby effective performance, each intervention varies in intensity, instructional method, monetary costs, time commitment, practicality, and potential negative transfer of training (Marks et al., 2002).

Because the use of each type of cross training is associated with specific up and downside factors, it is important to cultivate a clear understanding of the effects of these three interventions as separate techniques. This caveat is especially pertinent in light of meta-analytic research findings that provided meager support for the contribution of cross training to the overall

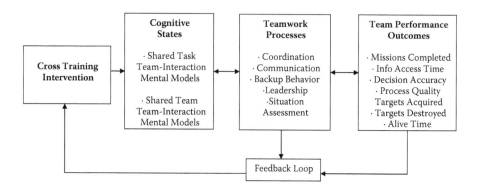

Figure 6.3 Heuristic framework of cross-training effects.

effectiveness of team training when it was conceptualized as a monolithic intervention (Salas et al., 2002). It is only through illuminating the unique benefits and costs of each of these three techniques that organizations can leverage that understanding to tailor the use of a given type of cross training to meet the specific training objectives in a particular context.

The following sections begin to delineate the proximal and distal effects of cross training by discussing the findings of prior research initiatives in light of an advanced heuristic framework (Figure 6.3). The discussion that follows of the empirical research studies investigating cross training is structured around the three techniques that comprise cross training (i.e., positional clarification, positional modeling, positional rotation). Best practices are extracted from these initiatives and advanced within each section. These best practices can be used by organizations conducting cross training to benchmark their current interventions or create new ones.

6.5.1.1   Positional clarification

Positional clarification relies upon written or verbal statements presented via lecture or discussion to convey IPK to team members about their teammates' positions. The instructor–trainee interactions that occur during a positional clarification intervention are conducted to raise the trainee's awareness about the content and timing of the taskwork and teamwork processes his/her teammates enact.

6.5.2   Best practice #9. Use positional clarification cross training to impart shared team-interaction mental models

To date, two research studies have investigated the positional clarification approach to cross training (see Cooke et al., 2003; Marks et al., 2002). The first of these two initiatives examined the effects of positional clarification on shared cognition in two laboratory settings using two different computer-based military simulations (Marks et al., 2002). This line of research also gauged the relationships between shared team-interaction mental models and both performance processes and performance outcomes. The results of the first of the two experiments conducted by Marks et al. (2002) suggested that teams exposed to positional clarification cross training developed a higher percentage of shared team-interaction knowledge than control group teams. The results of this first experiment also suggested that mental model similarity was correlated with backup behavior quantity, backup behavior quality, and team performance outcomes. Moreover, both backup behavior quantity and backup behavior quality were strongly correlated with team performance outcomes. The results of Marks et al.'s (2002) second experiment supported no such linkage between the positional clarification form of cross training and shared team-interaction mental models. The variable effects of positional clarification cross training on shared cognition as found in these two experiments is discussed in greater detail below.

6.5.3 Best practice #10. Conceptualize and measure the proximal effects of positional clarification cross training to gauge its impact

The second initiative undertaken to examine positional clarification was also conducted in a laboratory setting and similarly used a computer-based military simulation (Cooke et al., 2003). This study compared the usefulness of a conceptual version of cross training to a full version of positional clarification. While the results of this endeavor provided little support for conceptual cross training, they did suggest that positional clarification was an effective means of imparting both taskwork IPK and teamwork IPK. Although positional clarification did not directly affect team performance outcomes, taskwork IPK was predictive of performance outcomes in terms of mission completion rate. This latter finding is similar to the results of Marks et al. (2002), who found that the positional clarification affected proximal phenomena, such as shared cognition rather than directly impacting more distal team performance outcomes.

As noted, positional clarification is the least intensive of the three cross-training techniques. While positional clarification requires the least commitment of organizational resources in terms of time and money, some evidence suggests that it can be a useful means of imparting shared team-interaction models and, therefore, performance processes and performance outcomes (Marks et al., 2002). In fact, in the first experiment of a two-experiment research study, Marks and colleagues found that positional clarification was as effective as positional modeling in creating shared mental models. Given the additional resources required to conduct positional modeling, these results could be interpreted to suggest that positional clarification is a superior technique. Unfortunately, however, the picture is not so clear. This is because the results of the second experiment of this dual experiment research study suggested positional modeling was more effective at engendering shared team-interaction models than was positional clarification.

6.5.4 Best practice #11. Take into account the additional ambiguities inherent to distributed performance arrangements when choosing between positional clarification and more intensive forms of cross training

One explanation advanced by Marks et al. (2002) for the superiority of positional clarification over positional modeling in their first experiment, and its inferiority in their second experiment, is team member collocation or proximity. In their first experiment, ad hoc team members navigated a single helicopter so that they were always collocated. Marks and colleagues' reasoned that positional clarification was sufficient for creating interrole knowledge in this context because members worked closely together in real time on the same tasks. In contrast, members in the second experiment operated one of several tanks, which coordinated at a distance to accomplish a mission and, thus, a more intensive form of cross training was required to create shared cognition. It seems that as the coordination demands within a team increase (e.g., when a team is distributed) so too does the degree of cross training required (Travillian et al., 1993).

6.5.4.1 Positional modeling

The second form of cross training is positional modeling. The defining characteristics of positional modeling as a cross-training technique are demonstration and observation. More specifically, positional modeling includes the presentation of instructional content via an information technology-based medium (i.e., video, DVD) or by the efforts of a training instructor who demonstrates actions or strategies. Both of these mediums allow trainees to observe either desirable or undesirable behaviors.

6.5.5 Best practice #12. Use positional modeling cross training to impart shared team-interaction mental models

To date, two research studies have been conducted to examine the effects of positional modeling on team cognition, processes, and performance outcomes (see Duncan et al., 1996; Marks et al., 2002). The aforementioned Marks et al. (2002) research study consisted of two experiments, the first of which contrasted positional modeling with positional clarification and a control condition. The results of this first experiment suggested that positional modeling was an effective technique for fostering shared team-interaction mental models. Moreover, mental model similarity was correlated with backup processes and team performance outcomes. The results from Marks and colleagues' second experiment lent additional support to the effects of positional modeling on shared cognition. Specifically, the results of their second experiment suggested positional modeling was a useful technique for developing shared team-interaction knowledge and, in turn, this knowledge was correlated with coordination processes and team performance outcomes.

6.5.6 Best practice #13. Use positional modeling cross training to impart performance processes, such as leadership, backup behavior, and situation assessment

Duncan et al. (1996) also examined the effects of positional modeling. This research used Team Model Training (TMT), a 1-hour, computer-based, cross-training program, to examine the effects of cross training on proximal shared team-interaction mental models and more distal teamwork processes (i.e., backup behavior, leadership, situation assessment). The results of this initiative suggested that teams exposed to TMT developed richer shared mental models and engaged in less communication in terms of asking questions. Those teams exposed to TMT were also more effective at engaging in performance processes, such as backup behavior, leadership, and situation assessment, than control group teams.

6.5.6.1 Positional rotation

Positional rotation is the third type of cross training. This technique combines and extends the main elements of positional clarification (i.e., information presentation) and positional modeling (i.e., information demonstration and observation) by including actual practice in the tasks team members are charged with completing. A treatment is provided in this section of the five initiatives which have investigated positional rotation.

6.5.7 Best practice #14. Use positional rotation cross training to promote emergent cognitive states, such as IPK, and team performance processes, such as leadership

One early empirical examination of cross training was conducted with military teams who performed tasks in a simulated combat information center environment (Travillian et al., 1993). Travillian and colleagues used positional rotation to develop the IPK of 120 naval recruits randomly assigned to 40 three-person teams. The results of this study indicated that cross-trained teams were more effective at engaging in teamwork (i.e., coordination, leadership, orientation) and in generating desired team performance outcomes (i.e., accuracy of decisions) than control group teams who only received training on their own specific responsibilities.

Interestingly, the results of this initiative indicated the use of positional rotation was not related to mutual performance monitoring, backup behavior, providing feedback, and communication. The researchers interpreted these findings as suggesting that teams with less IPK spent more time monitoring, communicating, and backing up their teammates in order to understand the team's interactional requirements and, thereby, ascertain the "big picture" (Travillian et al., 1993).

6.5.8   Best practice #15. Revise performance measurement systems to account for the nuanced effects of cross training on team inputs and processes

The above results suggest that cross training serves to increase the frequency of some teamwork processes (i.e., coordination) while concurrently suppressing the use of other processes (i.e., communication). These findings have important implications for training evaluation as well as the appraisal of team performance in the workplace. Performance measurement systems used for training evaluation, team development, and/or compensation decisions must be tweaked to acknowledge that positional rotation leads to increases in some performance processes, while diminishing the use of other processes. For example, teams assessed in a maximum performance context after being exposed to cross training may display less communication. This does not necessarily suggest the team is performing poorly, as members may have drawn from their shared knowledge to switch to an implicit coordination strategy that minimizes the amount of overt communication required to complete a set of tasks (Entin and Serfaty, 1994). Moreover, as is suggested by the results of studies discussed later in this text, research suggests that some facets of communication (i.e., volunteers information) are enhanced by cross training, while other facets are unaffected (i.e., requests information) (Cannon-Bowers et al., 1998; Volpe et al., 1996). Thus, the variable effects of cross training on various criterion constructs and facets of criteria are also important considerations for organizational stakeholders.

A second use of positional rotation is found in research investigating the effects of cross training on the processes and performance of crew dyads engaging in a F-16 aircraft simulation (Volpe et al., 1996). Prior to the performance simulation, crew members received: (1) a 10-minute recorded message explaining how the simulation equipment operated, (2) written instructions for the task they were about to perform, and (3) a description of the roles and responsibilities of their teammates. This kind of information is identical to the instructional content that could be provided in positional clarification cross training. In addition to this information, each activity the crewmembers were expected to perform (e.g., aircraft maneuvering, radar locking, weapon selection, weapon firing) was demonstrated by the instructor using the flight simulation to depict appropriate actions. This type of demonstration is akin to positional modeling. Finally, each crewmember practiced his/her own tasks as well as the tasks assigned to his/her teammate (i.e., positional rotation).

6.5.9   Best practice #16. Use positional rotation cross training to reduce the latency with which targeted team performance outcomes are realized

The results of Volpe et al.'s (1996) research suggested that cross-trained crews exhibited more effective teamwork processes and produced more effective performance outcomes than their control group counterparts. For example, those dyads that received positional rotation cross training were more effective at communicating in terms of volunteering more information without being asked. Crews exposed to positional rotation were also more effective at maneuvering enemy targets into range, locking on to those targets, and destroying them in an expeditious manner. Cross-trained crews also received higher instructor ratings on team competency and overall team quality.

6.5.10   Best practice #17. Consider the workload typically experienced by a team, as positional rotation cross training is most effective when tasks are demanding

In a replication and extension of Volpe et al.'s (1996) research, Cannon-Bowers et al. (1998) examined the effects of positional rotation cross training on team processes and performance outcomes under varying workload conditions (i.e., low and high). Using the Tactical Naval Decision Making System to simulate Combat Information Center operations, positional rotation cross training enhanced both IPK and the amount of information volunteered by teams. Cannon-Bowers and colleagues' results also supported their expectations about the interaction of positional rotation and workload. Specifically, those teams that received positional rotation were able to leverage that advantage during the high workload condition to enact higher-quality processes, engage more targets correctly, and do so in less time than teams who did not receive cross training. In contrast, there was no difference between cross-trained and control group teams in the low workload condition in terms of process quality or performance outcomes.

The effects of a variation of positional rotation cross training, labeled experiential cross training in a team context, on team performance have also been investigated using a simulated naval surveillance task (McCann et al., 2000). McCann and colleagues sought to examine whether training team members, while performing the task as a team, was a viable training technique. This approach is in contrast to prior research that used a cross-training intervention to train individual team members and only later captured the performance of intact teams comprised of trained members. In addition, teams were reconfigured (i.e., each member was shifted to a new position) during the experiment to mimic the turnover and absenteeism of personnel, which teams in the workplace often experience as part of their ongoing operations.

6.5.11   Best practice #18. Train intact teams via experiential positional rotation cross training to minimize the performance decrements that result from turnover

Surprisingly, McCann et al.'s (2000) research results suggest that the only difference between cross-trained and untrained teams in terms of team processes was the fact that the former made slower decisions than the later. Interestingly, however, McCann and colleagues' results suggested that teams exposed to experiential cross training were more effective at coping with personnel reconfigurations than teams who were not exposed to positional rotation. In fact, teams in the experiential cross-training condition experienced no performance decrements as a result of the shift in personnel.

6.5.12   Best practice #19. Use positional rotation cross training to foster team processes, such as coordination, and thereby valued team performance outcomes.

The most recent examination of the effects of positional rotation cross training was conducted by Marks et al. (2002). These researchers examined the effects of positional rotation on shared cognition and the relationships between shared cognition and team processes and performance outcomes. The results of this experiment suggested that teams exposed to positional rotation cross training developed mental models with a higher percentage of shared team-interaction knowledge than those teams exposed to positional clarification. Moreover, team-interaction mental models were positively correlated with team coordination and performance outcomes in terms of the number of enemy installations captured and rebuilt on the battlefield.

6.6   Summary

There are many approaches to developing human capital that practitioners can utilize to develop teams and their members. This chapter reviewed the conceptual basis and supporting empirical evidence for one of these techniques, cross training. Cross training can be leveraged to develop shared team-interaction mental models or emergent cognitive states, which provide a structure for collecting and organizing information that is subsequently utilized to anticipate and execute the core processes of team performance. As performance unfolds and compiles over time, valued outcomes are achieved and stakeholders, both internal and external to a team, form judgments of effectiveness concerning a team's success and its need for further exposure to developmental opportunities.

The totality of empirical research findings reviewed herein strongly supports the use of cross training to develop shared team-interaction mental models. In fact, cross training is a useful means of imparting both teamwork and taskwork models (Figure 6.3). The evidence reviewed suggested the quantity and quality of these shared cognitive reservoirs were positively correlated with a wide range of performance processes, such as coordination, communication (volunteering information, but not requesting information), backup behavior, leadership, and situation assessment. In turn, these episodic processes were positively correlated with myriad performance outcomes, such as the number of missions completed, decision accuracy, and process quality (see Figure 6.3). Moreover, the use of cross training reduced decision latency and information access time.

Given the findings reviewed in this chapter, it seems cross training is a valuable approach for turning a team of experts into an expert team. It is important to remember, however, that cross training is a suite of interrelated techniques rather than a single intervention, and each of these approaches is uniquely grounded in the science of team training. In an effort to help practitioners maximize their limited resources, this chapter advanced 20 best practices, which can be leveraged by stakeholders when instituting positional clarification, positional modeling, and positional rotation programs in their organizations. Diligently following the advanced practices will help ensure the integrity, viability, and ultimately the utility of capital allocations for creating expert teams.

Acknowledgments

The views expressed in this work are those of the authors and do not necessarily reflect official Army policy. This work was supported in part by funding from the Army Research Laboratory's Advanced Decision Architecture Collaborative Technology Alliance (Cooperative Agreement DAAD19-01-2-0009). This work was also supported in part by the DoD Multidisciplinary University Research Initiative (MURI) program administered by the Army Research Office under grant DAAD19-01-1-0621.

References

American Society of Training and Development (ASTD). (2003). *STD 2003 State of the Industry Report Executive Summary*, Retrieved June 3, 2004, from http://www.astd.org/NR/rdonlyres/6EBE2E82-1D29-48A7-8A3A-357649BB6DB6/0/SOIR_2003_Executive_Summary.pdf.

Ancona, D., and Chong, C.L. (1999). Cycles and synchrony: the temporal role of context in team behavior. *Research on Managing Groups and Teams*, 2, 33–48.

Baker, C.V. (1991). The Effects of Inter-Positional Uncertainty and Workload on Team Coordination Skills and Task Performance. Unpublished Ph.D. dissertation, University of South Florida, Tampa, FL.

Bal, M. (1997). *Narratology: Introduction to the Theory of Narrative*. Toronto: University of Toronto Press.

Blickensderfer, E., Cannon-Bowers, J.A., and Salas, E. (1998). Cross-training and team performance. In J. A. Cannon-Bowers, and E. Salas (Eds.), *Making decisions under Stress: Implications for Individual and Team Training* (pp. 299–311). Washington, D.C.: APA.

Boudreau, J.W. (1983). Economic considerations in estimating the utility of human resource productivity improvement programs. *Personnel Psychology*, 36, 551–576.

Brigham, F., and Laios, L. (1974). Operator performance in the control of a laboratory processing plant. *Ergonomics*, 18, 53–66.

Burke, C.S., Stagl, K.C., Salas, E., Pierce, L., and Kendall, D. (2006). Understanding team adaptation: a conceptual analysis and model. *Journal of Applied Psychology* 91(6), 1189–1207.

Campbell, J.P., and Kuncel, N.R. (2002). Individual and team training. In N. Anderson, D.S. Ones, H.K. Sinangil, and C. Viswesvaran (Eds.), *Handbook of Industrial, Work and Organizational Psychology* (pp. 272–312). London: Sage Publications.

Cannon-Bowers, J.A., and Salas, E. (1990). Cognitive psychology and team training: shared mental models in complex systems. In K. Kraiger (Chair), *Cognitive Representations of Work*. Symposium conducted at the annual conference of the Society for Industrial and Organizational Psychology, Miami, FL, April.

Cannon-Bowers, Salas, E., and Converse, S.A. (1991). Cognitive psychology and team training: shared mental models of complex systems. *Human Factors Society Bulletin*, 37, 1–4.

Cannon-Bowers, Salas, E., and Converse, S.A. (1993). Shared mental models in expert team decision making. In N.J. Castellan, Jr. (Ed.), *Current Issues in Individual and Group Decision Making* (pp. 221–246). Hillsdale, NJ: Erlbaum.

Cannon-Bowers, J.A., Salas, E., Blickensderfer, E., and Bowers, C.A. (1998). The impact of cross-training and workload on team functioning: a replication and extension of initial findings. *Human Factors*, 40, 92–101.

Carnevale, A.P., Gainer, L.J., and Villet, J. (1990). *Training in America*. San Francisco: Jossey-Bass.

Cooke, N.J., Cannon-Bowers, J.A., Kiekel, P.A., Rivera, K., Stout, R.J., and Salas, E. (2000). Improving teams' interpositional knowledge through cross-training. *Proceedings of the IEA 2000/HFES 2000 Congress* (pp. 390–393). San Diego.

Cooke, N.J., Kiekel, P.A., Salas, E., Stout, R., Bowers, C., and Cannon-Bowers, J. (2003). Measuring team knowledge: a window to the cognitive underpinnings of team performance. *Group Dynamics: Theory, Research, and Practice*, 7, 179–199.

Day, D.V., and Halpin, S.M. (2001). *Leadership Development: A Review of Industry Best Practices* (Technical Report Number 1111). Alexandria, VA: U.S. Army Research Institute for the Behavioral and Social Sciences.

Duncan, P.C., Rouse, W.B., Johnston, J.H., Cannon-Bowers, J.A., Salas, E., and Burns, J.J. (1996). Training teams working in complex systems: a mental model-based approach. In W.B. Rouse (Ed.), *Human/Technology Interaction in Complex Systems* (vol. 8, pp. 173–231). Greenwich, CT: JAI Press.

Fiore, S.M., Johnston, J.H., and McDaniel, R. (2005). *Applying the Narrative Form and XML Metadata to Debriefing Simulation-Based Exercises*. Proceedings of the Human Factors and Ergonomics Society 49th annual meeting (pp. 2135–2139). Orlando, FL.

Fiore, S.M., Salas, E., Cuevas, H.M., and Bowers, C.A. (2005). Distributed Coordination Space: Toward a theory of distributed team process and performance. *Theoretical Issues in Ergonomic Science*, 4(3–4), 340–364.

Hemphill, J.K., and Rush, C.H. (1952). *Studies in Aircrew Composition: Measurement of Cross-Training in B-29 Aircrews* (AD B958347). Columbus: Ohio State University, Columbus Personnel Research Board.

Hollenbeck, J.R., DeRue, D.S., and Guzzo, R. (2004). Bridging the gap between I/O research and HR practice: Improving team composition, team training and team task design. *Human Resource Management*, 43, 4, 353–366.

Kahn, R.L., Wolfe, D.M., Quinn, R.P., Snoek, J.D., and Rosenthal, R.A. (1964). *Organizational Stress: Studies in Role Conflict and Ambiguity.* New York: John Wiley & Sons.

Kieras, D.E., and Bovair, S. (1984). The role of a mental model in learning to operate a device. *Cognitive Psychology,* 8, 255–273.

Klein, G.A. (1989). Recognition-primed decisions. In W.B. Rouse (Ed.), *Advances in Man-Machine Systems Research* (vol. 5, pp. 47–92). Greenwich, CT: JAI Press.

Klimoski, R., and Mohammad, S. (1994). Team mental model: construct or metaphor? *Journal of Management,* 20, 403–437.

Kozlowski, S.W.J. (1998). Training and developing adaptive teams: theory, principles, and research. In J.A. Cannon-Bowers and E. Salas (Eds.), *Making Decisions under Stress: Implications for Individual and Team Training* (pp. 91–114). Washington, D.C.: APA.

Kozlowski, S.W.J., Brown, K., Weissbein, D., Cannon-Bowers, J., and Salas, E. (2000). A multilevel approach to training effectiveness: enhancing horizontal and vertical transfer. In K. Klein, and S.W.J. Kozlowski (Eds.), *Multilevel Theory, Research and Methods in Organization* (pp. 157–210). San Francisco, CA: Jossey-Bass.

Kozlowski, S.W.J., Gully, S.M., Salas, E., and Cannon-Bowers, J.A. (1996). Team leadership and development: theory, principles, and guidelines for training leaders and teams. In M. Beyerlein, S. Beyerlein, and D. Johnson (Eds.), *Advances in Interdisciplinary Studies of Work Teams: Team Leadership* (vol. 3, pp. 253–292). Greenwich, CT: JAI.

Kozlowski, S.W.J., and Klein, K. (2000). A multilevel approach to theory and research in organizations: contextual, temporal, and emergent processes. In *Multilevel Theory, Research, and Methods in Organizations: Foundations, Extensions, and New Directions* (pp. 3–90). San Francisco: Jossey-Bass.

Kozlowski, S.W.J., and Salas, E. (1997). A multilevel organizational systems approach for the implementation and transfer of training. In., J.K. Ford (Ed.). *Improving Training Effectiveness in Work Organizations* (pp. 247–287). Mahwah, NJ: Lawrence Erlbaum Associates.

Kraiger, K., Ford, J.K., and Salas, E. (1993). Application of cognitive, skill-based, and affective theories of learning outcomes to new methods of training evaluation. *Journal of Applied Psychology,* 78, 311–328.

Liang, D., Moreland, R., and Argote, L. (1995). Group versus individual training and group performance: the mediating factor of transactive memory. *Personality and Social Psychology Bulletin,* 21, 384–393.

Marks, M.A., Mathieu, J.E., and Zaccaro, S.J. (2001). A temporally based framework and taxonomy of team process. *Academy of Management Review,* 26, 356–376.

Marks, M.A., Sabella, M.J., Burke, C.S., and Zaccaro, S.J. (2002). The impact of cross-training on team effectiveness. *Journal of Applied Psychology,* 87, 3–13.

Marks, M.A., Zaccaro, S.J., and Mathieu, J.E. (2000). Performance implications of leader briefings and team-interaction training for team adaptation to novel environments. *Journal of Applied Psychology,* 85, 971–986.

Mathieu, J.E., Heffner, T.S., Goodwin, G.F., Salas, E., and Cannon-Bowers, J.A. (2000). The influence of shared mental models on team process and performance. *Journal of Applied Psychology,* 85, 273–283.

Mathieu, J.E., and Leonard, R.L. (1987). Applying utility concepts to a training program in supervisory skills: a time-based approach. *Academy of Management Journal,* 30, 316–335.

Mathieu, J.E., and Schulze, W. 2006. The influence of team knowledge and formal plans on episodic team process performance relationships. *Academy of Management Journal* 49(3), 605–619.

McCann, C., Baranski, J.V., Thompson, M.M., and Pigeau, R.A. (2000). On the utility of experiential cross-training for team decision making under time stress. *Ergonomics,* 43, 1095–1110.

McIntyre, R.M., and Salas, E. (1995). Measuring and managing for team performance: emerging principles from complex environments. In R. Guzzo and E. Salas (Eds.), *Team Effectiveness and Decision Making in Organizations* (pp. 149–203). San Francisco: Jossey-Bass.

Moreland, R.L., and Myaskovsky, L. (2000). Exploring the performance benefits of group training: transactive memory or improved communication? *Organizational Behavior and Human Decision Processes,* 82, 117–133.

Morris, N.M., and Rouse, W.B. (1985). The effects of type of knowledge upon human problem solving in a process control task. *IEEE Transactions on Systems, Man, and Cybernetics,* SMC-15, 698–707.

Nembhard, D.A., Nembhard, H.B., and Qin, R. (2005). A real options model for workforce cross-training. *The Engineering Economist,* 50, 95–116.

Orasanu, J., and Salas, E. (1993). Team decision making in complex environment. In G. Klein, J. Orasanu, R. Calderwood, and C.E. Zsambok (Eds.), *Decision Making in Action: Models and Methods* (pp. 327–345). Norwood, NJ: Ablex.

Rouse, W.B., and Morris, N.M. (1986). On looking into the black box: Prospects and limits in the search for mental models. *Psychological Bulletin,* 100, 349–363.

Saavedra, R., Earley, P.C., and Van Dyne, L. (1993). Complex interdependence in task-performing groups. *Journal of Applied Psychology,* 1, 61–72.

Salas, E., Burke, C.S., and Stagl, K.C. (2004). Developing teams and team leaders: strategies and principles. In D. Day, S.J. Zaccaro, and S.M. Halpin (Eds.), *Leader Development for Transforming Organizations.* Mahwah, NJ: Erlbaum and Associates, Inc.

Salas, E., and Cannon-Bowers, J.A. (1997). Methods, tools, and strategies for team training. In M.A. Quinones and A. Ehrenstein (Eds.), *Training for a Rapidly Changing Workplace: Applications of Psychological Research* (pp. 249–280). Washington, D.C.: APA.

Salas, E., and Cannon-Bowers, J.A. (2001). The science of training: a decade of progress. *Annual Review of Psychology,* 52, 471–499.

Salas, E., Cannon-Bowers, J.A., and Johnston, J.H. (1997). How can you turn a team of experts into an expert team? Emerging training strategies. In C.E. Zsambok, and G. Klein (Eds.), *Naturalistic Decision Making* (pp. 359–370). Mahwah, NJ: Lawrence Erlbaum Associates.

Salas, E., Dickinson, T.L., Converse, S.A., and Tannenbaum, S.I. (1992). Toward and understanding of team performance and team training. In R. W. Swezey and E. Salas (Eds.), *Teams: Their Training and Performance.* Norwood, NJ: Ablex Publishing.

Salas, E., Mullen, B., Nichols, D.R., and Driskell, J.E. (2002). What contributes to the effective design of team training? In A. Towler (Chair), *Current Issues in Training Design and Evaluation.* Symposium presented at the 17th annual conference of the Society for Industrial and Organizational Psychology, (April) Toronto.

Salas, E., Priest, H.A., Stagl, K.C., Sims, D.E., and Burke, C.S. (2006a). Work teams in organizations: a historical reflection and lessons learned. In L. Koppes (Ed.), *Historical Perspectives in Industrial/Organizational Psychology* (chap. 17). Mahwah, NJ: Lawrence Erlbaum Associates.

Salas, E., Stagl, K.C., Burke, C.S., and Goodwin, G.F. (2006b). Fostering team effectiveness in organizations: toward an integrative theoretical framework of team performance. In J.W. Shuart, W. Spaulding, and J. Poland, (Eds.), *Modeling Complex Systems: Nebraska Symposium on Motivation, 52*, Lincoln: University of Nebraska Press.

Schmidt, F.L., Hunter, J.E., and Pearlman, K. (1982). Assessing the economic impact of personnel programs on work-force productivity. *Personnel Psychology, 35*, 333–347.

Stagl, K.C., Salas, E., and Burke, C.S. (2007). Best practices in team leadership: what team leaders do to facilitate team effectiveness. In J.A. Conger and R.E. Riggio (Eds.), *The Practice of Leadership*. New York: John Wiley and Sons.

Stagl, K.C., Salas, E., Rosen, M.A., Priest, H.A., Burke, C.S., Goodwin, G.F., and Johnston, J.H. (in press). Distributed team performance: a multilevel review of distribution, demography, and decision-making. In F. Yammarino and F. Dansereau (Eds.), *Multi-Level Issues in Organizations*,Amsterdam/New York: Elsevier.

Sundstrom, E., de Meuse, K.P., and Futrell, D. (1990). Work teams: applications and effectiveness. *American Psychologist, 45*, 120–133.

Sundstrom, E., McIntyre, M., Halfhill, T., and Richards, H. (2000). Work groups: from the Hawthorne studies to work teams of the 1990s and beyond. *Group Dynamics, 4*, 44–67.

Tannenbaum, S.I., Beard, R.L., and Salas, E. (1992). Team building and its influence on team effectiveness: an examination of conceptual and empirical developments. In K. Kelley (Ed.), *Issue, Theory, and Research in Industrial/Organizational Psychology* (pp. 117–153). Amsterdam: Elsevier.

Travillian, K.K., Volpe, C.E., Cannon-Bowers, J.A., and Salas, E. (1993). *Cross-Training Highly Interdependent Teams: Effects on Team Process and Team Performance*. Proceedings of the 37th annual Human Factors and Ergonomics Society conference (pp. 1243–1247). Seattle, Washington.

Volpe, C., Cannon-Bowers, J.A., Salas, E., and Spector, P.E. (1996). The impact of cross-training on team functioning: an empirical investigation. *Human Factors, 38*, 87–100.

Wenger, D.M. (1995). A computer network model of transactive memory. *Social Cognition, 13*, 319–339.

chapter 7

Team training in complex environments

Hari Thiruvengada and Ling Rothrock

Contents

7.1 Introduction

Training a workforce to work efficiently and effectively has always been the primary objective of training program designers, developers, and coordinators. Currently, a variety of training methodologies exist that focus on improving a multitude of characteristics related to both individual and team performance. Although such training methodologies generally improve individual and team performance, not all types of training are adequate and beneficial in every instance. Specifically, in high stress environments (such as military command and control or emergency response in hospitals), it is important that each team member understands the expectations of other teammates without overt communication, and implicitly coordinates his/her activities with other teammates to help maintain sustained team performance. Therefore, simply training individuals in their taskwork skills may be insufficient and lead to diminished performance. Additionally, appropriate team training (such as cross training or team coordination training) may be required in such situations to improve coordination and interaction between individual and team functions that yield better performance. Some existing team training methodologies that are popular (Cannon-Bowers and Salas, 1998) include cross training, team adaptation and coordination training (TACT), team dimensional training (based on guided team self correction), assertiveness training, and team leader training. These team-training methodologies are the outcome of the research performed by the U.S. Navy as part of its TADMUS (Tactical Decision Making under Stress) program (Cannon-Bowers and Salas, 1997). These methodologies emphasize various aspects of a team and are based on factors, such as team composition, task demands, organizational context, individual roles, and responsibilities that adversely affect the team structure and functioning.

Cross training is one of the most effective forms of training that is applicable to instructing employees within a workforce. Volpe et al. (1996) defined cross training as "an instruction strategy in which each team member is trained in the duties of his or her teammates." This definition is applicable to workforce cross training where the team size is large and multiple team members are assigned to similar tasks. Blickensderfer et al. (1998) classified cross training into three types: *positional clarification, positional modeling,* and *positional rotation*. Positional clarification involves the process of verbally acquainting the team member with the responsibilities of his/her teammates through lecture and discussion methods. Positional modeling involves both verbal discussion and observation of team members' roles. In positional rotation, each team member receives an on-the-job-training concerning the other team members' responsibilities. Positional clarification is the most intense form of cross training that involves long training periods. On the other hand, positional clarification is least intense and involves relatively short training periods.

There is a new type of cross training, known as Task Delegation Training (TDT), that can be used to augment the efficacy of teams by allowing delegation of tasks to its individual members. TDT is a positional clarification type of cross training where every team member has an understanding of the other team members' responsibilities and expectations. In TDT, the team members are trained and encouraged to understand the expectations and responsibilities of their teammates. Teams that are trained using TDT are able to compensate for the absence of a team leader through prior task delegation of the team leader's responsibilities. Such prior task delegation leads to nonoverlapping responsibilities for each team member and elevated team performance through efficient use of available resources. As a result, team members focus on specific tasks that relate to their primary taskwork as well as teamwork (backing up and error correcting other teammates) skills without inherent conflict in their role responsibilities.

This chapter is divided into eight major sections. In the introductory section, the need for team training is explained. Section 7.2 provides a discussion of the current practices in team training methodologies. Prior attempts to train fundamental team behaviors (namely backup and error correction behavior) are discussed in Section 7.3. Section 7.4 explains the relation between team training and team performance and in Section 7.5, an empirical investigation of team training effectiveness is introduced. The next section (Section 7.6) explains the results and analysis of comparison between team coordination training (TCT) and task delegation training (TDT) and a no training condition. Table 7.1 shows the basic characteristics of each of these training interventions.

The comparison allows us to assess the effectiveness of training on team performance. Section 7.7 summarizes the lessons learned from this research, and in the final section (Section 7.8), we provide concluding remarks about the training effectiveness on team performance. Empirical results indicate that teams that received TCT and TDT had comparable performance but outperformed teams that received no training.

7.1.1 Need for workforce training

Brannick and Prince (1998) define a team to be "two or more people with different tasks who work together adaptively to achieve specified and shared goals." Cannon-Bowers and Salas (1998) argue that a team is a set of two or more individuals who interact interdependently and adaptively toward a common goal or objective. The team members have specific roles and functions that relate to the task environment. Members of a team are bound together by a high level of interdependency, where every team member is interdependent and must coordinate his/her own activities with his/her teammates to achieve common goals. Coordination, the central feature

Table 7.1 Types of Training

No training (NT)	Team coordination training (TCT)	Task delegation training (TDT)
No specific training is imparted Team members are provided with information on the definition of team coordination and task delegation No specific tasks are delegated to each operator role	Team coordination is emphasized during training Team members are instructed on how to achieve effective coordination via demonstration of good and bad practices No specific tasks are delegated to each operator role	Task delegation is emphasized during training The radar scope on the operator's Graphical User Interface is split into two distinct areas and is designated to each of the two roles; operators monitor and perform actions within the designated area, while passing information pertaining to the other area onto their team mate Specific tasks are delegated to each operator role based on KSA competencies and operator capabilities.

of teamwork, requires adjustments on behalf of one or more of the team members so that the team goal can be reached. A workforce consists of small work teams (Sundstrom et al., 1990) that satisfy the requirements of the larger organization and still maintain adequate independence to perform specialized functions. External integration into a larger system and external differentiation through specialization, independence, or autonomy are critical to defining group-organization boundaries that partly specify what constitutes team effectiveness. Consequently, training plays a critical role in facilitating improved overall performance among members of a smaller work team or a larger workgroup. The existing principles and methodologies involved in team training, such as cross training or coordination training, could be applied to workforce training in order to improve productivity.

7.2   Current practices in team training

As mentioned earlier, there are several methodologies employed in team training that are geared toward improving specific team functions. In this section, we identify team competencies that govern the effectiveness of teams, discuss some existing team training practices, and analyze the benefits of cross training. We focus on the positional clarification form of cross training and on how to select an appropriate form of cross training.

7.2.1 Team competencies

The desired effect of team training is the development of team competencies. Cannon-Bowers et al. (1995) argued that three competencies are required for a team member to be effective: *knowledge, skills,* and *attitudes.* If team members are deficient in any of these competencies then their performance degrades substantially. Knowledge competencies correspond to the team members' knowledge about the tasks, the environment, and other team members. This leads to the development of interpersonal knowledge (IPK) and a shared mental model among the team members. Implicit coordination of effective teams is an outcome of high level of IPK and shared team mental model. Skills and competencies relate to each team member's adeptness at performing their primary taskwork and teamwork. Besides being adept at their primary taskwork, effective teams have an advanced understanding of teamwork skills that enable better communication, coordination, and adaptability.

Task-related attitudes or attitude competencies of teams influence a team member's choices and decisions to act in a certain way under particular circumstances (Dick and Carey, 1990). It determines the overall effectiveness and the way the team members feel about the task, the environment, and their impact on teamwork. Being proactive instead of reactive is helpful in improving team performance, as effective teams are able to reallocate tasks efficiently and anticipate the requirements of their teammates through implicit coordination (Kleinman and Serfaty, 1987). Such implicit coordination is critical for successful team performance in high stress environments. Orasanu (1990) observed that effective teams used the periods of low workload to plan strategies and prepare for high stress (crisis) situations. Therefore, maintaining better attitude competencies toward team tasks helps promote good overall team performance. A detailed definition of various team competencies, their implications on training design, and propositions linking situational or task characteristics to team competencies is provided in Cannon-Bowers et al. (1995). Members of a workforce (large teams) must exhibit appropriate team competencies in order to be efficient and productive.

7.2.2 Team-training techniques

The most notable forms of team training that are currently used are cross training, team coordination and adaptation training (TCT), team dimensional training (TDT), assertiveness training, and team leader training. While other team-training techniques exist, we restrict our discussion to forms that can be applied in complex environments.

Cross training (Cannon-Bowers et al., 1998; Volpe et al., 1996) is a form of training in which team members are trained on the tasks, duties, and responsibilities of other teammates. It is generally achieved through positional clarification, positional modeling, or positional rotation (Blickensderfer et al., 1998; Marks et al., 2002) and the level of knowledge that is imparted to individual team members in each cross-training technique is different. Cross training is beneficial for interdependent teams where the overall team performance depends on each team member's understanding of other team-mates' responsibilities and expectations.

Serfaty et al. (1998) introduced a model-driven form of team coordination training (TCT) known as team adaptation and coordination training (TACT). The focus of TACT is on the training of team coordination strategies through dynamic adaptation of teamwork processes to changing internal and external conditions. In the TACT form of training, team members were trained to utilize their low workload periods to plan task strategies for chaotic situations and task demands. Empirical results suggest that expert teams were able to anticipate other teammate's needs through implicit coordination and maintain superior performance during high stress scenarios.

A third form of training is known as team dimensional training through guided team self-correction (Smith-Jentsch et al., 1998). In this training, a team leader or instructor guides a team to self-diagnose problems in team processes and their interactions to develop effective solutions. The team members identify team processes that result in inferior performance through self-evaluation and work toward improving these processes during subsequent training. Effective feedback among team members is vital to the process of identifying areas of improvement in guided team self-correction.

The fundamental theory of assertiveness training (Smith-Jentsch et al., 1996) is that team members who are trained to be assertive when communicating feedback, stating opinions, offering a solution to a problem, initiating an action, offering or requesting assistance or backup can avert potential disaster in some environments. Team members who underwent assertiveness training are able to improve their attitudes toward being assertive in a team, even though there is no improvement in team performance. Assertiveness training is critical in highly sensitive environments, such as nuclear power plants or emergency room in hospitals, in order to prevent loss of life.

Team leader training (Tannenbaum et al., 1998) focuses on improving the behavior of a team leader as a facilitator, thereby improving the team climate for individual team members. Effective team leaders are able to maintain a good team climate during adverse conditions and facilitate the proper functioning of their teams. Researchers (Adelman et al., 1986; Fleishman and Zaccaro, 1992) suggest that teams exhibit effective decisionmaking in complex situations when they adopt implicit, flexible, and adaptive coordination (such as loading, sharing, and mutual performance monitoring) in the absence of explicit directives. Therefore, it is the responsibility of the team leader to prepare the team for future events based on past experience to help the team cope with adversities. One particular instance where this presumption fails is the case of reduced teams with no team leader where the team leader is unavailable to issue directives. This could lead to adverse consequences for the entire team if the individual members do not exhibit effective team coordination and adaptation. We contend that TDT is an effective team training methodology that helps maintain optimal team performance under such circumstances.

In this chapter, we focus on comparing the training effectiveness of TCT, TDT (positional clarification type of cross training), and no training methodologies on overall team performance in high-stress situations where cross training is required. We contend that the TDT condition provides more choices in accomplishing common goals than TCT or no training conditions and, thus, enables reduced teams to maintain superior performance under high-stress situations.

7.2.3 Cross training

Cross training is an instructional form of team training in which each team member is trained in the duties of their teammates (Volpe et al., 1996). It promotes the advancement of the knowledge required to perform the job functions of other team members and the overall team framework. Team communication is critical to the development of a shared understanding of the team functions, which is known as the *mental model of the team* (Blickensderfer et al., 1998). The extent to which the mental model of each team member overlaps determines the level of shared understanding or shared mental model for the team (Cannon-Bowers et al., 1993; Orasanu, 1990). As individuals gain experience on the task, they develop a shared understanding of the team functions and exhibit adaptability in adverse circumstances. Empirical research (Marks et al., 2000; Mathieu et al., 2000) indicates that increased similarity of a shared mental model improves coordination process leading to enhanced team performance.

Cross training is an essential strategy employed for training teams to understand the expectations of all its members. Although individuals in a team may be adept at performing their primary tasks, the team taken collectively may exhibit low team-level performance. Such a lowered performance is a consequence of the mismatch, which occurs in each team member's expectations of his/her teammates' task requirements. Team performance may be deterred further in dynamic and complex environments, where there is additional pressure induced due to temporal constraints and task complexity. In order to perform well in such adverse situations, the team must exhibit adaptability (Blickensderfer et al., 1998). Teams that are effective are able to adapt themselves to adverse conditions through implicit coordination (involving no overt communication), based on the information available in the task environment, and provide backup to their teammates by dynamically reallocating task responsibilities. Cross training enables the creation of interpositional knowledge (Baker et al., 1992; Volpe et al., 1996), also known as IPK, which includes each team member's understanding of appropriate task behavior of his/her teammates. Such understanding relates to each individual's task work skills, job function requirements, and context-dependent temporal relationships that explain the cause-and-effect of the underlying tasks.

Blickensderfer et al. (1998) provide the classification scheme for cross-training strategies and describe the level of interdependence and type of IPK required for different types of teams. This classification scheme is used to identify the appropriate type of cross-training methodology that could be adopted for enhancing team performance. Teams with low interdependence, such as advisory groups or review panels, require a basic level of IPK, as there is minimal interactive communication and coordination among its members. Backup behavior or team leadership is not critical to functioning of such teams; therefore, positional clarification seems to be an appropriate cross-training strategy for such teams. On the other hand, mining teams, task forces, and research teams may require a medium level of interdependence with some interactive communication and coordination, monitoring and feedback, team leadership, and backup behavior. The IPK in teams with a medium level of interdependence consists of a working knowledge of overall team structure, general dynamics, each member's responsibilities, and a basic understanding of the interaction process. Positional modeling is regarded as an appropriate cross-training methodology for such teams. Finally, teams that involve a high level of interdependence, such as surgical teams, emergency response teams, and combat information center teams, communicate and coordinate interactively. Monitoring, feedback, team leadership, and orientation are critical for optimal performance in such teams. These teams require thorough knowledge of specific job activities and interactions of each team member along with a shared understanding of the expectations of the entire team. Positional rotation is recommended as an appropriate cross-training methodology for interdependent teams.

7.2.4 Task delegation training (TDT)

Task delegation training is a positional clarification form of cross training where each team member is imparted with knowledge about the job functions, his/her interactions, and expectations of other teammates. Earlier studies in team cross training have focused on how a specific type of cross training could help improve overall team performance within the setting of a full team (teams where all team members are active and perform their task functions). However, it is plausible that a full team could evolve into a reduced team during the course of task execution. For instance, consider the case of a Special Weapons and Tactics (SWAT) team where the team commander is wounded and is no longer available to lead the team effectively. In such a situation, the other team members must coordinate and perform the duties of the team commander in order to survive. Teams must be trained to adapt to such a situation through additional training.

We submit that team coordination and TDT are capable of producing optimal team performance in such situations where additional clarification is required for the reallocation of tasks to each individual role. TDT is a positional clarification-type of cross training that involves relatively less training time and resources. In TDT, each team member is verbally instructed about the role responsibilities handled by his/her teammates and are provided with prior delegation of a subset of these responsibilities in case of his/her absence during the course of task execution. This avoids an overlap of among team tasks for each role and helps deploy the available resources effectively. We provide a comparison of the effectiveness of the TDT, TCT, and no training on team performance in this chapter and argue that TDT provides more effective paths to achieve good team performance in reduced teams.

7.3   Training for backup and error correction behavior

Performing secondary duties, such as providing backup and error correcting other teammates, is an essential component of teamwork. As teams are interdependent in nature, it is vital that every member of the team must provide appropriate backup and error correction for overloaded teammates in order to maintain superior performance. In congruence to Smith-Jentsch et al. (1998), we define the behaviors relating to providing backup and error corrections for other teammates as supporting behavior. There are other dimensions to teamwork, such as information exchange, communication, and initiative/leadership as defined by Smith-Jentsch et al. However, in this study, we strictly analyzed the effectiveness of training methodology on improving information exchange, team initiative and leadership, and supporting behavior. We did not consider communication as a dimension for analysis because teams exhibited implicit coordination without overt communication due to the high stress level of the task (Serfaty et al., 1998; Orasanu, 1990). Similarly, team initiative/leadership is used as a dimension for teamwork as reduced teams are still required to carry out their team leader's tasks during his/her absence.

Team coordination training attempts to enhance supporting behavior by training members to facilitate high levels of implicit coordination during high-stress scenarios. TCT aims at educating team members to utilize their periods of low workload to plan for chaotic situations and develop contingent plans based on teammate's expectations. Team dimensional training utilizes feedback on the supporting behavior dimension to train the team members on strategies to improve this dimension. Team members are encouraged to be assertive and forthcoming while providing backup and error correction. In TDT (positional clarification cross training) team members provide backup and error correction by performing their delegated tasks (tasks of the missing team member) in addition to fulfilling their primary responsibilities.

7.4   Relation between team training and team performance

In order to discuss the effectiveness of team training on team performance, it is necessary to understand the relationship between team processes, outcomes, and performance. Marks et al. (2002) argue that teamwork or team processes are the mediating links that link the relationship between team training and corresponding team outcome (performance) within the setting of input–process–outcome models. Coovert et al. (1990) suggest that team processes relate to the activities, strategies, responses, and behaviors employed in task accomplishment within teams. Team outcomes, on the other hand, pertain to the outcome of the various team processes. Any team performance measure (TPM) (Cannon-Bowers and Salas, 1997) must address the process as well as outcome measures in an appropriate manner.

Team outcome is usually not a diagnostic indicator, as it does not indicate the principal causes for observed performance. However, team process measures describe performance in a more direct manner. For example, let us consider the task where human subjects are required to judge the correlation between two datasets. They may either use a mathematical representation involving regression or a graphical plot of the datasets to identify the correlation. The predicted outcomes are positive, zero, or negative correlation, but it does not specify the underlying processes used by the subjects in their decision-making. Therefore, the subject's performance must be measured in terms the decision-making processes as well as outcomes (judgments) to help identify effective training methods. It is essential to identify and isolate the specific team processes that lead to lowered performance and reinforce processes that yield better performance. Figure 7.1 explains the relation between training, processes, outcomes, metrics, and performance in teams.

Team training is the input parameter to the team under training, which can be controlled in order to train teamwork and knowledge, skills, and attitudes (KSA) competencies that affect team processes, such as communication and coordination. These team processes have direct impact on the team performance measures (outcomes), which, in turn, influence overall team performance.

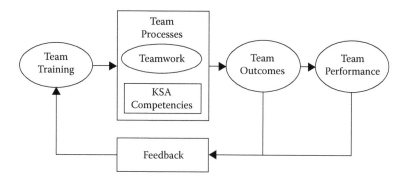

Figure 7.1 Relational diagram between team training and team performance.

7.4.1 Measurement instruments for individual and team performance

Cannon-Bowers and Salas (1997) state that the TPMs must consider multiple levels of measurement that are grounded both at individual and team levels. The justification provided for such a multilevel analysis is that both teamwork and taskwork skills influence team performance. Additionally, TPMs must include measures that address process as well as outcome. The process measures describe the activities, strategies, responses, and behaviors relevant to the human who is used to accomplish a task. On the other hand, outcome measures assess the quantity and quality of the end result. The process measures are usually referred to as *measures of performance* (MOP), and outcomes measures represent the *measures of effectiveness* (MOE). The distinction between MOP and MOE is that MOP relates to the human factor involved in the complex systems, while MOE includes other elements that are external to the human performance, such as equipment, environment, and fortune. TPMs that are used in training intervention must be able to provide a description, evaluation, and diagnosis of performance and form a basis for the remediation process.

Cannon-Bowers et al. (1995) propose that team competencies are held at different levels and relate to the KSA required to fulfill individual and team task requirements. They argue that knowledge, skills, and attitudes exist as individual competencies at individual level, as team competencies at individual level, and as team competencies at team level. Therefore, the team performance measure must adequately capture these competencies to act as a diagnostic indicator for training interventions.

In the past, researchers have used several instruments to assess and measure process and outcomes measures for operator actions at both individual and team level. Smith-Jentsch et al. (1998) provide a list of some of these instruments, which include Sequenced Actions and Latencies Index (SALI), Behavioral Observational Booklet (BOB), Antiair Teamwork Performance Index (ATPI), and Anti-Air Teamwork Observation Measure (ATOM). While SALI and BOB are measures used to evaluate individual level outcomes and process, ATPI and ATOM are used to evaluate team level outcomes and performance. In the SWAT example mentioned above, while SALI and BOB measure each SWAT team member's individual performance, ATPI and ATOM indicate team level performance. These instruments are used by experts in the field to provide subjective ratings for MOP and MOE at individual and team levels, and provide an indication of the expert's judgment of operator performance. Therefore, these ratings are subject to problems, such as inter-rater bias and reliability of the ratings. Additionally, the subjective ratings provided by the experts are decoupled from the objective measures of team performance. A sample ATOM for the supporting behavior dimension of teamwork is shown in Figure 7.2.

The ATOM shown in Figure 7.2 is used to elicit expert ratings for the supporting behavior dimension, specifically error correction. These expert ratings are subjective and are prone to inter-rater bias and require additional

Supporting Behavior:

Error correction - Instances where a team member points out that an error has been made and either corrects it him/herself or see that it is corrected by another team member

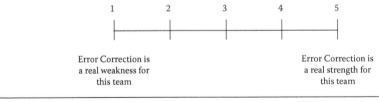

Figure 7.2 Sample of an antiair team observation measure (ATOM) for supporting behavior dimension.

validation. For example, in a case where one team member is observed to verbally point out a mistake, but is ignored by his teammate one rater may score the behavior a 4 (for mentioning the mistake), while another rater may score a 1 (for the lack of acknowledgment).

In contrast to existing MOPs and MOEs, we describe a measure called the relative accuracy index (RAI). RAI is an instrument that provides an objective assessment of process and outcome measures based on a framework known as a time window (Rothrock, 2001; Rothrock et al., 2005). Time windows do not prescribe what action should occur, but rather provide objective bounds on what action could be taken based on the operator's environment. These bounds represent a window in time, within an appropriate and related action would result in successful completion of the task. A more detailed explanation of a time window and the classification of its outcomes are provided in the following section. RAI avoids inter-rater bias problems, as it does not involve expert ratings, and evaluates outcomes based on opportunities that exist for an operator to perform an action within the environment. RAI also avoids the problem associated with reliability, as the outcomes are distinctly classified into six categories based on the operator's response to the environmental constraints. Therefore, RAI bridges the gap between performance and expert ratings, as it provides an indication of performance that is relevant to the process and outcome measures at individual and team level.

7.4.2   *Team performance measurement (TPM)*

Cannon-Bowers and Salas (1997) argue that any team performance measures, or TPM, must relate to the process as well as outcome measures in an appropriate manner in order to be treated as a diagnostic indicator. Team outcome measures are not diagnostic as they are affected by conditions external to the team members' control. For instance, during a forest fire, the efficacy of the fire-fighting team depends on the training provided to the individual members, their coordination, placement of each fire-fighting team (unit) as well as external conditions, such as wind conditions and weather. Therefore, it is critical to relate the appropriate team outcome measure with team process measures in order to diagnose the causal factors for current team performance.

7.4.2.1 Team process measures

Smith-Jentsch et al. (1998) used the seven original ATOM teamwork dimensions to define a simpler subset of four teamwork dimensions that avoids redundancy across the definitions of the seven dimensions. They suggest that teamwork could be appropriately captured and related to team process through four dimensions: information exchange, communication, team leadership/initiative, and supporting behavior. We use these four dimensions as a measure of team processes that influence teamwork in this study.

7.4.2.2 Team outcome measures

The team outcome measures used in our study are based on a construct called a time window (Rothrock, 2001; Rothrock et al., 2005), which was developed to evaluate the correctness and timeliness of operator actions based on the environment. A time window is based on the principles of signal detection theory (Green and Swets, 1988; Swets, 1996) and specifies temporal bounds for operator actions. It does not prescribe a correct action that needs to be taken but indicates whether an operator's action would lead to the required situation based on the current circumstances that exist within the environment. A time window is said to be open when an opportunity exists to execute an action that relates to that time window, or is otherwise closed. The possible classifications of a time window fall under six outcomes as shown in Figure 7.3, based on operator response (action) and environment.

An "on time" correct action — area marked (2) — is defined as an action executed by the operator that is relevant to the time window and results in

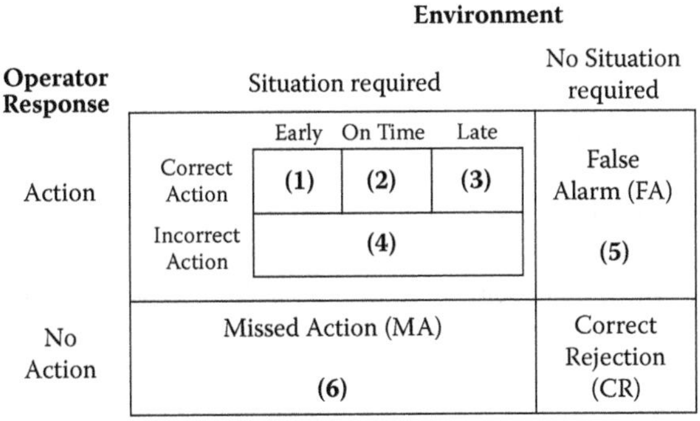

Figure 7.3 Classification of time window outcomes.

the required situation. An "early" correct action — area marked (1) — is defined as an operator action that is relevant to the time window and results in the required situation but is executed before the time window is opened. A "late" correct action — area marked (3) — on the other hand, is defined as an operator action that is relevant to a time window and leads to the required situation but is executed after a time window is closed. An "incorrect" action — area marked (4) — is defined as an action that is relevant to a time window, but does not result in the required situation. A "false alarm" action — area marked (5) — is an operator action that has no relevant time window. A "missed" action — area marked (6) — on the other hand, is an operator action that was not executed, but the time window for that action exists. The readers are referred to Rothrock (2001) for the mathematical representations of these six time window outcomes. A *truth maintenance system* (TMS) (Doyle, 1979) is used to keep track of information relating to all the time windows and operator actions. The TMS periodically records the appropriate information as needed in a systematic way for extracting meaningful inferences during data analysis. We describe a measure (RAI) for our data analysis. RAI is a cumulative measure of the relative accuracy of the operator's actions with respect to all the available opportunities to perform those actions. RAI can be expressed as the ratio of the number of "on time" correct actions executed by an operator for a class of time windows to the total number of time windows that are opened in that class for that specific operator role. The mathematical formulation for RAI is shown below.

Relative Accuracy Index (RAI) =

$$\frac{\text{Number of "On Time" correction actions for a class of time windows}}{\text{Total number of time windows that are opened in that class}}$$

Time windows that relate to a specific teamwork dimension, such as information exchange, are grouped together, and are said to belong to the same class of time windows for calculating RAI.

7.5   Example: Empirical investigation of team training effectiveness

In this section, we describe a study designed to enable direct measurement of, not only the performance outcomes, but also the requisite KSA competencies associated with each trial. The task used in this study is a command and control task known as the Team Aegis Simulation Platform (TASP). TASP (Bhandarkar et al., 2004; Bolton et al., 2004; Thiruvengada et al., 2004) is a team-in-the-loop computer simulation of three operator roles on board a warship. The roles considered in this study include the Aircraft Information Coordinator (AIC) and the Sensor Operator (SO). The type of team used in this study is an action team (Sundstrom et al., 1990) where tasks are dynamic but structured, and the team members have common goals with specialized individual tasks. The teams, however, are reduced in nature due to the absence of the Antiair Warfare Coordinator, serving as a team leader, who has a need for supporting behavior (backup and error correction). Each team consisting of an AIC and SO role engaged in 10- to 15-minute scenarios during the course of the task.

7.5.1 Study

Participants in this research were graduate and undergraduate students at The Pennsylvania State University. A total of 78 students (39 two-person teams), between the ages of 18 and 25, participated in this study. Of the total, 46 were male and 32 were female. They were skilled computer users and did not have any disabilities that restricted them from adequate use of mouse/keyboard interface. Additionally, the participants did not have any prior experience with the simulation environment. The participants engaged in a single session that lasted for about 3.5 hours on average and were provided with monetary compensation at the end of the study.

The two independent variables used in the study include training and workload. No training (NT), team coordination training (TCT), and task delegation training (TDT) are used as the training conditions. Workload stress levels are controlled by setting them at low and high levels. Different scenarios were developed for setting the low and high stress levels for workload. The nested design that was used in this study is shown in Table 7.2.

7.5.2 Procedure

Thirteen teams (one third of 39 total teams) randomly received one of the three training conditions. The team members were randomly assigned to the AIC or SO role. Each team was subjected to scenarios with both low and high workload stress levels. The teams' overall participation lasted for about 3.5 hours. Figure 7.4 shows the testing protocol that was used in this study.

The participants underwent an initial training of specific skills, which lasted for about 1 hour. This initial training enabled them to acquire skills that are necessary to accomplish tasks that are specific to their current roles. Four practice sessions (practice sessions 1 to 4) of 10 minutes' duration each were provided to the participants to hone their role-specific skills. During these practice sessions, the participants were given feedback on their performance relating to taskwork skills and were encouraged to ask any clarification questions. At the

Table 7.2 Nested Design for the Research Study

No training (NT)		Team coordination training (TCT)		Task delegation training (TDT)	
Low stress	High stress	Low stress	High stress	Low stress	High stress
Task 1	Task 2	Task 3	Task 4	Task 5	Task 6

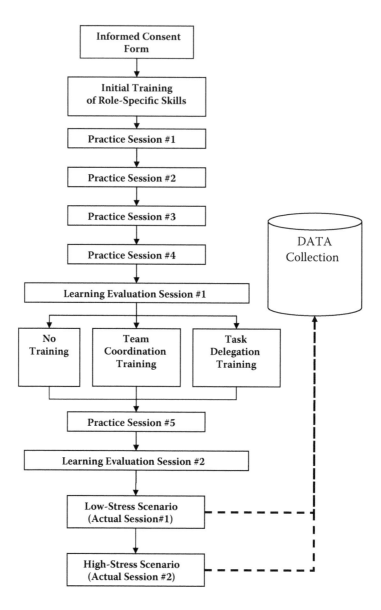

Figure 7.4 Protocol for the research study.

end of the four practice sessions, the teams were subjected to the first learning evaluation session (learning evaluation session 1) for a duration of 10 minutes, which assessed their learning on taskwork skills. During this session, each team member was assigned specific tasks that would require him/her to use his/her taskwork skills. Performance feedback was provided at the end of the session. After taskwork skills training, the teams were randomly exposed to one of the three team-training interventions.

In "no training" intervention, there was no hands-on training provided to the team regarding teamwork. Instead, they were instructed to read articles that explained the importance of teamwork and coordination. In "team coordination training" intervention, the teams were presented with instances of good and poor team coordination policies and were exposed to a video that demonstrated the same. In "task delegation training" intervention, the teams were provided with a presentation of different tasks that were delegated to their roles as part of the training intervention and were also shown a video that demonstrated teamwork associated with task delegation. After the appropriate training intervention was provided, the teams were given an opportunity to practice teamwork skills through a 10-minute practice session (practice session 5). Then, the teams were exposed to a second learning evaluation session (learning evaluation session 2) that assessed their teamwork skills. The teams were instructed to perform tasks that required the effective use of taskwork and teamwork skills. The teams were then subjected to two sessions (of 10 minutes duration each) with low and high stress levels of workload where data relating to their overall team performance were collected for further analysis.

7.5.3   Team process measures in TASP

Smith-Jentsch et al. (1998) defined four dimensions of teamwork for team dimensional training that are critical to overall team performance as information exchange, supporting behavior, communication, and team leadership/initiative. We use these dimensions to classify our team outcome measures (time windows) into team process measures. Communication is not used in this study as a dimension to classify time window outcome since teams are expected to exhibit implicit coordination due to the nature of the complexity involved in the Combat Information Center (CIC) task. Several time windows are opened and closed for each operator role, based on the environmental conditions. These time windows are summarized in Table 7.3 and are classified into the teamwork dimensions.

The information exchange dimension relates to the process of gathering information and effectively exchanging them to develop a shared mental model for the team. Therefore, the AIC must gather visual identification information for any unknown aircraft that approaches the vicinity of the ownship (ship on which the CIC team is located). In a similar fashion, the SO must detect sensor signal emissions and evaluate the intent of the signal as either friendly or hostile. The information that is gathered by both operators must be effectively exchanged between all team members. Therefore, time windows that are opened for visual identification and sensor evaluation process belong to this teamwork dimension. The tasks involved in this process are primary to the corresponding operator roles. Communication is external to the scope of this research as both operators exhibit implicit coordination without any overt communication.

The Anti-air Warfare Coordinator (AAWC) operator is primarily responsible for issuing directives to the AIC operator regarding defensive counter

Table 7.3 Teamwork Dimension Classification of Operator Responsibilities
for AIC and SO Roles

Task type	Teamwork dimension	Responsibilities for operator roles	
		Aircraft information coordinator (AIC)	Sensor operator (SO)
Primary	Information Exchange	Request Visual Identification (VID) report and pass it to other teammates	Evaluate incoming sensor signals Correlate sensor signal to a particular aircraft Transmit the correlated sensor signal
Backup	Communication	Operators did not use speech channels for communication (not considered)	Operators did not use speech channels for communication (not considered)
Primary	Team Initiative/ Leadership	Vector Defensive Counter Air (DCA) within 256 nautical miles (NM) from ownship Vector DCA outside 20 NM from ownship Vector DCA outside danger zones; (vectoring of DCA is done by changing its speed, course, and altitude)	Issue Level-1 warning to hostile aircrafts Issue Level-2 warning to hostile aircrafts Issue Level-3 warning to hostile aircrafts
Backup		Assign identification to unknown aircrafts Assign missiles to hostile aircrafts	Assign identification to unknown aircrafts
	Supporting Behavior	Engage missiles upon hostile aircrafts	
Error Correction		Change the identification of incorrectly identified aircrafts	Change the identification of incorrectly identified aircrafts

aircraft (DCA) flight objectives. Additionally, the AAWC is also responsible
for ordering SO to issue level warnings for incoming hostile aircraft. However, in a reduced team setting, the AAWC is unavailable to issue these team
directives and individual operators (AIC and SO) must exhibit team initiative
and leadership for the team's survival. Due to the absence of the AAWC,
time windows pertaining to flying DCA out of potential threats and issuing
warnings to approaching hostile aircrafts are classified under team initiative/
leadership dimension for AIC and SO, respectively. Finally, part of the supporting behavior dimension for AIC and SO roles is to identify the unknown
aircraft and error-correct any erroneous identifications. The AIC is also
responsible for supporting the AAWC role by assigning and engaging a

missile on hostile aircrafts that pose a high threat within close proximity to the ownship.

7.5.4 Team outcome measures in TASP

Each operator has a defined set of responsibilities based on the rules of engagement (ROE) and is expected to perform an action that satisfies their specific ROE. The operators are trained to perform their primary tasks and encouraged to provide backup and error correction to other operators whenever it is appropriate. Table 7.4 summarizes the rules of engagement and responsibilities of AIC and SO roles.

The AIC role is responsible for managing, controlling, and coordinating friendly assets known as defensive counter aircraft (DCA), identifying an unknown aircraft, and assigning and engaging a hostile aircraft with a missile.

Table 7.4 Rules of Engagement for AIC and SO Roles

AIC	SO
Engage a hostile or assumed hostile aircraft within 20 nautical miles (NM) from ownship (hostile or assumed hostile aircraft only). (AAWC RESPONSIBILITY BACKUP)	Issue Level-3 warning to hostile or assumed hostile only when it is within 20 to 30 nautical miles (NM)
Assign a missile to a hostile or assumed hostile aircraft within 30 NM from ownship (hostile or assumed hostile aircraft only). (AAWC RESPONSIBILITY BACKUP)	Issue Level-2 warning to hostile or assumed hostile only when it is within 30 to 40 NM
Maintain safety of DCA (e.g., keep DCA away from danger zones of hostile aircraft; don't let DCA run out of fuel, etc.)	Issue Level-1 warning to hostile or assumed hostile only when it is within 40 to 50 NM
Keep DCA within 256 NM from ownship	Make a primary identification of air contact (i.e., friendly, hostile, assumed hostile/friendly)*. (AAWC RESPONSIBILITY: BACKUP AND ERROR CORRECTION)
Keep DCA at least 20 NM away from ownship	Evaluate, correlate, and transmit all sensor value emissions that appear on the SO interface
Make a primary identification of air contact (i.e., friendly, hostile, assumed hostile/friendly)*. (AAWC RESPONSIBILITY: BACKUP AND ERROR CORRECTION)	

* Once an aircraft has come within 50 NM from ownship, it should be identified before it travels an excess of 10 NM. If an aircraft "pops up" within 50 NM, it should be identified before it travels an excess of 10 NM.

Note: Two overarching rules:

1. Defend ownship and ships in battle group.
2. Do not engage friendly or civilian aircraft.

The SO role responsibilities include detecting incoming sensor signal emissions, evaluating and transmitting them to other roles, identifying unknown aircraft, and issuing warnings to hostile aircraft that enter the vicinity of the ownship. The time windows corresponding to each teamwork dimension described in the next section was used to evaluate the RAI for that dimension.

7.6   Results

7.6.1   Training and information exchange

We obtained an average RAI by averaging the individual RAI outcomes of all time windows in information-exchange class: requesting visual identification (AIC), evaluating, and correlating and transmitting sensor signals (SO). There was no significant difference in information exchange dimension between teams that received either TCT or TDT intervention and no training (F (2, 72) = 6.298, p = 0.084, = 0.05). This lack of significance is attributed to the short duration of the training period involved for TCT and TDT and the CIC task complexity in TASP. Additionally, there were only 13 replicates (one per team) for each training condition and stress level of workload, which is a small sample size for producing statistical significance.

7.6.2   Training and team initiative/leadership

The AIC and SO roles exhibited team initiative and leadership by complying with time windows that pertain to vectoring DCA away from potential threats (AIC), keeping the DCA within the range of 20 nautical miles (NM) and 256 NM (AIC), and issuing different level warnings to hostile aircrafts that approach the vicinity of the ownship (SO). Therefore, the outcomes of time windows belonging to this category were classified as team initiative/leadership time windows and used to obtain the average RAI for the team along this dimension. There was no significant difference in team initiative/leadership between teams that received no training and TCT or TDT intervention (F (2, 72) = 0.617, p = 0.596, = 0.05). We believe that the absence of a team leader leaves a psychological effect on the attitudes of the team regardless of training interventions. The training program adopted in this study was inadequate in enabling the teams to cope with the loss of their team leader. Additionally, in the absence of a guiding authority, team members are not able to effectively carry out the duties of the team leader as well as their primary responsibilities in highly complex task environments.

7.6.3   Training and supporting behavior

Team members (AIC and SO) supported the AAWC operator role by executing backup and error correction actions. Time windows that were classified under this category are (1) assigning new identification to unknown aircraft (AIC and SO), (2) changing incorrectly assigned identification (AIC and SO), and (3) assigning and engaging a hostile aircraft with a missile (AIC). The outcomes of time windows mentioned above were used to obtain the average RAI along the supporting behavior teamwork dimension. There was no significant difference due to training on the supporting behavior dimension between TCT and TDT intervention and no training condition (F (2, 72) = 0.593, p = 0.607, = 0.05). The reliance of the teams on an explicit coordination mechanism through communication, need for adequate level of training, small sample size, and ineffective coordination schemes were responsible for the lack of significance along this dimension. Earlier studies report that supporting behavior could be improved through a shared mental model, which can develop as a result of appropriate level of team coordination training (Serfaty at al. 1998) and cross training (Blickensderfer et al., 1998).

7.6.4 Factor analysis

We performed a factor analysis by grouping the time windows into five categories as shown in Table 7.5.

The first variable, "sVID," represents the ratio between the total number of unique "on time" correct visual identification action and total number of unique visual identification time windows that were created and expressed on a standardized percentage scale. The second variable, "sSNR," represents the ratio between the total number of unique "on time" correct sensor evaluation, correlation, and transmit action, and total number of unique sensor evaluation, correlation, and transmit time windows that were created, expressed on a standardized percentage scale. The third variable, "sID," represents the ratio between the total number of unique "on time" correct track identification action and total number of assign and correct track

Table 7.5 Five Variables Used in Factor Analysis

Name	Role	Task type	Description
sVID	AIC	Primary	Standardized % of unique ONTIME correct actions for OPEN *visual identification* time windows
sSNR	SO	Primary	Standardized % of unique ONTIME correct actions for OPEN *sensor evaluation, correlation, and transmit* time windows
sID	AIC, SO	Backup	Standardized % of unique ONTIME correct actions for OPEN *identification* time windows
sLWRN	SO	Backup	Standardized % of unique ONTIME correct actions for OPEN *level 1 warning, level 2 warning and level 3 warning* time windows
sAE	AIC	Backup	Standardized % of unique ONTIME correct actions for OPEN *assign and engage with a missile* time windows

identification time windows that were created, expressed on a standardized percentage scale. The fourth variable, "sLWRN," represents the ratio between the total number of unique "on time" correct level one, level two, and level three warning action, and total number of unique level one, level two, and level three warning time windows that were created, expressed on a standardized percentage scale. The fifth variable, "sAE," represents the ratio between the total number of unique "on time" correct assign and engages with a missile action and total number of unique assign and engage missile time windows that were created, expressed on a standardized percentage scale.

The scree test is a graphical method used to determine factors which are significant based on eigenvalues (Cattell, 1966). The scree plot for NT intervention indicates that one factor could be extracted, which accounts for 31% of the total variance. A path analysis was conducted on the data in order to confirm the benefits due to the NT. A single path for the AIC operator's actions was extracted based on the first factor as shown in Table 7.6.

The path includes obtaining visual identification report from DCA (sVID) → assigning aircraft identification (sID) → issuing level warnings to hostile aircrafts (sLWRN) → assigning and engaging hostile aircraft with a missile (sAE). All AIC operators followed this single path to complete their tasks. This indicates that the AIC operators relied only on VID information and did not use sensor information that was transmitted by the SO role. This leads to a failure in information exchange dimension as the AIC could have achieved higher performance had he/she used the sensor information to avoid overkill in VID. This shows that the operators lack knowledge, skills, and attitudes pertaining to information exchange dimension at team level. Therefore, the target training should focus on enhancing these team competencies.

Table 7.6 Extracted Paths for NT, TCT, and TDT Interventions Based on Factor Analysis

Training type	Path (from factor)	Five variables used in factor analysis				
		sVID	sSNR	sID	sLWRN	sAE
NT	Path 1 (from Factor #1)					
TCT	Path 1 (from Factor #1) Path 2 (from Factor #2)					
TDT	Path 1 (from Factor #1) Path 2 (from Factor #2) Path 3 (from Factor #3)					

The analysis of the scree plot for TCT intervention shows that two factors could be extracted that account for 42.5 and 25.2% of the total variance, respectively. A path analysis was conducted to identify the effects of TCT on teamwork dimensions. Two paths for operator actions were identified based on the variables used in the factor analysis as shown in Table 7.6. The first path (for SO role) was evaluating, correlating, and transmitting sensor values (sSNR) →assigning aircraft identification (sID) → issuing level warnings to hostile aircrafts (sLWRN) →assigning and engaging hostile aircraft with a missile (sAE). The second path (for AIC role) was obtaining visual identification report from DCA (sVID) or evaluating, correlating, and transmitting sensor values (sSNR) → assigning aircraft identification (sID) → issuing level warnings to hostile aircrafts (sLWRN) → assigning and engaging hostile aircraft with a missile (sAE). This suggests that the operators had a choice of two paths to complete their tasks.

As expected, the teams followed implicit coordination as they accomplished tasks that related to their primary and backup responsibilities. There was no overt communication, but both operators were able to accomplish identification tasks based on VID and sensor information. There appears to be a shared mental model as operators were able to use both VID and sensor information for accomplishing identification tasks. This indicates that the operators possessed the knowledge, skills, and attitudes relevant to the information exchange and supporting behavior dimensions of teamwork. Additionally, the team members were able to achieve implicit coordination through a shared mental model. The target training should focus on enhancing team initiative and leadership skills for the team members.

A scree plot analysis for TCT intervention shows that three factors could be extracted that explain a variance of 25.3, 21, and 20.8%, respectively. Three paths for operator actions were identified based on the variables used in the factor analysis as shown in Table 7.6. The first path is evaluating, correlating, and transmitting sensor values (sSNR) → assigning aircraft identification (sID) → issuing level warnings to hostile aircrafts (sLWRN) → assigning and engaging hostile aircraft with a missile (sAE). The second path is obtaining visual identification report from DCA (sVID) or evaluating, correlating, and transmitting sensor values (sSNR) → assigning aircraft identification (sID) → issuing level warnings to hostile aircrafts (sLWRN) → assigning and engaging hostile aircraft with a missile (sAE). The third path is obtaining visual identification report from DCA (sVID) or evaluating, correlating, and transmitting sensor values (sSNR) → assigning and engaging hostile aircraft with a missile (sAE). The AIC operator followed the third path as a shortcut to achieve success because he/she felt urgency in assigning and engaging hostile tracks that were within 20 nautical miles, which violated the most important rule in the rules of engagement (RoE) and posed an imminent threat to the ownship. This result suggests that the operators had a choice of three paths to complete their tasks in case of TDT intervention. The AIC operators followed a longer path when there was no urgency and a shorter path (only assigning and engaging hostile aircrafts) when there was an imminent threat. Based on sensor information, the SO role was able to assign identification and issue level warnings. The team members were able to achieve implicit coordination through a shared mental model, which developed as a result of the employed task delegation in the TDT intervention. The teams that underwent TDT were able to compensate for the absence of their team member through knowledge, skills, and attitudes that they developed by participating in the training intervention.

Overall, TCT and TDT provided more paths to accomplish operator tasks than NT interventions. This reinforces the claim that TCT and TDT intervention helps augment overall situation awareness and leads to implicit coordination among team members.

7.7   Lessons learned

Lesson 1: *Team training must address both team processes and team outcomes in order to be effective* — The team process measures are a better diagnostic of team performance than outcomes because simple team outcomes exclude factors that are external to training. Therefore, a mapping between team process and team outcome measures is necessary to isolate the effectiveness of training on overall team performance. The training that is adopted must focus on improving team processes, thereby causing better team outcomes and performance.

Lesson 2: *Teams require both taskwork and teamwork skills to maintain superior performance* — It is important to note that even though individual members of a team may be skilled at performing their taskwork, the team may not exhibit superior performance due to the lack of teamwork. Being adept at taskwork does not necessarily transform into better team performance. Teams require both taskwork and teamwork skills to accomplish their common goals.

Lesson 3: *In case of reduced teams, the various components of training should adequately address team coordination, adaptation, and task reallocation (delegation) in order to be effective* — Team training should also relate to how teams would evolve within the environment. Reduced teams should be trained sufficiently enough to help them cope with the reduction in their team size through effective team coordination, adaptation, and task reallocation and readjustment. Training that focuses on improving teamwork processes would yield better team performance.

7.8   Conclusions

This research study focused on the analysis of the effectiveness of team coordination training and task delegation training (positional clarification form of team cross training). A need for training teams and workforce was established to demonstrate the potential effect of training on overall team performance. An explanation of team competencies and current practices in team training were explained to identify the link between team competencies and training methodologies. In addition to explaining team coordination training and task delegation training (positional clarification form of cross training), we addressed the selection procedure for an appropriate form of cross training. Prior research that focused on training backup and error correction behavior and the relation between team training and team performance were also explained.

We conducted an initial study within the setting of a command and control task to understand the effectiveness of team coordination training and task delegation training on teamwork among reduced teams. Initial findings did not provide any statistically significant improvements in information exchange, team initiative/leadership, and supporting behavior. This is attributed to the lack of adequate level of training and the short training period. However, existing literature in team training suggests that substantial improvements in overall team performance occurred when teams were subjected to coordination training and team cross training. Lessons learned from this research provide cautionary recommendations for future research and experimentation. Ideally, team training should focus on improving team processes and outcomes, thereby enhancing team performance. A work team usually has to satisfy requirements of a larger system (workforce) and still maintain enough independence to perform specific functions. The effectiveness of team training methodologies to enable smaller work teams to be effective functional units for larger workforce should be further investigated.

7.9   Future work

The research work described in this chapter focuses on providing a tool to transform team performance metrics into a form that is analyzable in terms of accuracy and latency of an individual as well as team performance. It also informs researchers and designers of training programs effectively about the relationships between team performance, team processes, and outcomes. In the future, performance metrics described in this research can be used as effective quantitative measures of team performance in teamwork and team training studies.

Acknowledgments

The views expressed herein are those of the authors and do not necessarily reflect those of any other organizations. This research was conducted in part with a funding grant from NAVAIR-TSD, Orlando, FL, under the METT (Modeling Effectiveness for Team Training) program. The authors wish to express their gratitude to Dr. Gwendolyn Campbell and all other individuals who were helpful and supportive in varying capacities for conducting the research. The authors would also like to thank D. L. McGann for editing a previous draft of this document.

References

Adelman, L., Zirk, D.A., Lehner, O.E., Moffet, R.J., and Hall, R. (1986). Distributed tactical decision making: conceptual framework and empirical results. *IEEE Transactions on Systems, Man, and Cybernetics*, 16: 794–805.

Baker, C.V., Salas, E., Cannon-Bowers, J.A., and Spector, P. (1992). The Effects of Interpositional Uncertainty and Workload on Team and Task Performance. Paper presented at the annual meeting of the Society for Industrial and Organizational Psychology, Montreal, Canada.

Bhandarkar, D., Thiruvengada, H., Rothrock, L., Campbell, G.E., and Bolton, A.E. (2004). TASP: a toolkit to analyze team performance in a complex task environment. In *Annual Conference of the Institute of Industrial Engineers*, 15–19 May 2004, Houston, TX.

Bolton, A.E., Campbell, G.E., Buff, W.L., Rothrock, L., Bhandarkar, D., Thiruvengada, H., and Kukreja, U. (2004). TASP: a flexible alternative for team training and performance research. Proceedings of *Interservice/Industry Training, Simulation and Education Conference (I/ITSEC)*, 6–9 December 2004, Orlando, FL.

Blickensderfer, E., Cannon-Bowers, J.A., and Salas, E. (1998). Cross training and team performance. In J.A. Cannon-Bowers and E. Salas (Eds.), *Making Decisions under Stress: Implications for Individual and Team Training*. Washington, D.C.: American Psychological Association, 299–311.

Brannick, M., and Prince, C. (1998). An overview of team performance measurement. In M.T. Brannick, E. Salas and C. Prince (Eds.), *Team Performance Assessment and Measurement: Theory, Methods, and Applications*. Mahwah, NJ: Lawrence Erlbaum Associates, 3–16.

Cannon-Bowers, J.A., and Salas, E. (1997). A framework for developing team performance measures in training. In M.T. Brannick, E. Salas, and J.A. Cannon-Bowers (Eds.), *Team Performance Assessment and Measurement: Theory, Methods, and Applications*. Mahwah, NJ: Lawrence Erlbaum and Associates, 45–62.

Cannon-Bowers, J A., and Salas, E. (1998). Team performance and training in complex environments: recent findings from applied research. *Current Directions in Psychological Science*, 7(3): 83–87.

Cannon-Bowers, J.A., Salas, E., Blickensderfer, E., and Bowers, C.A. (1998). The impact of cross-training and workload on team functioning: a replication and extension of initial findings. *Human Factors*, 40(1): 92–101.

Cannon-Bowers, J.A., Salas, E., and Converse, S.A. (1993). Shared mental models in expert team decision making. In N.J. Castellan, Jr. (Ed.), *Individual and Group Decision Making: Current Issues*. Hillsdale, NJ: Erlbaum, 221–246.

Cannon-Bowers, J.A., Tannenbaum, S.I., Salas, E., and Volpe, C.E. (1995). Defining competencies and establishing team training requirements. In R.A. Guzzo and E. Salas (Eds.), *Team Effectiveness and Decision Making in Organizations*. San Francisco: Jossey-Bass Publishers, 333–380.

Cattell, R.B. (1966). The Meaning and Strategic Use of Factor Analysis. In R.B. Cattell (Ed.), *Handbook of Multivariate Experimental Psychology*, Chicago: Rand McNally.

Coovert, M.D., Cannon-Bowers, J.A., and Salas, E. (1990). Applying mathematical modeling technology to the study of team training and performance. *Proceedings of the 12th Annual Interservice/Industry Training Systems Conference*.

Dick, W., and Carey, L. (1990). *The Systematic Design of Instruction* (3rd ed.). Glenview, IL: Scott, Foresman.

Doyle, J. (1979). Truth maintenance system. *Artificial Intelligence*, 12: 231–271.

Fleishman, E.A., and Zaccaro, S.J. (1992). Toward a taxonomy of team performance functions. In R.W. Swezey and E. Salas (Eds.), *Teams: Their Training and Performance*. Norwood, NJ: Ablex, 31–56.

Green, D.M., and Swets, J.A. (1988). *Signal Detection Theory and Psychophysics*. Los Altos, CA: Peninsula Publishing.

Klienman, D.L., and Serfaty, D. (1987). Team performance assessment in distributed decision making. In R. Gilson, J.P. Kincaid and B. Goldiez (Eds.), *Proceedings of the Symposium on Interactive Networked Simulation for Training*. Orlando: University of Central Florida, 22–27.

Marks, M.A., Sabella, M.J., Burke, C.S., and Zaccaro, S.J. (2002). The impact of cross-training on team effectiveness. *Journal of Applied Psychology*, 87(1): 3–13.

Marks, M.A., Zaccaro, S.J., and Mathieu, J.E. (2000). Performance implications of leader briefings and team-interaction training for team adaptation to novel environments. *Journal of Applied Psychology*, 85(6): 971–986.

Mathieu, J.E., Heffner, T.S., Goodwin, G.F., Salas, E., and Cannon-Bowers, J.A. (2000). The influence of shared mental models on team process and performance. *Journal of Applied Psychology*, 85(2): 273–283.

Orasanu, J. (1990). Shared mental models and crew performance. Paper presented at the *34th Annual Meeting of the Human Factors Society*, Orlando, FL.

Rothrock, L. (2001). Using time windows to evaluate operator performance. *International Journal of Cognitive Ergonomics*, 5(2): 95–119.

Rothrock, L., Harvey, C., and Burns, J. (2005). A theoretical famework and quatitative architecture to assess team task complexity in dynamic environments. *Theoretical Issues in Ergonomics Science*, 6(2): 157–171.

Serfaty, D., Entin, E.E., and Johnston, J.H. (1998). Team coordination training. In J.A. Cannon-Bowers and E. Salas (Eds.), *Making Decisions under Stress: Implications for Individual and Team Training*. Washington, D.C.: American Psychological Association, 221–245.

Smith-Jentsch, K.A., Salas, E., and Baker, D. (1996). Training team performance related assertiveness. *Personnel Psychology*, 49: 909–936.

Smith-Jentsch, K.A., Zeisig, R.L. Acton, B., and McPherson, J.A. (1998). Team dimensional training: a strategy for guided team self-correction. In J.A. Cannon-Bowers and E. Salas (Eds.), *Making Decisions under Stress: Implications for Individual and Team Training*. Washington, D.C.: American Psychological Association, 271–297.

Sundstrom, E., DeMeuse, K.P., and Futrell, D. (1990). Work teams: applications and effectiveness. *American Psychologist*, 45(2): 120–133.

Swets, J.A. (1996). *Signal Detection Theory and ROC Analysis in Psychology and Diagnostics: Collected Papers*. Mahwah, NJ: Lawrence Erlbaum Associates.

Tannenbaum, S.I., Smith-Jentsch, K.A., and Behson, S.J. (1998). Training team leaders to facilitate team learning and performance. In J.A. Cannon-Bowers and E. Salas (Eds.), *Making Decision under Stress: Implications for Individual and Team Training*. Washington, D.C.: American Psychological Association, 247–270.

Thiruvengada, H., Bhandarkar, D., Kukreja, U., and Rothrock, L. (2004). TASP: a toolkit for human experimentation in a synthetic environment with a hybrid human-agent team. In *Behavior Research in Modeling and Simulation (BRIMS) Conference*, 17–20 May 2004, Crystal City, VA.

Volpe, C.E., Cannon-Bowers, J., Salas, E., and Spector, P.E. (1996). The impact of cross-training on team functioning: an empirical investigation. *Human Factors*, 38(1): 87–100.

section III

Applications of workforce cross training

chapter 8

Workforce cross training in call centers from an operations management perspective

O. Zeynep Aksin, Fikri Karaesmen, and E. Lerzan Örmeci

Contents

8.1   Introduction

Call centers (also known as telephone, customer service, technical support, or contact centers) constitute a large industry worldwide, where the majority of customer–firm interactions occur. There are thousands of call centers in the world, with full-time employees ranging from a few to several thousand.

Modern call centers assume many different roles. While the telephone is the basic channel for call centers, contact centers incorporate contacts with customers via fax, e-mail, chat, and other web-based possibilities. Agents in multichannel contact centers have the possibility of responding to customer requests via several different media (Armony and Maglaras, 2004a, 2004b). Inbound call centers answer customer calls, whereas outbound centers make telephone calls to customers or potential customers typically for telemarketing or data collection purposes. Many call centers combine these two features using what is known as *call blending*, where agents who normally take inbound calls also perform outbound calls during times of low call volume (Bhulai and Koole, 2003; Gans and Zhou, 2003). As durable goods and technology companies globalize, technical support needs to be provided to customers who buy the same products around the globe. This support is given in several different languages in multilanguage call centers located in hubs such as Ireland, the Benelux countries (Belgium, The Netherlands, and Luxembourg), and Eastern Europe. In these centers, agents who possess several technical skills or speak several different languages respond to customer queries (Aksin and Karaesmen, 2002).

Today, call centers in mature industries, like financial services, are in the process of transforming into revenue or profit centers. To do this, these centers incorporate sales and *cross selling* into their processes, thus requiring agents to become experts in both service and sales (Aksin and Harker, 1999; Günes and Aksin, 2004). Many companies outsource their call center needs to third parties. An agent working in one of these outsourcing firms may be responding to calls originating from customers of different firms, all clients of the outsourcing company. Such agents need to possess know-how for products, promotions, and/or practices of different companies (Wallace and Whitt, 2005). Multichannel contact centers, call blending call centers, multilanguage technical support centers, service and sales centers, and call center outsourcing are all examples of the growing diversity and complexity of call center jobs. Workforce cross training has emerged as an important practice for companies that need to deal with this diversification in call center jobs.

There is a delicate balance between quality and costs in the management of call centers. As a direct and important point of contact with customers, the call center needs to provide good call content and high accessibility. Call content quality hinges extensively on human resource practices, such as staff selection, training, and compensation, since it occurs at the point of interaction between agents and customers. High accessibility implies, among other things, that a calling customer will be answered with a minimum wait time. Determining the appropriate number of agents to have in the presence of uncertain call volumes constitutes one of the most important challenges of call center management because adding staff implies adding costs for a call center. Indeed, 60 to 70% of call center costs are associated with staffing (Gans et al., 2003). To cope with the growing diversity of calls while keeping staffing costs to a minimum, a higher level of flexibility in answering calls is required. Structures with multiskill agents are becoming more prevalent, as the following studies on multichannel contact centers indicate:

> *Seventy-eight percent of call centers surveyed by ICMI used agents skilled in both phone and nonphone transactions to handle inbound calls, 17% used only agents dedicated to the phone channel, and 5% said they used some multiskilled agents.*

(Incoming Calls Management Institute, 2002)

> *When handling transactions from multiple channels (both service level and response time objective transactions), just over half (52%) of call center managers participating in a recent ICMI web-based seminar said they use blended agent groups for the different channels as the workload allows. Another 29% use separate agent groups for each channel, while 8% said they relied on multimedia queues and multiskilled agents who handled whatever type of transaction was next in queue.*

(Incoming Calls Management Institute, 2000)

Technology provides skills-based routing capability, enabling the routing of calls to the agents with the appropriate skills, thus allowing call centers to reap the benefits of flexibility. Through cross training, call centers can increase the flexibility of their staff. IDC, a market research firm, estimates revenue in the contact center training industry will grow from $415 million in 2001 to nearly $1 billion in 2006 (ICCM Weekly, 2002). Some of this growth will come from cross training.

This chapter will review the relatively young, however, growing literature on workforce cross training in call centers, making ties to related work in operations flexibility. In the subsequent section (Section 8.2), we discuss the costs and benefits of cross training that have been analyzed in the literature. This section also illustrates the multidisciplinary nature of the issue, on which we take a predominantly operational view. Section 8.3 establishes the importance of flexibility in call center operations. Viewing workforce cross training from this flexibility lens, we identify three basic questions pertaining to the design and control problems, which are introduced in Section 8.2. The first design problem is the skill set design, which we also label the *scope* decision. In terms of cross training, this refers to the questions of which skills should servers be cross-trained and how many skills should they have? Literature related to this problem is presented in Section 8.3. Having decided on the scope, the next issue is to decide what proportion of the workforce should be cross-trained. We label this design problem the *scale* decision and review related work on it at the end of Section 8.3. The design problems are closely related to the control problems of staffing and routing. Indeed, the value of different designs will interact with the subsequent staffing and routing decisions. Papers that focus on the control part of the problem are reviewed in Section 8.4. We end the chapter with a discussion of future directions in Section 8.5.

8.2   Costs and benefits of call center cross training

Like all human resource initiatives, there are benefits expected from cross-training practices. These are motivational benefits due to the enlargement or enrichment of jobs, cost benefits due to improved capacity utilization or improved speed, and quality benefits including improved customer service. These benefits come along with costs like training costs, loss of expertise and job efficiency, and mental overload. In this section, we review selected articles from the organizational behavior and operations management literature that identify different costs and benefits of cross training and then interpret these in the context of call centers.

8.2.1   The organizational behavior perspective

There is vast literature dealing with organizational behavior that explores the relationship between job design and performance. Grebner et al. (2003) focus on this problem in the call center context. The authors provide examples of earlier studies supporting the argument that call center jobs are predominantly specialized and simplified (Isic et al., 1999; Taylor et al., 2002), and as a result require a relatively short period of training (Baumgartner et al. 2002). In their article, they empirically test the premise that this division of labor and simplification, while reducing personnel costs, results in low variety and low job control, which leads to lower well-being and a higher turnover among call center workers. The authors argue that call center jobs need to be redesigned in order to improve autonomy, variety, and complexity. In general, it has been argued that cross training improves job enlargement (variety) and job enrichment (autonomy), which, in turn, has a positive impact on performance (Hackman and Oldham, 1976; Ilgen and Hollenbeck, 1991; Xie and Johns, 1995). However, in Xie and Johns, the authors demonstrate that there is a limit on the positive impact of job scope and, that beyond a threshold, job scope can become excessive and induce stress, which is dysfunctional for the organization. Thus, we find that the motivational benefits one expects from cross training in call centers can turn into costs if the resulting job scope is too high. Combined with Grebner et al.'s (2003) argument, this suggests that from a motivational standpoint, moderate job scope is superior to no variety or excessive scope job designs. An empirical investigation that determines the ideal region for job scope in different types of call centers remains to be done.

Campion and McClelland (1991, 1993) explore both the costs and benefits of job enlargement in service jobs. Taking an interdisciplinary perspective, they summarize four different models of job design coming from different disciplines: A *motivational* design that argues for job enlargement and enrichment, a *mechanistic* design recommending simplification and specialization, a *biological* model that advocates reduced physical stress and strain, and a *perceptual-motor* model recommending reduced attention and concentration requirements that lead to increased reliability. These models clearly point to some tradeoffs to be made between different benefits and costs involved. The authors identify the benefits of job enlargement as satisfaction (of individuals), mental underload, enhanced ability of catching errors, and improved customer service. The costs of job enlargement are stated to be higher mental overload, training requirements, higher basic skills requirements, more chance to make errors, decline in job efficiency, and more compensable factors, such as education and skills (leading to higher compensation). Job enlargement is classified into two types: *task enlargement*, which involves adding new tasks to the same job, and *knowledge enlargement*, which refers to adding requirements to the job that enhance understanding rules and procedures about other products of the organization. In their analysis, Campion and McClelland (1993) find that task enlargement mostly results in costs or negative benefits — more mental overload, greater chance of making errors, lower job efficiency, less satisfaction, less chance of catching errors, and worse customer service. On the other hand, knowledge enlargement is found to result in more satisfaction, less mental underload, greater chances of catching errors, better customer service, less mental overload, lesser chances of making errors, and higher efficiency.

Reinterpreting in a call center context, we could say that task enlargement involves completing greater portions of a customer request. For example, not just answering the initial part of a technical query but also completing the entire query, or not just attempting a sale but actually completing the entire sales transaction could be seen as examples of task enlargement. On the other hand, answering calls pertaining to different products, different languages, or different regions could be seen as examples of knowledge enlargement. Though the results remain to be tested in a call center setting, the Campion and McClelland results suggest that escalating calls to specialists in a technical support center, or separating service and sales roles may be desirable in order to avoid enlargement costs. Most call centers in the Evenson et al. (1999) study are reported to exhibit this type of service and sales separation, thus, in a way, providing a confirmation from practice. On the other hand, cross training that enhances flexibility through call blending of inbound and outbound or developing an expertise in several products seems to be beneficial from a job design perspective.

8.2.2   The operations management perspective

The organizational benefit derived from the flexibility of staff in the presence of uncertainty is not considered in the job design literature. Taking a modeling perspective and focusing on call centers, Pinker and Shumsky (2000) ask the following question: "Does cross training workers allow a firm to achieve economies of scale when there is variability in the content of work, or does it create a workforce that performs many tasks with consistent mediocrity?" The authors model the tradeoff between cost efficiency due to economies of scale resulting from cross-trained staff and quality benefits from experience-based learning in specialists. The authors find support for the well-known fact that a system with flexible servers can achieve the same throughput with fewer servers; however, they also demonstrate that higher flexibility results in lower quality and less customer satisfaction due to a decrease in experience in any given skill. This type of quality deterioration is also found to happen in a specialist system due to the possibility of low utilization, once again sacrificing experience-related quality performance. The authors conclude that an ideal design will consist of a mixture of specialized and flexible servers.

In a general context, taking a predominantly operations management perspective, Hopp and Van Oyen (2004) develop a framework to assess the appropriateness of cross-trained workers in different manufacturing and service settings. Called *agile workforce evaluation framework*, it identifies, by surveying an extensive literature, the links between cross training and organizational strategy as well as superior architectures and worker coordination mechanisms for workforce agility implementations. The links to organizational strategy are classified in two groups: direct and indirect. The direct links are in the form of labor cost reduction, time that captures efficiency, quality, and variety. Improvements in motivation, retention, ergonomic factors, experience-based learning, communication, and problem solving are the indirect links. Of these, variety is one that has not been directly considered in the job design literature. Based on the brief review of the literature we have presented so far, we note that all of the direct and indirect links play a role in evaluating worker cross training in call centers, emphasizing the importance of this practice for this industry.

The Hopp and Van Oyen research classifies workforce agility architectures in terms of skill patterns, worker coordination policies, and team structure. We provide examples of the first two features in the context of call centers. Skill patterns relate to the questions that we refer to under the scope decision mentioned in the introduction. It also resembles the task vs. knowledge enlargement of Campion and McClelland (1993). Skill patterns can be established taking skill types, entities (jobs or customers), or resources as a basis. In a call center setting, as mentioned by the authors, skill types can refer to calls pertaining to different products. Examples for entity-based tasks are similar to those identified under task enlargement mentioned above. Resource-based tasks could be, for example, cross training in a contact center in a way that servers are dedicated to particular channels (resources) but are cross trained to do all tasks within a channel.

Worker coordination policies determine how workers and tasks are matched over time. Call centers allow for a wide range of worker coordination policies:

- Fully cross-trained generalists who respond to all types of queries all the time.
- Partially cross-trained workers who can respond to queries about more than one type of product, but not all.
- Cross-trained workers who answer the same type of calls at certain times of the day (for example, outbound calls after 5 p.m., inbound before then).
- Cross-trained workers who help out different groups of specialists depending on system congestion (Örmeci, 2004).
- Worker groups whose skill sets are nested such that calls are escalated from the simplest level to more complex as the nature of the problem is explored (Shumsky and Pinker, 2003; Das, 2003; Hasija et al., 2005).

It is also possible to come across systems where each server has its own skill set and priorities defined over this set, such that calls are assigned to the next available worker possessing the required skills at the appropriate priority level. This occurs in systems that use skills-based routing to coordinate workers (Wallace and Whitt, 2005; Mazzuchi and Wallace, 2004).

The appropriateness of different design choices along the dimensions of skill pattern and worker coordination in different environments depends on *training efficiency* and *switching efficiency* (Hopp and Van Oyen, 2004). Of these, training efficiency is relevant for the skill pattern choice and captures the ease with which workers can be trained in different skills. This factor covers similar costs as those considered previously in the job enlargement context. Switching efficiency refers to the costs associated with changing tasks. These may be in the form of time lost between task changes or in switching from one resource to another or in terms of setup work that needs to be performed on each different entity (a customer, for example). Modern call center technology mitigates switching inefficiencies to some extent. Computer telephony integration and automatic call dispatching allow service representatives to view customer data and earlier history when calls need to be passed on. Since most call centers operate in a paperless environment, a physical switching time is nonexistent or minimal. Agents use the phone and their computers as the basic resource and this does not change from task to task. Since switching inefficiencies are relatively small, the main costs to be considered in the cross-training design for call centers seem to be those mentioned under training efficiency.

Our review on the costs and benefits of cross training suggests that apart from direct costs like the cost of training, compensation for additional skills, or quantifiable efficiency losses, costs associated with cross training are typically beyond the scope of operations management models. The latter often focuses on the benefits of cross training, particularly from a flexibility standpoint.

The organizational behavior literature that focuses on job design seems to recommend moderate scope in cross training. In Section 8.3, we review articles that explore the job scope question from a flexibility perspective, thus ignoring some of the human resource-related issues analyzed in the job design literature. Quality–efficiency tradeoffs modeled in Pinker and Shumsky (2000) imply that good designs mix specialized and flexible servers. Papers addressing the question of the appropriate mix of specialized and flexible servers are also reviewed in this section. Finally, in Section 4, we review related literature on worker coordination mechanism in call centers as problems of staffing and routing.

8.3   Cross training and its impacts on performance

This section focuses on the performance improvements brought about by cross training. In Section 8.3.1, the cross-training issue is related to the framework of operational resource flexibility, and the related literature is reviewed. Section 8.3.2 outlines some of the general guidelines for the scope of cross training suggested by the operational flexibility literature. Finally, Section 8.3.3 focuses on the issue of scale of cross training. This section introduces and discusses the general principles, but postpones a number of critical operational issues, such as staffing and routing in a call center context, to Section 8.4.

8.3.1   Cross training and operational flexibility literature

In this section, we present the impacts of cross-training structures on the performance of a call center focusing on operational issues. As already mentioned in Section 8.2, our discussion focuses on the benefits in terms of resource flexibility. In a survey article on manufacturing applications, Sethi and Sethi (1990) describe several different dimensions of flexibility. Our discussion will be based on a particular type of flexibility most closely related to cross training. Resource flexibility will be understood as the capability of a resource to perform several different tasks.

In the operations literature, resource flexibility has been studied extensively. A typical problem in that setting is to evaluate the performance of the system, such as a flexible manufacturing system, under a particular type of resource flexibility structure (see Sethi and Sethi, 1990). More recently, a number of papers have addressed operational issues in the presence of flexible resources. A relatively well-studied problem from that perspective is one of capacity investment. These papers assume a certain form of flexibility and then explore the question of the ideal level of this flexibility and how it relates to value in terms of throughput or revenues under uncertain demand (see, e.g., Fine and Freund, 1990; Netessine et al., 2002). Van Mieghem (1998) provides a review of the literature on this problem.

Our focus as the main design issue in this section is cross training or flexibility structure. There is a rich selection of operations literature on this problem. Starting with the influential work of Jordan and Graves (1995), this stream of work keeps all other design parameters fixed and investigates the isolated effects of varying the flexibility structure. Motivated by an automative sector example, Jordan and Graves explore the problem of assigning multiple products to multiple plants. The demand for each product is random and the flexibility of the plants determines which products they can handle. They develop general guidelines for this problem. One of their important findings is that well-designed limited flexibility is almost as good as full flexibility. We discuss some of these general guidelines in Section 8.3.2.

While the general flexibility principles of Jordan and Graves were established for a particular model, a number of subsequent papers have observed that these principles are robust to the assumptions of the particular model. For instance, the model of Jordan and Graves is a single-period model that ignores the dynamics of the system. Sheikzadeh et al. (1998) consider a flexible manufacturing system modeled by parallel queues and observe that the general flexibility principles hold in this case. Gurumurthi and Benjaafar (2004) also present a numerical investigation of the benefits of flexibility based on a queuing model. Jordan et al. (2004) also consider a queuing-based model, and confirm and extend some of the general principles.

Another interesting observation is that even though the model in Jordan and Graves (1995) is a parallel resource-type system, most general principles also hold for serial systems, such as production lines or assembly lines. Graves and Tomlin (2003) perform a similar analysis for a multistage version of the system in Jordan and Graves, and develop a flexibility measure for such systems. Hopp et al. (2004) explore the benefits of chaining in the context of cross training for production lines. Inman et al. (2004) investigate different cross-training structures for workers in an assembly line.

Despite the existing flexibility principles, it remains difficult to say whether one particular flexibility structure is better than the other in terms of operational performance when cross-training costs are disregarded. An interesting recent development has been the work of Iravani et al. (2005), who develop a *structural flexibility index* that quantifies the structural flexibility in production and service systems and allows a ranking based on performance of these systems. Numerical results confirm that the index is quite accurate for comparing system performance.

8.3.2 General flexibility guidelines

Let's examine an example in detail to demonstrate the impacts of different cross-training structures and assume that there are three different types of calls: *A, B,* and *C,* each one corresponding to a particular skill. Let us also assume that the call center is organized in three departments corresponding to the main skills demanded (*A, B,* or *C*). Figure 8.1 depicts a sample of eight different cross-training structures. In this figure, for each structure, the nodes on the left represent the different call types, and the nodes on the right represent resources (servers) and their skills. Structure 1 in Figure 8.1 depicts

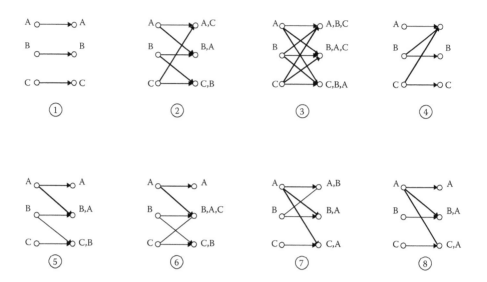

Figure 8.1 Different cross-training structures.

a system where there is no cross training and each group of servers can respond to a single type of call. While this system has obvious advantages in terms of hiring and training costs, it also has an important disadvantage. Since calls arrive randomly over time, it may be possible that one of the departments is completely busy, while in another one there may be idle servers. This causes customer service levels to fall even though capacity (i.e., total number of servers) is sufficient.

Let us now assume that the servers whose main skills are *A* are also trained in skill *C*, those whose main skills are *B* are trained in *A*, and those whose main skills are *C* are trained in *B*. The resulting structure is depicted as Structure 2 in Figure 8.1. We can view this new structure as consisting of three service departments *A*, *B*, and *C*, which are distinguished by their main skills. In this case, if department *A* is busy, arriving type *A* calls can be answered by cross-trained servers in department *B* (if they are available). Similar assignments may also take place for calls of type *B* and *C*.

Structure 3 in Figure 8.1 represents the case of completely multiskilled servers who are all able to respond to all three types of calls. In this structure, all servers are cross trained to have all the three skills. As can be seen, in this structure, the notion of different departments loses its significance. In a way, there is a single multiskilled pool of servers who do not differentiate between different types of calls. The system is as efficient as it gets in terms of customer service performance. In this structure, a customer waits only because all servers are busy, and not because of a mismatch between demand type and the skill-set of a server available. This is, however, at the expense of increased cross-training costs and possibly higher burnout rates for servers due to increased task complexity.

The problem of designing the right cross-training structure now appears clearly. The system with dedicated departments in Structure 1 of Figure 8.1 is cost effective in terms of cross-training requirements but may suffer in customer performance, whereas Structure 3 is at the opposite extreme. Is there a cross-training structure somewhere in between the two extremes, that is, satisfactory in terms of customer service performance but not too costly in terms of cross-training expenses? In other words, what is the right scope of resource flexibility? The answer to this question in a given situation depends on a number of complicated aspects of the problem, such as customer service performance objectives, the capacities of departments, the difficulty and the costs associated with cross training the servers, and the operational costs of using cross training effectively (i.e., by correctly routing the calls). On the other hand, there are certain general properties, which provide a guideline to these issues. Starting with the work of Jordan and Graves (1995), this question has received a lot of attention. Below, we outline some of the general principles concerning cross-training structure.

We begin with a basic property, motivated by a comparison between two systems where one has additional cross-trained servers. An example would be a comparison of Structures 1, 5, and 2 in Figure 8.1. If we leave aside problems of call routing, answer quality, and call duration, clearly the performance of Structure 5 is at least as good as that of Structure 1 but cannot be better than the performance of Structure 2. In fact, additional cross training can be seen as a possible call assignment option, which cannot degrade performance.

8.3.2.1   Property 1

Increased cross training increases (in a nonstrict sense) customer service performance — Despite its simplicity and generality, Property 1 is of limited use for a cross-training design because it does not enable us to make a useful statement on how much cross training is required given that cross training is costly. In order to obtain a better sense of the performance vs. cost tradeoff, more subtle results on flexibility structure are required. To this end, let us define the concept of chaining as introduced by Jordan and Graves (1995). A chain is a group of demand and resource types that are either directly or indirectly connected by demand–resource assignment decisions. Jordan and Graves argue that performance is enhanced by constructing chain structures in the demand–resource assignments. One of the main principles stated in Jordan and Graves is that longer and fewer chains are better than multiple shorter chains.

8.3.2.2   Property 2

Longer and fewer chains in the cross-training structure lead to improved customer service performance — As an example of Property 2, consider the two structures in Figure 8.2 corresponding to four different demand types A, B, C, and D. Structure 10 has two chains, whereas Structure 9 has a single but longer chain. In general, the performance of Structure 9 turns out to be better than that of

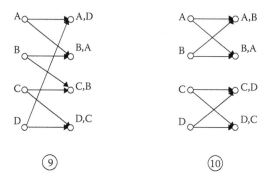

Figure 8.2 Two different structures: two chains vs. a single chain.

Structure 10. The intuition here is subtler. Structure 10 can absorb peaks in the demand of a single class but cannot absorb peaks of both A and B or both C and D type demands, whereas Structure 9 can actually handle peaks in A and B by shifting some of the demand to departments C and D.

Property 2 establishes that one should strive to construct long chains when designing the skill structure. Structure 2 in Figure 8.1 comprises a single long chain that connects departments *A*, *B*, and *C*. On the other hand, this property does not enable a comparison of different chains of the same length. In fact, Structures 2 to 7 are all single long chains. It turns out, however, that the performance of Structure 2 is, in general, better than all other structures, except the fully flexible Structure 3. Aksin and Karaesmen (2002, 2007) provide a formal justification of this general property for a particular model.

8.3.2.3 *Property 3*

If the call rates for different call types are balanced and the number of servers in each department is roughly identical, then balanced cross-training designs outperform less balanced ones — In order to interpret Property 3, note that all servers have two skills in Structure 2 of Figure 8.1, whereas in Structure 4, there are some servers with three skills and others with one skill. If demands and, therefore, staffing levels at different departments are roughly balanced, the three-skill department in Structure 4 may be heavily demanded and may not be able to absorb peaks from multiple demands as efficiently as in Structure 2.

By Property 3, Structure 2 is superior to structures 4, 5, 6, and 8. On the other hand, Structure 7 seems comparable to Structure 2 since all servers have two skills in both structures. It turns out that Structure 2, in general, is superior to Structure 7 because, in Structure 7, type *C* calls can only be handled by department *C*, which diminishes the effectiveness of cross-trained resources in that department (see Aksin and Karaesmen, 2007).

8.3.2.4 *Property 4*

If the call rates for different call types are balanced, then cross-training designs, enabling balanced routings, outperform designs that allow less- balanced routings — The final important design issue pertains to the question of combining Property 1, which states that more flexibility is better for performance with Properties 2, 3, and 4 that highlight the importance of balanced chain structures. Can we compare chain structures that may have different amounts of flexibility? One of the main results in flexibility design is that the performance Structure 2 is almost as good as Structure 3 (Figure 8.1) under a variety of assumptions and for different models. Jordan and Graves (1995), Aksin and Karaesmen (2002), Gurumurthi and Benjaafar (2004), and Jordan et al. (2004) all present numerical examples in different settings and Aksin and Karaesmen (2007) provide a theoretical justification for a particular model.

8.3.2.5 *Property 5*

Well-designed limited resource flexibility is almost as good as full resource flexibility in terms of performance — It is interesting that Property 5, like Properties 1 to 4, is robust to modeling assumptions and stays consistent in different settings ranging from single-period models introduced by Jordan and Graves (1995), to queuing-based models in manufacturing, such as in Sheikzadeh et al. (1998), Gurumurthi and Benjaafar (2004), Jordan et al. (2004), and Inman et al. (2004). More support for these findings come from the work of Iravani et al. (2005), whose structural flexibility index (SFI) is proposed as an indicator of the performance of the structure. First, SFI supports the above-stated properties. For instance, the SFI indices for Structures 2, 3, 5, and 8 of Figure 8.1 are, respectively, 6, 9, 3.73, and 4, implying that, regardless of the setting, Structure 2 is anticipated to be superior to Structures 5 and 8. Second, in a systematic study, Iravani et al. (2005) test the predictive performance of SFI on a variety of different models and find that the index is robust with respect to model structure and assumptions.

Most research discussed so far that proposes flexibility guidelines describes models and applications in manufacturing settings. Even though the queuing literature is relatively rich in the investigation of pooling issues (see Buzacott, 1996, and Mandelbaum and Reiman, 1998, for example), there is less work on systematic investigation of these structural properties in a call center setting. Aksin and Karaesmen (2002) propose an approximation for a call center based on an upper bound on system performance and show that this approximation possesses some of the general structural properties presented above.

A major challenge in the transition from the static Jordan and Graves (1995) framework to a queuing framework more appropriate for call centers is that the static framework assumes that calls (demand) will be allocated optimally to resources. In a queuing framework, this is not a trivial issue because calls have to be routed dynamically. It is known that careless routing policies may have a negative effect on performance despite a correct flexibility structure (see Gurumurthi and Benjaafar, 2004, and Jordan et al. (2004, for examples).

The routing issue will be discussed in detailed in Section 8.4, which also presents a review of some recent call center research.

8.3.3 Scale issues in cross training

The operational design issue discussed so far in Section 8.3.1 and Section 8.3.2 is essentially the problem of skill set design: How many different skills should the servers have and which ones? This issue can also be called the *scope* decision (Aksin and Karaesmen. 2002). Another important question is with regards to the *scale* decision: What is the right proportion of servers to cross train, given desirable skill-set structures? Going back to Structure 2 of Figure 8.1, it is plausible that department A could consist of a mixture of specialists (A skills only) and cross-trained servers (A, B skills). The question is to find the right tradeoff for the proportion of specialists and cross-trained servers. This question seems to have received less attention. Aksin et al. (2005) present some results for the Jordan and Graves framework and show that there is a diminishing returns property in terms of the scale of cross training; the marginal value of an additional cross-trained server is decreasing in the number of existing cross-trained servers. This implies that finding the right scale is an important question since initial increases in scale improve performance significantly, but the improvement drops as scale is added.

To our knowledge, only a few papers directly address cross-training scale design issues in call centers. Pinker and Shumsky (2000) investigated the mix of cross-trained workers vs. specialists taking into account the quality tradeoff. Chevalier et al. (2004) find that, for a number of different cost structures, cross training around 20% of the servers is the optimal tradeoff. Both of these papers investigate the scale issue for relatively simple cross-training structures, and it would be interesting to verify whether such results are robust to the scope of cross training. In a more recent paper, Jouini et al. (2004) describe a call center design problem where a transition occurs from a completely pooled structure to a dedicated team-based organization where each team is assigned to a group of customers. The dedicated organization is similar to the Structure 1 of Figure 8.1 and does not benefit from the demand pooling effect. The authors, however, find that this structure can be surprisingly efficient if the teams can accept only a small number of calls from customer groups other than their own. This suggests that a significant performance improvement can be obtained by cross training a limited number of servers in each team.

8.4 Staffing and routing

In the previous section, the aim has been to characterize preferable skill set designs in a multiskill call center, which gives possible cross-training levels. In this context, these decisions correspond to strategic level decisions. In this section, on the other hand, we will consider the tactical and operational decisions regarding cross training. More explicitly, we want to compute the number of operators from each skill set determined previously, such that the call center satisfies certain constraints due to quality of service and workforce scheduling, with a minimal cost. Hence, we start by describing the operations of a call center, which includes, among others things, the quality of service criteria, certain staffing issues, and routing rules for calls of different types. Then, we will concentrate on the relations between cross training and the operational characteristics of call centers.

Figure 8.3 shows the process of answering an incoming call, from which we can identify a number of factors that affect the performance of a call center:

- Arrival rates of calls referred to as call volumes
- Call types
- Routing policy
- Number of total trunk lines
- Abandonment behavior
- Number and capability (i.e., cross-training level) of operators
- Service rates of operators

Among these terms, only routing policy needs further explanation. Routing policy 1 routes an incoming call to one of the available servers, if any, or delays it; whereas routing policy 2 assigns one of the calls waiting in the line, if any, to an agent who has just finished service or keeps the agent idle. Call center management can completely control some of these factors, such as the routing policy, number of trunk lines, number and capability of operators; partially control some, such as service rates of operators, abandonment behavior; and, finally, cannot control some, such as call volumes and call types.

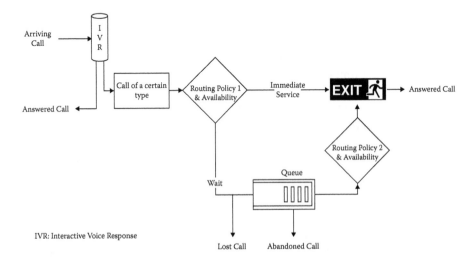

Figure 8.3 Call answering process.

This section will take number and capability of operators and the routing policy as the control variables, with the objective of operating with a minimal cost (which typically consists of staffing and communication costs) while certain constraints regarding customer satisfaction, which we refer to as quality of service (QoS) constraints, are satisfied. More explicitly, in Section 8.4.1 we will assume that there is only one class of calls, so that we need to optimize only the number of operators. The aim of this section is to introduce the basic concepts on staffing call centers, so the issue of cross training will not be mentioned here. The remaining sections, however, are devoted to discuss the effects of cross training on staffing and routing in call centers. Therefore, in Section 8.4.2 to Section 8.4.4, we consider call centers that give different kinds of services. We assume that their skill set designs have already been determined, i.e., the possible cross-training levels are fixed. Section 8.4.2 discusses call routing policies given the number of agents in each skill set. Section 8.4.3 aims to find optimal staffing levels for each skill set present in the given design, while assuming that a routing policy is chosen. At this stage, it is possible to assign no agents to a skill set, so that the given skill-set design may not be fully used. Hence, the capability of the operators (or the skill-set design) is fully determined only at this stage. Finally, in Section 8.4.4, we address the problem of determining the staffing levels and the routing policy together.

8.4.1   Staffing, shift scheduling, and rostering

We first define the problems to be introduced. The *staffing problem* seeks to find a minimal workforce level for short time intervals to guarantee a certain service level during these intervals. Usually, it is assumed that each of these short time intervals behaves independently of all other intervals, although this is generally not true (see Avramidis et al., 2004; Brown et al., 2005; Steckley et al., 2004). In the next level, a number of shifts are defined, such that lunchtime and, sometimes, coffee breaks, are known, as well as the starting and finishing time of the shift. Then, the *shift-scheduling problem* determines the number of employees to be scheduled in each shift in order to meet the minimal workforce levels. As a result of the "independence" assumption, the solution of the staffing problem becomes an input to the shift-scheduling problem. In fact, this assumption can be relaxed so that the two problems can be solved together, which will yield a better solution. However, due to its complexity, the combined problem has been considered only recently, as we will see below.

Finally, a call center has a certain number of employees, where each employee has certain rights regarding the number of days off per week, the number of weekends off, the number and sequence of day and night shifts, etc. The *rostering problem* aims to create a work schedule for each of these employees such that constraints due to staffing and shift scheduling problems are satisfied as well as these additional constraints. The *rostering* problem can either take the solution of the shift-scheduling problem as an input or solve the shift-scheduling and rostering problems together. In order to solve the rostering problem, feasible schedules are defined over a longer time horizon, so that each schedule satisfies all the constraints due to employee rights and/or company policies. Then the problem is to assign the employees to these schedules in such a way that the QoS constraints are also satisfied. Usually, similar methodologies are developed to solve both shift scheduling and rostering problems, so we will review both problems together. Each of these problems is difficult to solve in call centers, even if the additional complexity of cross training is not present. Hence, in this section, we describe these problems without referring to cross-training issues. In other words, this section assumes a call center operating with only one class of calls.

Call centers usually operate 24 hours a day, 7 days a week (24/7). Figure 8.4, Figure 8.5, and Figure 8.6 present typical call volume patterns for 1 day, for 6 weeks, and for 1 year, respectively. From these figures, we observe strong seasonality effects on both daily and weekly call volumes, which lead to significant differences on call volumes in relatively short intervals. This has a significant effect on the staffing policies. In order to convert the call volumes to staffing levels, we need to define the performance measures significant to call centers and QoS constraints, which build up on these measures. Here are the most common performance measures:

> The percentage of calls that are answered within a specified amount of time, say, w seconds.
> The percentage of abandonments (calls that leave the system before being answered).
> The percentage of lost calls (calls that do not find an available trunk line to an available operator and, so cannot enter the system).
> The average amount of waiting time in the queue for answered calls and/or for all calls that enter the system.

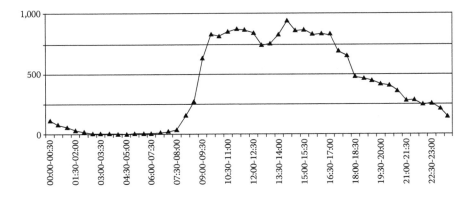

Figure 8.4 Half-hourly call volumes for a day.

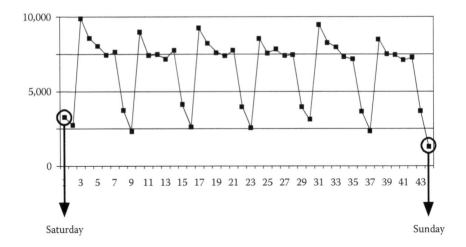

Figure 8.5 Daily call volumes for 6 weeks.

Each call center specifies its own QoS constraints by setting a minimum or maximum level for one or more of these performance measures. Call centers usually compute the minimum required staffing levels for each half-hour. Moreover, in practice, the primary QoS constraint builds on the first type of performance measure. Hence, the minimum required staffing levels are computed such that at least $p\%$ of all calls are answered within a specified amount of time, say, w seconds, in each half-hour. The parameters p and w change with the type of the call center and/or with the type of calls to be answered. Much of the workforce-scheduling software, commonly used in call centers, determine the minimum staffing levels by a numerical procedure based on the well-known Erlang-C or Erlang-delay formula. This method has two main drawbacks. The first one is its underlying assumptions:

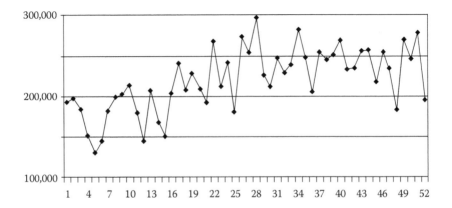

Figure 8.6 Weekly call volumes for a year.

Erlang-C formula gives the steady-state behavior of a queuing system with Poisson arrivals and exponential service times, where the parameters are assumed to be constant over the half-hour being considered. However, half-hours are not always sufficiently long for a system to move into steady state, especially because the arrival rates are not constant during the half-hours. Another crucial assumption is the independence of intervals. Moreover, the effects of a finite number of trunk lines (alleviates busy signals) and of call abandonments are completely ignored. The second drawback is due to the lack of intuition because this procedure, being numerical, does not provide information about how changes in the parameters may affect the staffing levels. Hence, different staffing schemes are proposed in the literature. Here, we review only the most recent work, and refer the interested readers to the references given in Section 4 of Gans et al. (2003).

The most commonly known methodology, both in academia and in practice, is the square-root safety staffing: If the arrivals were deterministic and constant over time, then it would be enough simply to have the number of servers exactly equal to the load of the system, say ρ. The square-root safety staffing rule proposes to have an additional number of servers, proportional to the square root of the arrival rate, $\sqrt{\rho}$, in order to compensate for the randomness in the arrival process. Then, the required number of servers will be $\rho + \beta\sqrt{\rho}$, where β can be considered as the target service level of the call center. There are a number of papers that analyze square-root safety staffing rule in different systems, such as Jennings et al. (1996); Garnett et al. (2002); Borst et al. (2004); Feldman et al. (2006); Armony and Mandelbaum (2004); and Green et al. (2006). More recent work models the effects of different factors on staffing, such as Whitt (2006a) considering uncertain arrival rate and absenteeism and Steckley et al. (2004) considering random arrival rates.

In the staffing problem, the specified constraint or constraints has to be satisfied in each half-hour, as opposed to 8 hours or 1 day. Then, the minimum required staffing levels, regardless of the procedure used, follow the half-hourly call volume patterns closely. However, the actual staff levels cannot match the minimum requirements exactly, since the operators work in shifts either full time, being present for 6 to 9 hours with a 1-hour break, or part-time, being present for a minimum of 3 hours.

The shift-scheduling problem, as mentioned earlier, aims at finding the number of employees to work in each shift by satisfying the minimum staffing requirements with a minimal cost. This problem has been considered by many authors, starting in the 1950s with Dantzig (1954) in. Here, we will review recent work closely related to call centers. Henderson and Mason (1998) develop a new technique for rostering in a call center, which combines simulation with integer programming by relaxing the QoS constraints from half-hours to adjacent half-hours. Atlason et al. (2004) and Ingolfsson et al. (2005) use similar techniques to solve the staffing and rostering problems together. Koole and van der Sluis (2003), on the other hand, develop a local search algorithm for a call center staffing problem with a global service level constraint (as opposed to half-hourly constraints).

8.4.2 Skill-based routing in call centers with different kinds of calls

In this section, we consider a call center serving different types of calls with possibly cross-trained operators. In section 8.4.2.1, we present examples to show the effects of different routing policies on the performance measures of a call center, which, in turn, affects the staffing levels. Section 8.4.2.2 reviews the literature on the routing policies of different systems.

8.4.2.1 Effects of routing policies on the staffing of call centers

We first consider the routing of calls in a simple system, which aims to minimize the total waiting costs. Assume that we have two types of calls, A and B, and two service stations, one dedicated to B calls, the other fully flexible, meaning that it can serve both A and B calls. The routing policy for incoming calls is as follows. All calls start receiving service if there is one available agent capable of answering the incoming call. An incoming B call randomly goes to one of the two stations whenever both have available servers. If there is no available server, then the call joins the queue. The routing policy for the agent who just finishes a service is straightforward for the dedicated station, since he/she just checks the queue for B calls and starts serving if there is at least one call in queue. On the other hand, an agent in the fully flexible station chooses a call randomly from the queue, whenever there is at least one call. The problem with this policy is obvious. The fully flexible station may choose to serve a B call with a significant probability, while there are A calls waiting in line; and afterwards an agent in the dedicated station may stay idle because he/she cannot serve A calls. In other words, this routing policy does not use the flexible capacity wisely, as it increases the probability of having A calls waiting in the queue, while the flexible station is busy with B calls and the dedicated station has idle capacity. In fact, the optimal policy for this type of a system is given by Xu et al. (1992): B calls are served in the dedicated station, while A calls are served in the fully-flexible station. The fully flexible station serves B calls only if the number of B calls in line exceeds a threshold. This shows that the routing policy should protect the flexible capacity for those who really need it.

Following, we introduce and discuss a number of issues in a call center that offers three different services. Then the set of all possible kinds of calls is (A, B, C). We assume that the call center has three stations, where all operators at each station are capable of answering two types of calls, with the skill set design given by $S = ([A,B], [B,C], [C,A])$. Finally, we take the number of agents in each station fixed.

The first effect of having a number of services is on the description of the QoS constraints. Although it is possible to have one global QoS constraint for all calls, as in Section 8.4.1, it is more common to set a different service level for each type of call, especially when the skills correspond to tasks bringing different returns, or to the capability of serving different customer segments. To describe explicitly, our example may have the following QoS goals: At least 90% of A calls are answered within 10 seconds, at least 80% of B calls are answered within

20 seconds, and at least 80% of C calls are answered within 30 seconds. This type of constraint is more difficult to evaluate because its evaluation drastically depends on the call routing policies, as we will discuss below.

If our call center has these QoS constraints, then the management should value A calls the most and C calls the least, while B calls should be valued in between the two. We have seen above that the performance of a call center strongly depends on "routing policies." Moreover, in the simple example described above, we know that a policy of "threshold" type is optimal. Now we will describe a threshold routing policy for our call center, which preserves the priorities of the QoS constraints. As mentioned in the beginning of this section, the routing policy has two functions, one to direct the incoming calls to an agent or delay them, the other to assign a call waiting in line to an agent who has just become available or keep him/her idle. Assume that our routing policy directs an incoming A call to any available agent with the right skill, an incoming B or C call to any available agent with skills (B,C), and delays them if all agents in pool (B,C) are busy. An agent in pool (A,B) or (C,A) who just finishes his/her service takes an A call if there is one in queue; if not, s/he serves a B or C call, respectively, only if the number of B or C calls is greater than a threshold. If an agent in pool (B,C) becomes idle, he/she first checks the B queue and starts serving a B call, if any; if not, the operator checks the C queue and serves a C call, if any. With such a routing policy, the QoS for A calls will always be satisfied given that there is sufficient capacity, but QoS constraints for B and C calls may not be satisfied at all, as they have such a low priority in routing. If the routing policy is not modified, the staffing levels may be very high to achieve the specified QoS constraints. Hence, the optimal staffing levels from each skill set depend strongly on routing policies. Here we also would like to note that the performance measures of a call center with a routing policy as described above do not have a closed-form solution. In general, there are two methods to understand whether the QoS constraints are satisfied: to simulate the whole system or to derive good approximations under general conditions, where the former is computationally intensive, and the latter is very difficult. There are several papers that approximate certain performance measures (see, e.g., Franx et al., 2006; Koole et al., 2003; Shumsky, 2004), but usually an easier routing policy is assumed.

8.4.2.2   *Review of routing policies*

In skill-based routing (SBR) problems, the performance measures of the call center are not restricted to the ones described in Section 8.4.1. Some of the common objectives used in SBR are to minimize total waiting costs of all calls, to maximize the revenue generated by all calls, and to maximize throughput of calls from a certain class, while satisfying certain QoS constraints for other classes. Section 8.5 of Gans et al. (2003) has a comprehensive review of SBR. Here, we will consider only the most recent work.

One line of research concentrates on call centers that have one dedicated station for each call type and one fully flexible station with agents who can serve all call types. Örmeci (2004) considers optimal dynamic admission control of such a call center with no waiting room and two kinds of calls. The objective is to maximize the total revenue generated. It is shown that a call should be routed to a dedicated station if possible, and the optimal admission policy to the fully flexible facility is of the threshold type. Chevalier et al. (2004) generalizes the first part of this result to call centers with more than two classes of calls. Koole and Pot (2005), on the other hand, consider a call center with an infinite waiting room and possibly more than two classes of calls. They use the technique of approximate dynamic programming to find good call routing heuristics. Finally, Bhulai (2005) considers a call center with no waiting room and several stations having general sets of skills, as opposed to call centers with one fully flexible and many dedicated stations. He uses approximate dynamic programming, as Koole and Pot (2005), to find dynamic call routing policies.

In some call centers, agents are being cross trained in call-back or e-mail response services, in addition to the traditional job of answering incoming phone calls. This leads to *call blending* problems. Several papers analyze the effects of call blending on the performance of call centers and develop efficient blending policies (Gans and Zhou, 2003; Bhulai and Koole, 2003; Armony and Maglaras, 2004a, 2004b).

Another growing issue in the call center industry is outsourcing. According to Datamonitor, the total value for U.S. outsourcing alone will be almost $24 billion by 2008, compared to the current $19 billion (Datamonitor, 2004). In this case, routing of calls between the outsourcing firm and in-house call centers becomes an important issue, analyzed by Gans and Zhou (2004).

8.4.3   *Staffing in call centers with cross training*

In this section, we still consider a call center serving different types of calls with a fixed skill set design, but now we take the routing policy given and aim at finding optimal staffing levels. We will continue using the same call center introduced in Section 8.4.2, so that our call center is offering three different services, i.e., the set of all possible kinds of calls is (A, B, C). We have been using the term *skill set* to give the general idea, but in this section, we need to be more precise. Hence, we explain it by using the example in the above section. The agents working in our call center may be trained for answering only one call type or they may be cross trained for two or three call types. We specify an agent's cross-training level with his/her skill set, where the skill set is defined as the set of call types that an agent can answer. For example, if an agent is trained to answer B and C calls, then his/her skill set is (B,C). Hence, all possible skill sets in our example are (A), (B), (C), (A,B), (B,C), (C,A), (A,B,C). In the previous section, we have identified methods to find a good skill set design. To illustrate the link to this section with the previous ones, we assume that we have analyzed this call center at the strategic level and we have chosen the skill set as $S = ([A], [B], [C], [A,B], [B,C], [C,A])$, i.e., all agents will have, at the most, two skills. Therefore, when we compute the number of required agents from each skill set, we will never consider having an agent with three skills.

In the remaining parts of this section, we assume that a skill-set design, S, is given. Let the number of elements in S be s, and S_k be the kth element of

S, where $k = 1, 2, \ldots, s$. Our aim in this section is to find the number of operators from each skill set S_k for short time intervals (e.g., for half-hours) to guarantee a certain service level during these intervals with a minimal cost.

Çezik and L'Ecuyer (2004) use the methodology of Atlason et al. (2004) to find optimal staffing levels for each half-hour with the objective of minimizing the staffing costs of a multiskill call center. The call center has an infinite waiting room, and the calls may or may not abandon the system. There is always a global QoS constraint so that at least $p\%$ of all calls has to be answered in w seconds; a QoS constraint for each different type of call can also be added. The routing policy is a nonidling policy with simple priorities: an incoming i call checks for an available server in skill sets, $\{S_k : i \in S_k\}$, by a fixed numeric order; and when an agent in skill set S_k becomes idle, he/she checks the waiting line for calls that he/she can answer according to a given numeric order. As they note, this kind of routing policy makes the system highly unbalanced, and the service level of low-priority call types tends to be very low. An iterative cutting-plane algorithm on an integer program, where QoS constraints are estimated by simulations, solves this problem. Due to computational complexities, especially in large problems, finding an optimal solution is not always possible, so they also propose practical heuristics.

Bhulai et al. (2005) consider the same problem with Çezik and L'Ecuyer (2004). The main difference is that the QoS constraints are estimated by approximations based on steady-state behavior of Markovian queuing systems. This reduces the computational time needed to solve the problem, which allows them to consider the shift-scheduling problem in addition to staffing.

Wallace and Whitt (2005) use simulation to find optimal staffing levels as well as optimal numbers of trunk lines, where the objective is to minimize the total number of agents first and to minimize total trunk lines next, while still satisfying certain QoS constraints. The routing policy is still a simple priority policy, but they also differentiate the skill level of agents. For example, an agent with skills (B,C) has a primary skill of B and a secondary skill of C, whereas an agent with skills (C,B) has a primary skill of C and a secondary skill of B. Hence, in general, they assume that each agent has a primary skill (a level of 1) and may have skills at levels 2, 3, etc. When we let the skill sets depend on the order of skills, our framework represents this situation as well. They use an Erlang formula to find an initial value for the total number of servers, then square root staffing to allocate these servers to different primary skills. Finally they describe a rule to add skills to these servers. After specifying the initial solution, they use simulation to improve the solution. They numerically show, via simulation, that a call center with all agents having at most two skills performs very closely to a call center with fully flexible servers. Mazzuchi and Wallace (2004) conduct an experimental design on this call center, where they investigate the effects of call volume and of the number of skills of each agent on certain performance measures. They reach the same conclusion as Wallace and Whitt (2005). Sisselman and Whitt (2006), on the other hand, use this framework to incorporate the expected value of certain call types generated by certain kinds of agents, in particular the preference of agents to answer certain types of calls in call routing.

The conclusions of these papers confirm the long-time observation of "little flexibility goes a long way," and what they suggest is to cross train the agents in, at most, two skills. However, implementing this suggestion in all call centers may not be possible. Certain types of call centers, such as technical support call centers and call centers of certain banks, require the skill sets to be nested. The entry-level agents know only the basics, and adding one more skill means "teaching the subject one level deeper." Then, we can label the skills as $1, \ldots, n$, where level 1 is the entry level and level n is the "guru" level. In this kind of a system, the agents will, eventually, be cross trained in more than two skills. Hence, we probably need different tools to analyze this kind of a system. Finally, we would like to note another issue that brings a nested skill set into picture: The necessity of offering a career plan to the employees. When the cross-training levels are set to two, there is no career path for the agents, which will take them to a "better" position in the organization.

Chevalier et al. (2004), as mentioned above, consider a call center, which has one dedicated station for each call type and one fully flexible station with agents who can serve all call types. They assume that an incoming call is served in its dedicated station, if possible; if not, it is directed to an available agent in the full-flexible station, and if that station is also full, then the call is lost (due to no waiting room). We note that directing calls to the dedicated stations first is shown to be optimal. They show, through a numerical study, that spending 80% of the staffing budget on the dedicated agents (and the remaining 20% spent on the fully flexible station) works well over a wide range of parameters in systems with unlimited waiting space as well as in those with no waiting room.

8.4.4 Routing and staffing in call centers with cross training

In Section 8.4.2 and Section 8.4.3, we have seen that the routing policies have a strong effect on the performance measures and, thereby, on the staffing levels. Hence, if the staffing problem is solved in combination with the optimal routing problem, significant improvements can be achieved. In this section, we still consider a call center serving different types of calls with a fixed skill set design, S, but now we aim at finding an optimal routing policy as well as optimal staffing levels. The combined problem of routing and staffing is considerably difficult, so the studies on this subject are based on fluid approximations, which ignores stochastic fluctuations observed in queuing systems.

Harrison and Zeevi (2005) and Bassamboo et al. (2004) formulate the problem as a two-stage stochastic linear program with recourse. At the first stage, the staffing levels are determined, while in the second stage the routing problem is optimized by using a fluid approximation of the call center.

The second stage observes the random features of the system, so that the corresponding expected value is approximated via Monte Carlo simulation. In this way, the daily call patterns (see Figure 8.4) can be incorporated in the model. In the routing problem, the objective is to minimize total penalty due to abandonment, whereas the overall objective is to minimize the staffing costs and the expected abandonment penalty. Bassamboo et al. (2004), on the other hand, derive an asymptotic lower bound on the expected total cost for the same problem and propose a staffing and routing method, which is shown to achieve this asymptotic lower bound. Whitt (2006b) is another study that proposes a fluid model to solve routing and staffing problems simultaneously, where two optimization problems are constructed based on the fluid approximations and a discussion on how to implement stochastic fluctuations and dynamic routing is presented.

8.5 Future directions

As described in Section 8.3, the literature exploring the skill set design question assumes given or identical capacity in the different skill types. It is described how properties of superior cross-training structure are quite robust to underlying system characteristics. So, irrespective of the operating details of their systems, managers can follow some of these guidelines in determining cross-training policies. However, this analysis is performed taking a benefits perspective. Determining the appropriate structure for a given call center requires understanding costs in addition to benefits. The operational costs of a flexibility design, however, depend on capacity that is deployed within the designed skill-set structure. Section 8.4 illustrates that a call center performance is closely tied to capacity and explains the way it is eventually exploited through routing decisions. Linking skill-set design to capacity choice in call centers or other systems is an important direction for future research, which would enable systematically making a tradeoff between the benefits from flexibility and the costs of staffing. For example, the flexibility literature recommends structures having balanced flexibility with capacity pools that have the same number of skills. Are such structures still superior to others when the cost of capacity is incorporated in the analysis? These costs will be characterized more precisely as research focusing on staffing and routing develops. In particular, approaches that are capable of jointly optimizing staffing and routing will enable a better assessment of costs. This is another important area for future research.

Another direction for investigation is one that adopts an interdisciplinary approach to cross-training design and combines motivational, biological, and perceptual-motor aspects of the issue with operational ones. Empirical research that explores the motivational impact of the superior structures in Section 8.3 or that helps define better routing policies as those that improve operational as well as motivational performance is needed for different call center environments. Other questions lying at the interface of operations and human resource management are: What is the relationship between these flexibility structures and career paths? How can career paths be formulated such that skill sets come closer to designs that are known to be operationally superior?

Acknowledgments

The authors would like to thank Tolga Çezik, Gü Gürkan, Ger Koole, Robert Shumsky, and Ward Whitt for their comments on an earlier version.

References

Aksin, O.Z., and Harker, P.T. (1999). To Sell or Not to Sell: Determining the Tradeoffs Between Service and Sales in Retail Banking Phone Centers, *Journal of Service Research*, 2:1 19–33.

Aksin, O.Z., and Karaesmen, F. (2002). Designing Flexibility: Characterizing the Value of Cross-Training Practices, Working paper, Koç University, Istanbul.

Aksin, O.Z., and Karaesmen, F. (2007). Characterizing the Performance of Process Flexibility Structure, Forthcoming, Operations Research Letters.

Aksin, O.Z., Karaesmen, F., and Örmeci, E.L. (2005). On the Interaction Between Resource Flexibility and Flexibility Structures, in *Proceedings of the Fifth International Conference on Analysis of Manufacturing Systems — Production Management*.

Armony, M., and Maglaras, C. (2004a). On Customer Contact Centers with a Call-Back Option: Customer Decisions, Routing Rules, and System Design, *Operations Research*, 52:2, 271–292.

Armony, M., and Maglaras, C. (2004b). Contact Centers with a Call-Back Option and Real-Time Delay Information, *Operations Research*, 52:4, 527–545.

Armony, M., and Mandelbaum, A. (2004). Design, Staffing and Control of Large Service Systems: The Case of a Single Customer Class and Multiple Server Types', Working paper, Israel Institute of Technology Haifa.

Atlason, J., Epelman, M.A., and Henderson, S.G. (2004). Call Center Staffing with Simulation and Cutting Plane Methods, *Annals of Operations Research*, 127, 333–358.

Avramidis, A.N., Deslauriers, A. and L'Ecuyer, P. (2004). Modeling daily arrivals to a telephone call center, *Management Science*, 50:7, 896–908.

Bassamboo, A., Harrison, J.M., and Zeevi, A. (2004). Design and Control of a Large Call Center: Asymptotic Analysis of an LP-based Method.

Baumgartner, M., Good, K. and Udris, I. (2002). Call Centers in der Schweitz, *Psychologische Untersuchungen in 14 Organisationen*, (*Call Centers in Switzerland, Psychological Investigations in 14 Organisations*), Zurich, Switzerland.

Bhulai, S., and Koole, G. (2003). A Queuing Model for Call Blending in Call Centers, *IEEE Transactions on Automatic Control*, 48:8, 1434–1438.

Bhulai, S., Koole, G., and Pot, A. (2005). Simple Methods for Shift Scheduling in Multi-skill Call Centers, Working paper, Free University, The Netherlands.

Bhulai, S. (2005). Dynamic Routing Policies for Multi-Skill Call Centers, Working paper, Free University, The Netherlands.

Borst, S., Mandelbaum, A., and Reiman, M. (2004). Dimensioning Large Call Centers, *Operations Research*, 52:1, 17–34.

Brown, L., Gans, N., Mandelbaum, A., Sakov, A., Shen, H., Zeltyn, S., and Zhao, L. (2005). Statistical analysis of a telephone call center: A queuing-science perspective, *Journal of the American Statistical Association*, 100, 36–50.

Buzacott, J.A. (1996). Commonalities in Reengineered Business Processes: Models and Issues, *Management Science*, 42:5, 768–782.

Campion, M.A., and McClelland, C.L. (1991). Interdisciplinary Examination of the Costs and Benefits of Enlarged Jobs: A Job Design Quasi-experiment, *Journal of Applied Psychology*, 76, 186–198.

Campion, M.A., and McClelland, C.L. (1993). Follow-up and Extension of the Interdisciplinary Costs and Benefits of Enlarged Jobs, *Journal of Applied Psychology*, 78:3, 339–351.

Çezik, T., and L'Ecuyer, P. (2004). Staffing Multiskill Call Centers via Linear Programming and Simulation, Technical Report, Université de Montréal, Quebec, Canada.

Chevalier, P., Shumsky, R.A., and Tabordon, N. (2004). Routing and Staffing in Large Call Centers with Specialized and Fully Flexible Servers, Working paper.

Dantzig, G.B. (1954). A Comment on Edie's 'Traffic Delays at Toll Booths,' *Operations Research*, 2:3, 107–138.

Das, A. (2003). Knowledge and Productivity in Technical Support Work, *Management Science*, 49:4, 416–431.

Datamonitor (2002), Cited in *Call Center Management Review*, May 2002: http://www.ccmreview.com

Datamonitor (2004), 4-26-2004, Cited in: http://www.incoming.com/statistics

Datamonitor (2005), 4-14-05, Cited in: http://www.incoming.com/statistics

Feldman, Z., Mandelbaum, A., Massey, W.A., and Whitt, W. in press. Staffing of Time-Varying Queues to Achieve Time-Stable Performance, *Management Science*, forthcoming.

Fine, C.H., and Freund, R.M. (1990). Optimal Investment in Product-Flexible Manufacturing Capacity, *Management Science*, 36:4, 449–466.

Evenson, A., Harker, P.T., and Frei, F.X. (1999). Effective Call Center Management: Evidence from Financial Services. Working paper 99–25–B, Wharton Financial Institutions Center, University of Pennsylvania, Philadelphia.

Franx, G.J., Koole, G.M., and Pot, S.A. (2006). Approximating Multi-Skill Blocking Systems by Hyperexponential Decomposition, *Performance Evaluation*, 63:8, 799–824.

Gans, N., Koole, G.M., and Mandelbaum, A. (2003). Telephone Call Centers: Tutorial, Review, and Research Prospects, *Manufacturing & Service Operations Management*, vol. 5, 97–141.

Gans, N., and Zhou, Y. (2003). A Call-Routing Problem with Service-Level Constraints. *Operations Research*, 51, 255–271.

Gans, N., and Zhou, Y. (2004). Overflow Routing for Call Center Outsourcing, Working paper, The Wharton School, University of Pennsylvania, Philadelphia.

Garnett, O., Mandelbaum, A., and Reiman M. (2002). Designing a Call Center with Impatient Customers, *Manufacturing and Service Operations Management*, 4:3, 208–227.

Graves, S.C., and Tomlin, B.T. (2003). Process Flexibility in Supply Chains, *Management Science*, 49, 907–919.

Grebner, S., Semmer, N.K., Lo Faso, L., Gut, S., Kalin, W., and Elfering, A. (2003). Working Conditions, Well-Being, and Job-Related Attitudes among Call Center Agents, *European Journal of Work and Organizational Psychology*, 12:4, 341365.

Green, L.V., Kolesar, P.J., and Whitt, W. (2006). Coping with Time-Varying Demand When Setting Staffing Requirements for a Service System, *Production and Operations Management (POMS)*, forthcoming.

Güne, E.D., and Aksin, O.Z. (2004). Value Creation in Service Delivery: Relating Market Segmentation, Incentives, and Operational Performance, *Manufacturing and Service Operations Management*, 6:4, 338–357.

Gurumurthi, S., and Benjaafar, S. (2004). Modeling and Analysis of Flexible Queuing Systems, *Naval Research Logistics*, 51, 755–782.

Hackman, J.R., and Oldham, G.R. (1976). Motivation through the Design of Work: Test of a Theory, *Organizational Behavior and Human Performance*, 16, 250–279.

Harrison, J.M., and Zeevi, A. (2005). A Method for Staffing Large Call Centers Based on Stochastic Fluid Models, *Manufacturing & Service Operations Management*, 7:1, 20–36.

Hasija, S., Pinker, E., and Shumsky, R. (2005). Staffing and Routing in a Two-Tier Call Center, *International Journal of Operational Research*, vol. 1, no. 1/2, 8–29.

Henderson, S., and Mason, A. (1998). Rostering by Iterating Integer Programming and Simulation. In *Proceedings of the 1998 Winter Simulation Conference*, Washington, DC, 677–683.

Hopp, W.J., Tekin, E., and Van Oyen, M.P. (2004). Benefits of Skill Chaining in Production Lines with Cross-Trained Workers, *Management Science*, 50, 83–98.

Hopp, W.J., and Van Oyen, M.P. (2004). Agile Workforce Evaluation: A Framework for Cross-Training and Coordination, *IIE Transactions*, 36:10, 919–940.

ICCM Weekly (2002), 4-3-2002, Cited in: http://www.incoming.com

Ilgen, D.R., and Hollenbeck, J.R. (1991). The Structure of Work: Job Design and Roles, In M.D. Dunnette and L.M. Hough, Eds., *Handbook of Industrial and Organizational Psychology*, Consulting Psychologists Press, Palo Alto, CA, 2, 165–207.

Incoming Calls Management Institute (ICMI, 2000). 9-1-2000, Cited in *Call Center Management Review*, September 2000: http://www.ccmreview.com

Incoming Calls Management Institute (ICMI, 2002). 7-1-2002 Multichannel Call Center Study Final Report, cited in *Call Center Management Review*, July 2002: http://www.ccmreview.com

Ingolfsson, A., Cabral, E., and Wu, X. (2005). Combining Integer Programming and the Randomization Method to Schedule Employees, Technical Report, University of Alberta, Edmonton, Canada.

Inman, R.R., Jordan, W.C., and Blumenfeld, D.E. (2004). Chained Cross-Training of Assembly Line Workers, *International Journal of Production Research*, 42:10, 1899–1910.

Iravani, S.M., Van Oyen, M.P., and Sims, K.T. (2005). Structural Flexibility: A New Perspective on the Design of Manufacturing and Service Operations, *Management Science*, 51, 151–166.

Isic, A., Dormann, C., and Zapf, D. (1999). Belastungen und Resources an Call Center Arbeitsplatzen" (Job stressors and resources among call center employees), *Zeitschrift fur Arbeitswissenschaft*, 53, 202–208.

Jennings, O., Mandelbaum, A., Massey, W., and Whitt, W. (1996). Server Staffing to Meet Time-Varying Demand, *Management Science*, 42:10, 1383–1394.

Jordan, W.C., and Graves, S.C. (1995). Principles on the Benefits of Manufacturing Process Flexibility, *Management Science*, 41:4, 577–594.

Jordan, W.C., Inman, R.R., and Blumenfeld, D.E. (2004). Chained Cross-Training of Workers for Robust Performance, *IIE Transactions*, 36, 953–967.

Jouini, O., Dallery, Y., and Nait-Abdallah, R. (2004). Analysis of the Impact of Team-Based Organizations in Call Center Management, Working paper, Ecole Centrale Paris.

Koole, G.M., and Pot, S.A. (2005). Approximate Dynamic Programming in Multi-Skill Call Centers, In *Proceedings of the 2005 Winter Simulation Conference*, Orlando, FL.

Koole, G.M., Pot, S.A., and Talim, J. (2003). Routing Heuristics for Multi-Skill Call Centers, In *Proceedings of the 2003 Winter Simulation Conference*, New Orleans, 1813–1816.

Koole, G., and van der Sluis, E. (2003). Optimal Shift Scheduling with a Global Service Level Constraint, *IIE Transactions*, 35, 1049–1055.

Mandelbaum, A., and Reiman, M.I. (1998). On Pooling in Queuing Networks, *Management Science*, 44:7, 971–981.

Mazzuchi, T.A., and Wallace, R.B. (2004). Analyzing Skill-Based Routing Call Centers Using Discrete-Event Simulation and Design Experiment, in *Proceedings of the 2004 Winter Simulation Conference washington, DC*, R.G. Ingalls, M.D. Rossetti, J.S. Smith, and B.A. Peters, Eds.

Netessine, S., Dobson, G., and Shumsky, R. (2002). Flexible Service Capacity: Optimal Investment and the Impact of Demand Correlation, *Operations Research*, 50:2, 375–389.

Örmeci, E.L. (2004). Dynamic Admission Control in a Call Center with One Shared and Two Dedicated Service Facilities, *IEEE Transactions on Automatic Control*, 49:7, 1157–1161.

Pinker, E.J., and Shumsky, R.A. (2000). The Efficiency-Quality Trade-Off of Cross-Trained Workers, *Manufacturing & Service Operations Management*, 2:1, 32–49.

Sethi, A.K., and Sethi, S.P. (1990). Flexibility in Manufacturing: a Survey, *International Journal of Flexible Manufacturing Systems*, 2, 289–328.

Sheikzadeh, M., Benjaafar, S., and Gupta, D. (1998). Machine Sharing in Manufacturing Systems: Total Flexibility Versus Chaining, *International Journal of Flexible Manufacturing Systems*, 10, 351–378.

Shumsky, R.A. (2004). Approximation and Analysis of a Queuing System with Flexible and Specialized Servers, *OR Spectrum*, 26:3, 307–330.

Shumsky, R.A., and Pinker, E.J. (2003). Gatekeepers and Referrals in Services, *Management Science*, 49:7, 839–856.

Sisselman, M.E., and Whitt, W. (2006). Value-Based Routing and Preference-Based Routing in Customer Contact Centers. *Production and Operations Management*, forthcoming.

Steckley, S.G., Henderson, S.G., and Mehrotra, V. (2004). Service System Planning in the Presence of a Random Arrival Rate, Technical Report, Cornell University, Ithaca, NY.

Taylor, P., Mulvey, G., Hyman, J., and Bain, P. (2002). Work Organization, Control and the Experience of Work in Call Centers, *Work, Employment, and Society*, 16, 122–150.

Van Mieghem, J.A. (1998). Investment Strategies for Flexible Resources, *Management Science*, 44, 1071–1078.

Van Oyen, M.P., Gel, E.G.S., and Hopp, W.J. (2001). Performance Opportunity for Workforce Agility in Collaborative and Noncollaborative Work Systems, *IIE Transactions*, 33:9, 761–777.

Wallace, R.B., and Whitt, W. (2005). A Staffing Algorithm for Call Centers with Skill-Based Routing, *Manufacturing and Service Operations Management*, 7, 276–294.

Whitt, W. (2006a). Staffing a Call Center with Uncertain Arrival Rate and Absenteeism, *Production and Operations Management*, 15:1, 88–102.

Whitt, W. (2006b). A Multi-Class Fluid Model for a Contact Center with Skill-Based Routing. *International Journal of Electronics and Communications (AEU)*, 60:2, 95–102.

Xie, J.L., and Johns, G. (1995). Job Scope and Stress: Can Job Scope be too High? *Academy of Management Journal*, 18, 1288–1309.

Xu, S.H., Righter R., and Shanthikumar J.G. (1992). Optimal Dynamic Assignment of Customers to Heterogeneous Servers in Parallel, *Operations Research*, 40, 1126–1138.

chapter 9

Partial pooling in tandem lines with cooperation and blocking

Nilay Tanik Argon and Sigrún Andradóttir

Contents

Abstract

For a tandem line of finite, single-server queues operating under the production-blocking mechanism, we studied the effects of pooling several adjacent stations and the associated servers into a single station with a single team of servers. We assumed that the servers are cross trained (so that they can work at several different stations) and that two or more servers can cooperate on the same job. For such a system, we provided sufficient conditions on the service times and sizes of the input and output buffers at the pooled station under which pooling would decrease the departure time of each job from the system (and, hence, increase the system throughput). We also showed that pooling decreases the total number of jobs in the system at any given time and the sojourn time of each job in the system if the departure time of each job from the system is decreased by pooling and there is an arrival stream at the first station. Moreover, we provided sufficient conditions under which pooling will improve the holding cost of each job in the system incurred before any given time and extended our results to closed tandem lines and to queuing networks with either a more general blocking mechanism or probabilistic routing. Finally, we presented a numerical study aimed at quantifying the improvements in system performance obtained through pooling and at understanding which stations should be pooled to achieve the maximum benefit. Our results suggest that the improvements gained by pooling may be substantial and that the bottleneck station should be among the pooled stations in order to obtain the greatest benefit.

9.1 Introduction

One of the main issues in the design and management of stochastic production and service systems is the division of the overall processing requirement into tasks assigned to different stations, workers, or machines. For example, consider a job shop with three workers. Assuming that each worker is capable of performing all tasks of the processing requirement, would it be more efficient to divide the processing requirement into three tasks and then assign each worker to one task, or to let all workers join forces and then complete the entire processing requirement together as a team? Obviously, there are more alternatives than these two. However, this simple example leads us to the concept of pooling in queuing networks, namely the replacement of several queuing stations by a single, functionally equivalent station (we define a station as a unit that covers a queue, a server, and a task).

9.1.1 Problem definition

In this chapter, we study the effects of pooling some of the stations in a system of $N \geq 2$ queuing stations in tandem numbered $1, \ldots, N$, where each station $j \in \{1, \ldots, N\}$ has one server (referred to as server j) and the jobs are served in the order that they arrive (i.e., according to the first-in-first-out, FIFO, queuing discipline). We assume that there are $0 \leq b_j \leq \infty$ buffers in front of station $j \in \{2, \ldots, N\}$, unlimited supply of jobs in front of the first station

$(b_1 = \infty)$, and infinite capacity buffer space following the last station $(b_{N+1} = \infty)$. Consequently, if all buffers in front of station $j \in \{2,...,N\}$ are full when station $j-1$ completes a job, then we assume that this job remains at station $j-1$ until the job in service at station j is moved to station $j+1$ or leaves the system (if $j = N$). This type of blocking is usually called *production blocking*. Since we assume that the output buffer space for station N is unlimited, station N will never be blocked. Such tandem queuing systems are observed commonly in manufacturing environments, e.g., in the form of asynchronous production flow lines (see Buzacott and Shanthikumar, 1993).

Throughout this chapter, $\mu_{\ell,j} \geq 0$ denotes the rate at which worker ℓ processes jobs at station j, for $\ell, j \in \{1,...,N\}$ (the workers are identical if $\mu_{\ell,j} = \mu_{k,j}$ for all $j, k, \ell \in \{1,...,N\}$). Also, we let $X_j(i)$ be the service time of job $i \geq 1$ at station $j \in \{1,...,N\}$ in the original line (if job i is in the system at time zero, then $X_j(i)$ denotes the remaining service time of job i at station j at time zero). Then $\mu_{j,j}X_j(i)$ represents the service requirement of job $i \geq 1$ at station $j \in \{1,...,N\}$.

We are interested in studying the departure process from the system described above when several adjacent stations $K \in \{1,...,N-1\}$ through $M \in \{K,...,N\}$ are pooled. By pooling several stations, we mean pooling the queues, servers, and tasks. Throughout this chapter, we refer to pooling all of the stations in the network as *complete pooling* (corresponding to $K = 1$ and $M = N$) and to pooling some of the stations in the network as *partial pooling*. When stations $K,...,M$ are pooled, then we let $X_j^{(K,M)}(i)$ be the service time of job $i \geq 1$ at station $j \in \{1,...,K-1,M+1,...,N\}$ in the pooled system and we let $X^{(K,M)}(i)$ denote the service time of job $i \geq 1$ at the pooled station. If pooling stations K through M do not affect the other stations in the network, then $X_j^{(K,M)}(i) = X_j(i)$ for all $j \in \{1,...,K-1,M+1,...,N\}$ and $i \geq 1$. Moreover, if the pooled workers $K,...,M$ form a team that completes the tasks at stations $K,...,M$ sequentially for each job $i \geq 1$ and if the nonnegative function $g_j^{(K,M)}(\mu_{K,j},...,\mu_{M,j})$ represents the rate at which the team performs task $j \in \{K,...,M\}$ (assuming that $g_j^{(j,j)}(\mu_{j,j}) = \mu_{j,j}$ for all $j \in \{1,...,N\}$), then

$$X^{(K,M)}(i) = \sum_{j=K}^{M} \frac{\mu_{j,j}X_j(i)}{g_j^{(K,M)}(\mu_{K,j},...,\mu_{M,j})}, \qquad (9.1)$$

for all $K \in \{1,...,N-1\}, M \in \{K,...,N\}$, and $i \geq 1$ (note, however, that most of our results do not require this particular structure for the pooled service times). In Equation (9.1), the worker rates are *additive* if $g_j^{(K,M)}(\mu_{K,j},...,\mu_{M,j}) = \sum_{\ell=K}^{M} \mu_{\ell,j}$ for all $j \in \{K,...,M\}$ (i.e., workers neither lose nor gain any efficiency by pooling). For some manufacturing, service, and computer systems, the additivity of the server rates is a valid assumption, e.g., when tasks involve many subtasks that can be done in parallel without loss of efficiency. Nevertheless, it is clear that the assumption that the server rates are additive is restrictive. For example, the number of workers working together on a job may be limited (because cooperation of more than a specific number of workers may be infeasible, dangerous, or undesirable). In such cases, several workers working together may not be as efficient as when they work alone, corresponding to $g_j^{(K,M)}(\mu_{K,j},\ldots,\mu_{M,j}) < \sum_{\ell=K}^{M} \mu_{\ell,j}$ for all $j \in \{K,\ldots,M\}$; we say that such systems have *subadditive* work force. On the other hand, there are other systems where the cooperation of workers increases their individual efficiency (because they work better in teams than individually), in which case $g_j^{(K,M)}(\mu_{K,j},\ldots,\mu_{M,j}) > \sum_{\ell=K}^{M} \mu_{\ell,j}$ for all $j \in \{K,\ldots,M\}$. Such systems have, what we call, *superadditive* work force.

9.1.2   Literature review

The early work on pooling starts with the resource sharing concept, i.e., pooling parallel queues without pooling the servers. For example, it is well known that an M/M/m queue with arrival rate $m\lambda$ and service rate μ for each server is superior to m parallel M/M/1 queues, each having an arrival rate of λ and service rate μ, in terms of the average waiting time in the system. (We refer the reader to Smith and Whitt, 1981, Calabrese, 1992, Section 8.4.1 of Buzacott and Shanthikumar, 1993, and Benjaafar, 1995, for further studies of resource sharing.) Moreover, it is also known that an M/M/1 queue with arrival rate λ and service rate $m\mu$ is superior to an M/M/m queue with arrival rate λ and service rate μ for each server, in terms of the average waiting time in the system.

Two important studies on pooling in queuing networks are due to Buzacott, 1996, and Mandelbaum and Reiman, 1998. More specifically, Buzacott (1996) studies two models for (complete) pooling of tasks in a tandem line with infinite buffer spaces and general service time distributions, namely parallel facilities and teams. In the parallel system, each worker picks a job from the arrivals queue and completes all process requirements for that job. Buzacott (1996) shows that high task processing time variability makes the parallel system attractive compared to the tandem system. Buzacott (1996) also considers the case where the workers work as a team and are allocated different tasks. The total processing time of a job is the maximum of the random durations of the subtasks completed by different team members. The conclusion is that pooling with teams is not superior to the (unpooled) tandem line unless factors, such as motivation, improve the performance of team members.

Mandelbaum and Reiman (1998) study the pooling of stations in a Jackson queuing network. As in this chapter, Mandelbaum and Reiman assume that when several stations are pooled, their quees and servers are also pooled. The authors mainly concentrate on the complete pooling of all the stations in the network and compare the mean steady-state sojourn times of the pooled and original systems under the assumptions of additive server rates and identical servers. For a tandem Jackson network, they show that complete pooling always helps, but for Jackson networks with more general routing, complete pooling becomes advantageous only when the service variability (which depends on the structure of the network) is low. Mandelbaum and Reiman also show that partial pooling of stations that are not neighbors in a tandem Jackson network may cause the network to become unstable (this follows from Theorem 1 of

Bramson (1994)). Here, the problem of instability arises due to the fact that the tasks of the pooled stations are not pooled (i.e., if a job requires several tasks to be completed by a pooled server, then that job rejoins the queue in front of the pooled server for each task). In this study, we only consider the partial pooling of neighboring stations in a line, so that we can pool the tasks without affecting the order of the tasks that each job goes through. Hence, we can assume that the pooled server does not start working on a new job until the entire service requirement for the current job is finished at the pooled station.

In order for it to be beneficial to pool several stations in a queuing network, it may be necessary to cross train the workers to be capable of performing multiple tasks. There are several studies of queuing systems with cross-trained workers, including Bischak (1996), Zadavlav et al. (1996), Bartholdi et al. (2001), Andradóttir et al. (2001), Andradóttir and Ayhan (2005), Van Oyen et al. (2001), and references therein. The first three of these studies assume that only one server can work on a job at a time, i.e., the work is not collaborative. Andradóttir et al. (2005, 2001, 2003) and Van Oyen et al. (2001), on the other hand, study queuing networks with cross-trained and collaborative workers. More specifically, Andradóttir et al. (2005, 2001, 2003) study the dynamic assignment of workers to stations with the goal of maximizing the throughput for queuing networks with either finite or infinite buffers. Moreover, Van Oyen et al. show that when all workers are identical in a tandem line, then complete pooling maximizes the throughput along all sample paths; this policy also minimizes the cycle time for each job and the average work-in-process (WIP) inventory. Note that all the studies discussed in this paragraph (except for Van Oyen et al.) discuss the dynamic assignment of workers to tasks, and that the policies under consideration do not involve permanent pooling of the servers, queues, or tasks. Moreover, the studies on systems with collaborative workers assume that the server rates are additive.

The studies on pooling reviewed above focus on the complete pooling of all the stations in the network. In this chapter, we study partial pooling, which we believe is more realistic than complete pooling. An important motivation for considering partial pooling is that it requires less cross training than complete pooling. For example, there may be some tasks that require skills that only a subset of the workers have, in which case, complete pooling leads to an inefficient use of the available workers. Moreover, with partial pooling it is reasonable to assume that the workers can be fully efficient while working together on a job. However, this assumption is unreasonable for complete pooling of a large number of workers.

The outline of this chapter is as follows. In Section 9.2 and Section 9.3, we study the effects of pooling on the departure time of each job from each station and on the steady-state throughput of the system; we discuss under what conditions pooling is beneficial in Section 9.2 and compare different pooling schemes in Section 9.3. In Section 9.4, we look at the effects of pooling on the WIP, sojourn times, and holding costs. In Section 9.5, we extend our model to tandem lines with a more general blocking mechanism than production blocking and to more general queuing networks than open tandem lines.

In Section 9.6, we use numerical results to quantify the benefits of pooling on the system throughput and WIP and to gain insights into how partial pooling should be done to obtain the most benefit. Finally, we provide our concluding remarks in Section 9.7.

Most of the material presented in this chapter was originally published in *Queueing Systems — Theory and Applications* (see Argon and Andradóttir, 2006). The interested reader is referred to Argon and Andradóttir for the proofs of our technical results.

9.2 When is partial pooling advantageous?

In this section, we determine conditions under which pooling stations $K \in \{1, \dots, N-1\}$ through $M \in \{K, \dots, N\}$ will yield earlier departures from the system. We start by considering the deterministic system in which the service time $X_j(i)$ of each job $i \geq 1$ at each station $j \in \{1, \dots, N\}$ is a constant X_j that depends only on the station and not on the job. For $i \geq 1$, let $D_j(i)$ be the departure time of job i from station $j \in \{1, \dots, N\}$ in the original line and $D_j^{(K,M)}(i)$ be the departure time of job i from station $j \in \{1, \dots, K-1, M, \dots, N\}$ when stations K through M are pooled. (We arbitrarily label the pooled station as station M, unless otherwise stated.) Then, it is easy to show that $D_N(i) \leq D_N^{(K,M)}(i)$, for all $i \geq 1$, if pooling stations K through M does not affect the other stations and

$$X^{(K,M)} \leq \min\left\{ \sum_{j=K}^{M} X_j, \max_{j \in \{1,\dots,N\}}\{X_j\} \right\}.$$

Defining the throughput of the pooled and unpooled systems as $T^{(K,M)} = \lim_{i \to \infty}\{i / D_N^{(K,M)}(i)\}$ and $T = \lim_{i \to \infty}\{i / D_N(i)\}$, respectively, we can conclude that as long as pooling stations K through M do not affect the other stations and $X^{(K,M)} \leq \max_{j \in \{1,\dots,N\}}\{X_j\}$, then $T^{(K,M)} \geq T$ (for conditions that guarantee that these limits exist almost surely, see, e.g., Proposition 4.8.2 in Glasserman and Yao, 1994).

Note that the size and allocation of buffer spaces do not affect whether pooling is beneficial in deterministic systems. By contrast, when the service times are stochastic, then the size and allocation of buffer spaces can affect whether pooling is beneficial or not. One problem is that there are $M - K$ more stations in the original system than in the pooled system. Since stations also act as storage spaces, it is clear that even when all buffers in the original system (including those between the pooled stations) are kept when pooling is performed, the original system has more storage spaces than the pooled system, which can lead to a better throughput for the unpooled system.

To illustrate, consider a tandem network with three stations and no buffers, and suppose that the service times at stations 1, 2, and 3 are independent and exponentially distributed with means 10, 0.2, and 2 minutes, respectively. Modeling this network as a continuous time Markov chain (CTMC), we find that the steady-state throughput is approximately 5.9566 jobs per hour. Now, consider the system where stations 2 and 3 are pooled, $\mu_{3,2} = \mu_{2,2}$, $\mu_{2,3} = \mu_{3,3}$, and the server rates are additive. It is easy to see that the steady-state throughput for this pooled line is approximately 5.9401 jobs per hour. The reason for the higher throughput of the original system relative to the pooled system may well be the one additional storage space that it has.

The above discussion suggests that for a fair comparison of the pooled and unpooled systems, the total number of storage spaces in both systems should be equalized. This is accomplished by providing $M - K$ additional buffer spaces for the pooled system. In the example of the previous paragraph, if we equalize the total number of storage spaces in both systems by adding one buffer space to the pooled line, then the steady-state throughput of the pooled system is approximately 5.9939 jobs per hour, which is larger than that of the original line. Note, however, that pooling without additional buffers does not necessarily degrade performance. This issue is explored in the numerical results given in Section 9.6.3.

Pooling for systems with stochastic service times is studied in Section 9.2.1 and Section 9.2.2. In Section 9.2.1, we study the benefits of pooling when the input and output buffers of the pooled station are larger than specified bounds. However, these bounds are such that the pooled system generally has more storage spaces than the unpooled system. In Section 9.2.2, we consider pooling when the total number of storage spaces in the unpooled and pooled systems are equal. In this case, it is not always obvious how to allocate the $M - K$ addi- tional buffers and the buffers between stations $K,\ldots,M$ of the original system around the pooled station to guarantee a decrease in the departure time of each job from the system. Rather than attempting to identify the best buffer allocation, we show that pooling is beneficial for one particular buffer allocation structure. (For research concerned with determining the buffer allocation that maximizes the throughput, see, e.g., Conway et al., 1988, Hillier et al., 1993, and Glasserman and Yao, 1994.)

9.2.1 Pooling with large buffers around the pooled station

In this section, we consider pooling stations $K \in \{1,\ldots,N-1\}$ through $M \in \{K,\ldots,N\}$ and placing large buffers around the pooled station in a tandem line with stochastic service times. Let $0 \le P^{(K,M)}, Q^{(K,M)} \le \infty$ be the sizes of the input and output buffers of the pooled station, respectively. (Throughout this chapter, we suppress the superscripts in $P^{(K,M)}$ and $Q^{(K,M)}$ when it is not likely to cause confusion.)

We next state our main result in this section whose proof is provided in Argon and Andradóttir (2006).

Theorem 1. For $1 \le K \le M \le N$, if $X_j^{(K,M)}(i) \le X_j(i)$ for all $j \in \{1,\dots,K-1, M+1,\dots,N\}$ and $i \ge 1$, and there exist $\alpha_k \in [0,1]$ for $k \in \{K,\dots,M\}$ such that:

(i) $X^{(K,M)}(i) \le \sum_{k=K}^{M} \alpha_k X_k(i)$ for all $i \ge 1$;

(ii) $\sum_{k=K}^{M} \alpha_k = 1$; and

(iii) $P \ge \sum_{k=K}^{L_{max}} b_k + L_{max} - K$ and $Q \ge \sum_{k=L_{min}+1}^{M+1} b_k + M - L_{min}$, where $L_{min} = \min \{k \in \{K,\dots,M\} : \alpha_k > 0\}$ and $L_{max} = \max\{k \in \{K,\dots,M\} : \alpha_k > 0\}$,

then we have that $D_M^{(K,M)}(i) \le \min\{D_{L_{max}}(i), \sum_{\ell=K}^{M} \alpha_\ell D_\ell(i)\}$ and $D_j^{(K,M)}(i) \le D_j(i)$ for all $j \in \{1,\dots,K-1, M,\dots,N\}$ and $i \ge 1$.

Theorem 1 implies that if the buffers around the pooled station are sufficiently large (relative to the buffers $b_K,\dots,b_{M+1}$) and the service time of each job at the pooled station is not too large (relative to the service times at the stations that are pooled), then pooling will result in smaller departure times from the system. Therefore, if both throughputs τ and $\tau^{(K,M)}$ are well defined and the conditions of Theorem 1 are satisfied, then $\tau^{(K,M)} \ge \tau$. Note, however, that when $b_2,\dots,b_N$ are all finite and $L_{min} < L_{max}$, then it is necessary to increase the total number of storage spaces in the line in order to satisfy the conditions on P and Q in Theorem 1. Hence, Theorem 1 is of most interest when either $L_{min} = L_{max}$ or at least one of $b_2,\dots,b_N$ is infinite.

We next discuss a special case under which conditions (i) and (ii) of Theorem 1 hold. Suppose that $\mu_{\ell,j} = \theta_\ell \eta_j$, for all $\ell, j \in \{K,\dots,M\}$, where $\eta_K,\dots,\eta_M$ reflect the difficulty of the tasks at stations $K,\dots,M$, while $\theta_K,\dots,\theta_M$ represent the rates at which workers $K,\dots,M$ operate. (For example, this is satisfied when the workers are identical.) Suppose also that Equation (9.1) holds with additive or superadditive server rates. Then, $\mu_{j,j} / g_j^{(K,M)}(\mu_{K,j},\dots,\mu_{M,j}) \le \theta_j / \sum_{\ell=K}^{M} \theta_\ell$ for all $j \in \{K,\dots,M\}$ and, hence, conditions (i) and (ii) of Theorem 1 hold with $\alpha_j = \theta_j / \sum_{\ell=K}^{M} \theta_\ell$ for all $j \in \{K,\dots,M\}$ (conditions (i) and (ii) in fact may hold even if the server rates are subadditive at some of the pooled stations, as long as they are sufficiently superadditive at some other pooled stations). We discuss other implications of Theorem 1 in Section 9.2.2.

Note that Theorem 1 requires that $X^{(K,M)}(i) \le \sum_{k=K}^{M} \alpha_k X_k(i)$ for all $i \ge 1$. We next extend Theorem 1 to cases where this inequality only holds stochastically. Let $\mathcal{Y} = \{Y(i)\}_{i \ge 1}$ and $\mathcal{Z} = \{Z(i)\}_{i \ge 1}$ be stochastic processes with state space $\mathbb{R}^d$, where $d \in \mathbb{N}$. Then, $\mathcal{Y}$ is smaller than $\mathcal{Z}$ in the usual stochastic ordering sense ($\mathcal{Y} \le_{st} \mathcal{Z}$) if and only if $E[f(\mathcal{Y})] \le E[f(\mathcal{Z})]$ for every nondecreasing functional $f : \mathbb{R}^\infty \to \mathbb{R}$, and $\mathcal{Y}$ is smaller than $\mathcal{Z}$ in the increasing convex ordering sense ($\mathcal{Y} \le_{icx} \mathcal{Z}$) if and only if $E[f(\mathcal{Y})] \le E[f(\mathcal{Z})]$ for every continuous (with respect to the product topology in $\mathbb{R}^\infty$), nondecreasing, and convex functional $f : \mathbb{R}^\infty \to \mathbb{R}$ (provided the expectations exist). (A functional $f : \mathbb{R}^\infty \to \mathbb{R}$ is nondecreasing if $f(\{y_1, y_2,\dots\}) \le f(\{z_1, z_2,\dots\})$ whenever $y_i \le z_i$ for all $i \ge 1$, and it is convex if $f(\alpha \mathbf{y} + (1-\alpha)\mathbf{z}) \le \alpha f(\mathbf{y}) + (1-\alpha)f(\mathbf{z})$ for all $\alpha \in [0,1]$ and $\mathbf{y}, \mathbf{z} \in \mathbb{R}^\infty$. A functional $\phi : \mathbb{R}^\infty \to \mathbb{R}^\infty$ is nondecreasing/convex if every component of ϕ is nondecreasing/convex.) For more information on the usual stochastic and increasing convex orders for stochastic processes, see, e.g., Shaked and Shanthikumar, 1994, Sections 4.B.7 and 5.A.4, respectively.

For any vector $(Z_1,\ldots,Z_n)$, where $n \geq 1$, define a subvector $\mathbf{Z}_{k,\ell} = (Z_k,\ldots,Z_\ell)$ for $1 \leq k \leq \ell \leq n$. Furthermore, for all $i \geq 1$, define $\mathbf{D}(i) = (\mathbf{D}_{1,K-1}(i), D_{L_{max}}(i), \mathbf{D}_{M+1,N}(i))$, $\mathbf{D}^{(K,M)}(i) = (\mathbf{D}_{1,K-1}^{(K,M)}(i), D_M^{(K,M)}(i), \mathbf{D}_{M+1,N}^{(K,M)}(i))$, and $\mathbf{X}^{(K,M)}(i) = (\mathbf{X}_{1,K-1}^{(K,M)}(i), X^{(K,M)}(i), \mathbf{X}_{M+1,N}^{(K,M)}(i))$. We are now ready to state the following result, whose proof is given in Argon and Andradóttir [6].

Proposition 1. *For $1 \leq K \leq M \leq N$, if there exist $\alpha_k \in [0,1]$ for $k \in \{K,\ldots,M\}$ such that conditions (ii) and (iii) of Theorem 1 hold and*

$$\{\mathbf{X}^{(K,M)}(i)\}_{i\geq 1} \leq_{st(icx)} \left\{ \mathbf{X}_{1,K-1}(i), \sum_{k=K}^{M} \alpha_k X_k(i), \mathbf{X}_{M+1,N}(i) \right\}_{i\geq 1}, \qquad (9.2)$$

then we have that $\{\mathbf{D}^{(K,M)}(i)\}_{i\geq 1} \leq_{st(icx)} \{\mathbf{D}(i)\}_{i\geq 1}$.

We conclude this section by showing how much improvement in throughput can be gained by pooling several stations in a tandem line. Consider a tandem line with $b_j = \infty$ for $j = 2,\ldots,N$ and independent and identically distributed (i.i.d.) service times at each station. Then, the system throughput is determined by the bottleneck station, i.e., $\tau = \min_{j \in \{1,\ldots,N\}} \{1/E[X_j(1)]\}$ (see, e.g., Muth, 1973). Also, the throughput of the pooled line will be given by $\tau^{(K,M)} = \min\{1/E[X^{(K,M)}(1)], \min_{j \in \{1,\ldots,N\}\setminus\{K,\ldots,M\}} \{1/E[X_j(1)]\}\}$ if the stations that are not being pooled are unaffected by pooling. Let $J \in \{1,\ldots,N\}$ be such that $E[X_J(1)] = \max_{j \in \{1,\ldots,N\}} \{E[X_j(1)]\}$. Then, if Equation (9.1) holds and $E[X^{(K,M)}(1)] \leq E[X_J(1)]$, it is easy to obtain that

$$1 \leq \frac{\tau^{(K,M)}}{\tau} \leq \frac{E[X_J(1)]}{E[X^{(K,M)}(1)]} = \left(\sum_{j=K}^{M} \frac{\mu_{J,j} E[X_j(1)]}{g_j^{(K,M)}(\mu_{K,j},\ldots,\mu_{M,j}) E[X_J(1)]} \right)^{-1}. \qquad (9.3)$$

(Note that the condition that $E[X^{(K,M)}(1)] \leq E[X_J(1)]$ is similar to the condition required for pooling to be beneficial in deterministic lines and is satisfied if conditions (i) and (ii) of Theorem 1 hold.) Hence, at least in the infinite buffer setting, pooling the bottleneck station J with some of its neighboring stations increases the throughput as long as the pooled station is faster than station J. When the workers are identical with additive rates, the right-hand side of inequality (9.3) becomes $(M - K + 1)E[X_J(1)]/\sum_{j=K}^{M} E[X_j(1)]$, which is less than or equal to $M - K + 1$ when $J \in \{K,\ldots,M\}$. Hence, in this special case, pooling the bottleneck station with a group of substantially faster adjacent stations may increase the throughput by a factor of up to $M - K + 1$.

9.2.2 *Pooling with equal total number of storage spaces*

Suppose that we pool stations $K \in \{1,\dots,N-1\}$ through $M \in \{K,\dots,N\}$ and place P buffers before and Q buffers after the pooled station. For a fair comparison, in this section we provide $M-K$ additional buffers to the pooled system to ensure that the total number of storage spaces is equal in the pooled and unpooled systems. The buffers are allocated in such a way that $P \ge b_K$, $Q \ge b_{M+1}$, and $P+Q = \sum_{j=K}^{M+1} b_j + M - K$ (see Section 9.2.1).

Under this assumption, when several stations at the beginning (end) of the line are pooled (i.e., when $K=1$ or $M=N$), then it is natural to place all the buffers between the pooled stations and the additional $M-K$ buffers right after (before) the pooled station, since we assume that the buffer spaces before the first station and after the last station are unlimited. Then, condition (*iii*) of Theorem 1 will be satisfied, so that *pooling several stations at the beginning or end of the line is beneficial* in terms of departure times (and, hence, throughput), as long as conditions (*i*) and (*ii*) of Theorem 1 are satisfied. In fact, Theorem 1 shows that pooling at the beginning or end of a tandem line may be beneficial even when the pooled system has fewer storage spaces than the unpooled system (i.e., when $K=1$ and $L_{min} > 1$ or when $M=N$ and $L_{max} < N$).

However, if neither $K=1$ nor $M=N$, then it is not clear how the $\sum_{j=K+1}^{M} b_j + M - K$ available buffer spaces should be split between the input and output buffers of the pooled station. To illustrate, suppose that the original line has four stations with $b_2 = 0$, $b_3 = 4$, and $b_4 = 0$, and assume that the service times at the successive stations are independent and exponentially distributed with rates 1, 100, 2, and 100 jobs per unit time, respectively. Consider pooling stations 2 and 3 assuming that Equation (9.1) holds, $\mu_{3,2} = \mu_{2,2}$, $\mu_{2,3} = \mu_{3,3}$, the server rates are additive, and $P+Q = 5$. By modeling the original and pooled lines as CTMCs, it is easy to see that pooling increases the system throughput if and only if $P \ge 2$ (see Argon, 2002, for the details). Hence, this simple example shows that the buffer allocation should be done carefully when several intermediate stations are pooled. We now specify sufficient conditions that guarantee a benefit from partial pooling when the total number of storage spaces in the pooled and unpooled lines are equal. More specifically, in the remainder of this section we assume that $P = \sum_{j=K}^{L} b_j + L - K$ and $Q = \sum_{j=L+1}^{M+1} b_j + M - L$, where $L \in \{K,\dots,M\}$. In other words, we assume that the pooled station is placed at the position of station $L \in \{K,\dots,M\}$ in the original (unpooled) line. For the example discussed in this paragraph, $L=2$ and $L=3$ correspond to $P=0$ and $P=5$, respectively.

Theorem 1 implies that if $X^{(K,M)}(i) \le X_L(i)$, $X_j^{(K,M)}(i) \le X_j(i)$ for all $j \in \{1,\dots, K-1, M+1,\dots,N\}$ and $i \ge 1$, then $D_j^{(K,M)}(i) \le D_j(i)$ for all $j \in \{1,\dots, K-1, L, M+1,\dots,N\}$ and $i \ge 1$. Similarly, Proposition 1 yields that if

$$\{\mathbf{X}^{(K,M)}(i)\}_{i\ge1} \le_{st(icx)} \{\mathbf{X}_{1,K-1}(i), X_L(i), \mathbf{X}_{M+1,N}(i)\}_{i\ge1}, \text{ then} \tag{9.4}$$

$$\{\mathbf{D}^{(K,M)}(i)\}_{i\ge1} \le_{st(icx)} \{\mathbf{D}_{1,K-1}(i), D_L(i), \mathbf{D}_{M+1,N}(i)\}_{i\ge1}. \tag{9.5}$$

Hence, given condition (9.4), pooling stations $K,\ldots,M$ and placing the pooled station at the position of station L in the original system is stochastically better than the unpooled system in terms of the departure times from the system. Note, however, that for a given tandem line and stations K and M, there need not exist a station $L \in \{K,\ldots,M\}$ such that condition (9.4) is satisfied (see Argon, 2002, for an example). We next discuss some special cases for which a station L that satisfies condition (9.4) under increasing convex orders $(\leq_{icx})$ can be found.

> *Proposition 2.* (i) *If* $\{X_j\}_{j=1}^n$ *are i.i.d. random variables, then* $\sum_{k=1}^n X_k / n \leq_{icx} X_j$, *for all* $j = 1,\ldots,n$.
> (ii) *Let* $X_1,\ldots,X_n$ *be independent exponential random variables with rates* $\mu_1,\ldots,\mu_n$, *respectively. If all* $\mu_1,\ldots,\mu_n$ *are distinct and* $\{L\} = \mathrm{argmin}_{j=1,\ldots,n}\{\mu_j\}$, *then* $\sum_{k=1}^n X_k / n \leq_{icx} X_L$.
> (iii) *Let* $X_1,\ldots,X_n$ *be independent normal random variables with means* $m_1,\ldots,m_n$ *and variances* $\sigma_1^2,\ldots,\sigma_n^2$, *respectively. Suppose that* $\alpha_1,\ldots,\alpha_n$ *are nonnegative scalars such that* $\sum_{k=1}^n \alpha_k = 1$. *If for some* $L \in \{1,\ldots,n\}$, *it is true that* $\sum_{k=1}^n \alpha_k m_k \leq m_L$ *and* $\sum_{k=1}^n \alpha_k^2 \sigma_k^2 \leq \sigma_L^2$, *then* $\sum_{k=1}^n \alpha_k X_k \leq_{icx} X_L$.

Proposition 2 is proved in Argon and Andradóttir (2006). Note that if $\{Y_i\}_{i\geq 1}$ and $\{Z_i\}_{i\geq 1}$ are sequences of independent random variables, then $Y_i \leq_{icx} Z_i$ for all $i \geq 1$ implies that $\{Y_i\}_{i\geq 1} \leq_{icx} \{Z_i\}_{i\geq 1}$ (see Shaked and Shanthikumar, (1994), Sections 5.A.2 and 5.A.4). Moreover, Equation (9.1) yields $X^{(K,M)}(i) = \sum_{k=K}^M X_k(i) / (M - K + 1)$, for all $i \geq 1$, when the workers are identical with additive rates. Hence, Proposition 2 can be used to identify cases for which condition (9.4) is satisfied.

Note that even if condition (9.4) cannot always be satisfied, it provides insights into which pooling strategies are likely to result in improved system performance. In particular, condition (9.4) suggests that *pooling several stations near a bottleneck station and placing the pooled station at the position of the bottleneck station is likely to be beneficial* as long as the pooled station is faster than the bottleneck station (similar results were obtained earlier for tandem lines with either deterministic service times or infinite buffers). We provide numerical results that are consistent with this insight in Section 9.6.2.

Finally, suppose that we divide the stations in a tandem line into $R \in \{2,\ldots,N\}$ nonoverlapping groups and pool stations in each group (where a group may consist of a single station) to obtain a tandem line with several pooled stations. We say that the rth group consists of stations $M_{r-1}+1,\ldots,M_r$ for $r = 1,\ldots,R$, where $M_0 = 0$, $M_R = N$, and $M_r \in \{M_{r-1}+1,\ldots,N-R+r\}$ for $r = 1,\ldots,R-1$. Then, it is easy to show that if all conditions of Theorem 1 hold for each group of pooled stations, then the departure time of each job from each station $r \in \{1,\ldots,R\}$ in the pooled system will be smaller than the departure time of that job from station M_r in the unpooled system.

Similarly, if conditions (*ii*) and (*iii*) of Theorem 1 hold for each group of stations and

$$\{X^{(M_{r-1}+1,M_r)}(i); r=1,\dots,R\}_{i\geq 1} \leq_{st(icx)} \left\{ \sum_{k=M_{r-1}+1}^{M_r} \alpha_k X_k(i); r=1,\dots,R \right\}_{i\geq 1},$$

then it can be shown that the departure times for the pooled system are stochastically smaller than the departure times for the unpooled system.

9.3 Adding stations into a group of pooled stations

In Section 9.2, we presented conditions under which partial pooling improves the departure times and throughput of a tandem line. We now go one step further and consider whether it is beneficial to add more stations to the group of pooled stations. Our motivation is to gain insights into how many stations should be pooled to maximize the system throughput. Also, in this section, we focus on pooling station $M+1$ with the already pooled stations $K,\dots,M$. Similar results on pooling station $K-1$ with the already pooled stations $K,\dots,M$ also can be obtained (see Argon, 2002).

In the following proposition, we show that pooling station $M+1$ with the already pooled stations $K,\dots,M$ yields earlier departures from the system if the service times at the pooled station are sufficiently small and the buffers around the pooled station are sufficiently large.

Proposition 3. For $1 \leq K \leq M \leq N-1$*, if* $X_j^{(K,M+1)}(i) \leq X_j^{(K,M)}(i)$ *for all* $j \in \{1,\dots,K-1,M+2,\dots,N\}$ *and* $i \geq 1$*, and if there exists* $\alpha \in [0,1]$ *such that:*

$$(i)\ X^{(K,M+1)}(i) \leq \alpha X^{(K,M)}(i) + (1-\alpha)X_{M+1}(i), \quad \forall i \geq 1;$$

$$(ii)\ P^{(K,M+1)} \geq P^{(K,M)} + \begin{cases} 0 & \text{if } \alpha=1, \\ Q^{(K,M)}+1 & \text{otherwise;} \end{cases}$$

$$(iii)\ Q^{(K,M+1)} \geq b_{M+2} + \begin{cases} 0 & \text{if } \alpha=0, \\ Q^{(K,M)}+1 & \text{otherwise,} \end{cases}$$

then we have that $D_j^{(K,M+1)}(i) \leq D_j^{(K,M)}(i)$ *for all* $j \in \{1,\dots,K-1,M+1,\dots,N\}$ *and* $i \geq 1$*.*

Proposition 3, which follows directly from Theorem 1, has several interesting implications for comparing two tandem lines with equal number of storage spaces. First, if condition (*i*) of Proposition 3 is satisfied with $\alpha \in \{0,1\}$, then conditions (*ii*) and (*iii*) will be satisfied automatically by placing the pooled station (obtained by pooling stations $K,\dots,M+1$) at the position of either station $M+1$ (so that $P^{(K,M+1)} = P^{(K,M)}+Q^{(K,M)}$ and $Q^{(K,M+1)} = b_{M+2}$) or the already pooled station (obtained by pooling stations $K,\ldots,M$, so that $P^{(K,M+1)} = P^{(K,M)}$ and $Q^{(K,M+1)} = Q^{(K,M)} + b_{M+2}$), respectively. Also, if $K = 1$, then conditions *(ii)* and *(iii)* will be satisfied, assuming that the station obtained by pooling stations $K,\ldots,M+1$ is placed at the beginning of the line. Finally, if $M = N - 1$, then conditions *(ii)* and *(iii)* will be satisfied, assuming that the group of pooled stations $K,\ldots,M+1$ is placed at the end of the line. However, if $\alpha \notin \{0,1\}$, $K \neq 1$, $M \neq N - 1$, and $Q^{(K,M)} < \infty$, then satisfying conditions *(ii)* and *(iii)* will require increasing the total number of storage spaces in the line.

Now consider the case where Equation (9.1) holds with $\mu_{\ell,j} = \theta_\ell \eta_j$ and $g_j^{(K,J)}(\mu_{K,j},\ldots,\mu_{J,j}) = \eta_j g_j^{(K,J)}(\theta_K,\ldots,\theta_J)$ for $\ell, j \in \{K,\ldots,J\}$ and $J \in \{M, M+1\}$. Also, suppose that $\theta_{M+1} + g^{(K,M)}(\theta_K,\ldots,\theta_M) \leq g^{(K,M+1)}(\theta_K,\ldots,\theta_{M+1})$, so that the additional worker contributes to the pooled rate at least in the amount of its rate when it works alone (this condition is satisfied by additive rates but not necessarily by superadditive rates). Then, it is clear that condition *(i)* of Proposition 3 is satisfied. Consequently, Proposition 3 implies that *pooling as many stations as possible at the beginning of a line is beneficial (implying that complete pooling is best)* if Equation (9.1) holds with additive server rates of the form $\mu_{\ell,j} = \theta_\ell \eta_j$ for $\ell, j \in \{K,\ldots,M+1\}$ (e.g., if the workers are identical) and the total number of storage spaces in the pooled and unpooled lines are equal. Similarly, using Proposition 2.6 in Argon (2002), we conclude that *pooling as many stations as possible at the end of a tandem line is beneficial* under reasonable conditions on the service times and buffer spaces.

Note that Proposition 3 can be generalized to stochastic orders using an argument similar to the one used to prove Proposition 1. A special case of such results follows:

Proposition 4.   *(i) For $1 \leq K \leq M \leq N - 1$, if the pooled station is placed at the position of the already pooled station (so that $P^{(K,M+1)} = P^{(K,M)}$ and $Q^{(K,M+1)} = Q^{(K,M)} + b_{M+2}$) and*

$$\{\mathbf{X}^{(K,M+1)}(i)\}_{i\geq 1} \leq_{st(icx)} \left\{ \mathbf{X}_{1,K-1}^{(K,M)}(i), X^{(K,M)}(i), \mathbf{X}_{M+2,N}^{(K,M)}(i) \right\}_{i\geq 1},$$

then we have that $\{\mathbf{D}^{(K,M+1)}(i)\}_{i\geq 1} \leq_{st(icx)} \left\{ \mathbf{D}_{1,K-1}^{(K,M)}(i), D_M^{(K,M)}(i), \mathbf{D}_{M+2,N}^{(K,M)}(i) \right\}_{i\geq 1}.$

(ii) For $1 \leq K \leq M \leq N - 1$, if the pooled station is placed at the position of station $M+1$ of the existing system (so that $P^{(K,M+1)} = P^{(K,M)} + Q^{(K,M)}$ and $Q^{(K,M+1)} = b_{M+2}$) and

$$\{\mathbf{X}^{(K,M+1)}(i)\}_{i\geq 1} \leq_{st(icx)} \left\{ \mathbf{X}_{1,K-1}^{(K,M)}(i), \mathbf{X}_{M+1,N}^{(K,M)}(i) \right\}_{i\geq 1},$$

then we have that $\{\mathbf{D}^{(K,M+1)}(i)\}_{i\geq 1} \leq_{st(icx)} \left\{ \mathbf{D}_{1,K-1}^{(K,M)}(i), \mathbf{D}_{M+1,N}^{(K,M)}(i) \right\}_{i\geq 1}.$

Proposition 4 in the current chapter and Proposition 2.8 in Argon (2002) (which is concerned with adding station $K - 1$ to the group of pooled stations

$K,\ldots,M$) suggest that *adding a station to an already pooled group of stations will generally be beneficial* as long as the pooled station is positioned at the right place (i.e., at the position of the slower station it replaces). Our final result in this section, which is proved in Argon and Andradóttir (2006), compares two pooled systems in which either stations $K,\ldots,M+1$ or stations $K-1,\ldots,M$ are pooled.

Proposition 5. For $2 \le K \le M \le N-1$, if in the pooled system where stations $K,\ldots,M+1$ are pooled, the pooled station is placed at the position of station $M+1$; in the pooled system where stations $K-1,\ldots,M$ are pooled, the pooled station is placed at the position of station $K-1$; and $\{X^{(K,M+1)}(i)\}_{i\ge1} \le_{st(icx)} (\ge_{st(icx)})$ $\{X^{(K-1,M)}(i)\}_{i\ge1}$, we have that $\{D^{(K,M+1)}(i)\}_{i\ge1} \le_{st(icx)}$ $(\ge_{st(icx)})\{D^{(K-1,M)}(i)\}_{i\ge1}$.

One implication of Proposition 5 is that if station $J \in \{2,\ldots,N-1\}$ is slower than one of its neighboring stations F and faster than its other neighboring station S, where $\{S,F\} \in \{J-1,J+1\}$, then pooling station J with S and placing the pooled station at the position of S is better than pooling it with F and placing the pooled station at the position of F. This implication agrees with our numerical results in Section 9.6.

9.4 Other performance measures

In this section, we observe the effects of pooling on other performance measures (besides departure times and throughput), namely the WIP, sojourn times, and holding costs. When two tandem lines are compared according to these criteria, then the total number of jobs that enter the two systems should be the same for a fair comparison. Hence, in this section, we assume that there is an arrival stream at the first station (recall that the size b_1 of the buffer in front of the first station is assumed to be infinite). We start by showing that our earlier results hold under this assumption. For all $1 \le K \le M \le N$, let $X_0^{(K,M)}(i)$ and $X_0(i)$ be the interarrival times between jobs $i-1$ and i for $i \ge 1$ at the pooled and unpooled systems, respectively. Proposition 6 is proved in Argon and Andradóttir (2006).

Proposition 6. Suppose that there is an arrival stream to station 1, which is independent of the service time processes for both the original and pooled systems. Then:

(i) Theorem 1 and Proposition 3 are still true, if for all $i \ge 1$, $X_0^{(K,M)}(i) \le X_0(i)$ and $X_0^{(K,M+1)}(i) \le X_0^{(K,M)}(i)$, respectively.

(ii) Propositions 1, 4, and 5 are still true, if $\{X_0^{(K,M)}(i)\}_{i\ge1} \le_{st(icx)} \{X_0(i)\}_{i\ge1}$, $\{X_0^{(K,M+1)}(i)\}_{i\ge1} \le_{st(icx)} \{X_0^{(K,M)}(i)\}_{i\ge1}$, and $\{X_0^{(K,M+1)}(i)\}_{i\ge1} \le_{st(icx)} (\ge_{st(icx)}) \{X_0^{(K-1,M)}(i)\}_{i\ge1}$, respectively.

Next, suppose that we are comparing two tandem lines of N stations, namely systems 1 and 2, and that the arrival processes to the two systems are exactly the same. Let $D_j^{(m)}(i)$ be the departure time of job i from station

j in system m, $W_j^{(m)}(i)$ be the total amount of time job i spends at stations $1,\ldots,j$ in system m, and $Q_j^{(m)}(t)$ the total number of jobs at stations $1,\ldots,j$ in system m at time t for $m=1,2$, $j=1,\ldots,N$, $i\geq1$, and $t\geq0$. It is easy to show that if $D_j^{(1)}(i)\leq D_j^{(2)}(i)$ for all $j=1,\ldots,N$ and $i\geq1$, then $W_i^{(1)}(i)\leq W_i^{(2)}(i)$ and $Q_j^{(1)}(t)\leq Q_j^{(2)}(t)$ for $j=1,\ldots,N$, $i\geq1$, and $t\geq0$. Similarly, if $\{\mathbf{D}_{1,N}^{(1)}(i)\}_{i\geq1}\leq_{st(icx)}$ $\{\mathbf{D}_{1,N}^{(2)}(i)\}_{i\geq1}$, then we have $\{\mathbf{W}_{1,N}^{(1)}(i)\}_{i\geq1}\leq_{st(icx)}\{\mathbf{W}_{1,N}^{(2)}(i)\}_{i\geq1}$ (see Tembe and Wolff [31] for a similar result on $W_N^{(m)}(i)$) and $\{\mathbf{D}_{1,N}^{(1)}(i)\}_{i\geq1}\leq_{st}\{\mathbf{D}_{1,N}^{(2)}(i)\}_{i\geq1}$ implies that $\{\mathbf{Q}_{1,N}^{(1)}(t)\}_{t\geq0}\leq_{st}\{\mathbf{Q}_{1,N}^{(2)}(t)\}_{t\geq0}$. These results, which are proved in Argon and Andradóttir (2006), can be used to compare pooled and unpooled systems when the arrival processes to both systems are exactly the same. Hence, we conclude that whenever pooling (stochastically) decreases the departure times from the system for each job, then it also (stochastically) decreases the sojourn time of each job in the system and the total number of jobs in the system at any given time if there is a stochastic arrival stream at the first station and the arrival process is independent of the service time processes.

In the remainder of this section, we observe the effects of pooling on the long-run average holding cost. Let $h_j\geq0$ be the holding cost per unit time of a job at station j and its input buffer for $j=1,\ldots,N$. We assume that the holding cost per unit time of a job at the pooled station and its input buffer is equal to h_K (see Remark 2 at the end of this section for a discussion on how this assumption can be relaxed). This is consistent with the way holding costs are charged at all other stations in the pooled and unpooled lines, and is reasonable if the holding costs represent the investment made in each job at the stations where the job has been processed so far. It is generally assumed that as jobs move along the production line, more value is added, yielding $h_1\leq h_2\leq\cdots\leq h_N$. However, there are several studies in the literature that consider the case where $h_1\geq h_2\geq\cdots\geq h_N$, see, e.g., Veatch and Wein (1994) and Ahn et al. (1999). (Veatch and Wein note that letting $h_1\geq h_2\geq\cdots\geq h_N$ intentionally may be used as a tool to model the benefits of just-in-time manufacturing where the goal is to minimize WIP.)

One might expect that pooling would decrease the average holding cost if it eliminates the largest holding costs in the system. However, this is not true in general, even if pooling yields earlier departures from the system (see Argon, 2002, for an example). This effect of pooling arises from the fact that by pooling several stations in the line, jobs may be pushed more quickly to stations where it is expensive to keep jobs. We next provide conditions under which pooling stations $K\in\{1,\ldots,N-1\}$ through $M\in\{K,\ldots,N\}$ decreases the holding cost of each job in the system incurred before any given time. Throughout this section, we assume that the arrival process is not affected by pooling, i.e., $D_0^{(K,M)}(i)=D_0(i)$ for all $i\geq1$, where $D_0^{(K,M)}(i)$ and $D_0(i)$ denote the arrival time of job i to the pooled and unpooled lines, respectively. Let $H(i,t)$ and $H^{(K,M)}(i,t)$ denote the total holding cost of job $i\geq1$ in the original and pooled systems, respectively, incurred by time $t\geq0$. Let also $h_\ell'=h_\ell-h_{\ell+1}$ for $\ell=1,\ldots,N$, where $h_{N+1}=0$. Hence, $h_j=\sum_{\ell=j}^{N}h_\ell'$ for all $j=1,\ldots,N$. The following result is proved in Argon and Andradóttir (2006).

Proposition 7. For $1 \leq K \leq M \leq N$, we have $H^{(K,M)}(i,t) \leq H(i,t)$ for all $i \geq 1$ and $t \geq 0$, if and only if

$$\sum_{\ell \in \{1,\ldots,K-1,M+1,\ldots,N\}} h'_\ell \left(\min\{t, D_\ell(i)\} - \min\left\{t, D_\ell^{(K,M)}(i)\right\} \right)$$

$$+ \sum_{\ell=K}^{M} h'_\ell \left(\min\{t, D_\ell(i)\} - \min\{t, D_M^{(K,M)}(i)\} \right) \geq 0, \quad \forall \, i \geq 1 \quad \text{and} \quad t \geq 0. \tag{9.6}$$

The first sum in inequality (9.6) of Proposition 7 is nonnegative if

$$h'_\ell \left(D_\ell(i) - D_\ell^{(K,M)}(i) \right) \geq 0, \forall \, \ell \in \{1,\ldots,K-1,M+1,\ldots,N\} \quad \text{and} \quad i \geq 1. \tag{9.7}$$

Inequality (9.7) is satisfied when either h'_ℓ or $D_\ell(i) - D_\ell^{(K,M)}(i)$ is equal to zero or when both have the same sign. More specifically, if $h_\ell \geq (\leq)h_{\ell+1}$, then inequality (9.7) is satisfied as long as $D_\ell(i) \geq (\leq) D_\ell^{(K,M)}(i)$ for all $\ell \in \{1,\ldots,K-1, M+1,\ldots,N-1\}$ and $i \geq 1$. This leads to an intuitive conclusion that *pooling decreases holding costs if the jobs are pushed faster (more slowly) toward downstream stations where it is cheaper (more expensive) to hold jobs.* Since h'_N is always nonnegative, inequality (9.7) is satisfied for $\ell = N$ only when $D_N(i) \geq D_N^{(K,M)}(i)$ for all $i \geq 1$. Finally, if $b_J = \infty$, where $J \in \{1,\ldots,K\}$, then inequality (9.7) would be automatically satisfied for stations $1,\ldots,J-1$ since in that case $D_\ell(i) = D_\ell^{(K,M)}(i)$ for all $\ell = 1,\ldots,J-1$ and $i \geq 1$.

We next study the second sum in inequality (9.6) of Proposition 7. One case under which this sum is nonnegative is when $h'_\ell(D_\ell(i) - D_M^{(K,M)}(i)) \geq 0$ for all $\ell = K,\ldots,M$ and $i \geq 1$. For example, this condition holds if $h_K \geq h_{K+1} \geq \cdots \geq h_{M+1}$ and the assumptions of Theorem 1 hold with $\alpha_K = 1$. Note also that the second sum in inequality (9.6) is nonnegative if and only if the following inequality holds for all $i \geq 1$ and $t \geq 0$:

$$(h_K - h_{M+1})\left(\min\{t, D_M(i)\} - \min\{t, D_M^{(K,M)}(i)\} \right) \geq \sum_{l=K}^{M} h'_l(\min\{t, D_M(i)\} - \min\{t, D_l(i)\})$$

$$= \sum_{\ell=K}^{M-1} h'_\ell \sum_{j=\ell+1}^{M} S_j(i,t) = \sum_{j=K+1}^{M} \sum_{\ell=K}^{j-1} h'_\ell S_j(i,t) = \sum_{j=K+1}^{M} (h_K - h_j) S_j(i,t). \tag{9.8}$$

The right-hand side of condition (9.8) is equal to $\hat{H}(i,t) - H(i,t)$, where $\hat{H}(i,t)$ is the holding cost of job $i \geq 1$ by time $t \geq 0$ in the original line if $h_j = h_K$ for $j \in \{K+1,\ldots,M\}$. Therefore, if $\hat{H}(i,t) \leq H(i,t)$ for all $i \geq 1$ and $t \geq 0$ (e.g., if $h_K \leq h_j$ for $j = K+1,\ldots,M$), then condition (9.8) will hold as long as $(h_K - h_{M+1})$ $(D_M(i) - D_M^{(K,M)}(i)) \geq 0$ for all $i \geq 1$, which is similar to condition (9.7). In other words, if replacing $h_{K+1},\ldots,h_M$ by h_K is beneficial (without any changes in the departure times), then the only concern at the pooled stations is whether pooling leads to a desirable change in the departure times from the pooled station.

If, on the other hand, $\hat{H}(i,t) \geq H(i,t)$ for some $i \geq 1$ and $t \geq 0$, then a possible benefit from the changes in the departure times should offset the cost associated with replacing $h_{K+1}, \ldots, h_M$ by h_K due to pooling.

Our final result in this section (which is proven in Argon and Andradóttir, 2006) provides conditions under which pooling decreases the total holding cost of each job in the system. Let $H(i)$ and $H^{(K,M)}(i)$ be the total holding cost of job $i \geq 1$ in the original and pooled systems, respectively.

Proposition 8. *For $1 \leq K \leq M \leq N$, we have $H^{(K,M)}(i) \leq H(i)$ for all $i \geq 1$, if*

(i) There exist $\alpha_j \in [0,1]$ for $j \in \{K, \ldots, M\}$ such that $\sum_{j=K}^{M} \alpha_j = 1$ and

$$\sum_{\ell \in \{1, \ldots, K-1, M+1, \ldots, N\}} h'_\ell \left(D_\ell(i) - D_\ell^{(K,M)}(i) \right) + (h_K - h_{M+1})$$

$$\times \sum_{\ell=K}^{M} \alpha_\ell \left(D_\ell(i) - D_M^{(K,M)}(i) \right) \geq 0, \quad \forall i \geq 1;$$

(9.9)

(ii) and, if $K < M$, then $h_j \geq h_{M+1} \sum_{\ell=K}^{j-1} \alpha_\ell + h_K \sum_{\ell=j}^{M} \alpha_\ell$ for all $j = K+1, \ldots, M$.

Note that if the conditions of Theorem 1 hold, so that $D_\ell(i) \geq D_\ell^{(K,M)}(i)$ for $\ell \in \{1, \ldots, K-1, M+1, \ldots, N\}$ and $\sum_{\ell=K}^{M} \alpha_\ell D_\ell(i) \geq D_M^{(K,M)}(i)$ for all $i \geq 1$, then condition (i) of Proposition 8 will hold under the assumption that $h_1 \geq h_2 \geq \cdots \geq h_K \geq h_{M+1} \geq \cdots \geq h_N$. Hence, if Theorem 1 holds, then Proposition 8 essentially replaces condition (9.6) of Proposition 7 with an assumption about $h_{K+1}, \ldots, h_M$ (i.e., condition (ii) of Proposition 8) and an assumption about $h_1, \ldots, h_K, h_{M+1}, \ldots, h_N$. However, this is achieved at the expense of having a weaker result in terms of $H(i)$ instead of $H(i,t)$ for all $i \geq 1$ and $t \geq 0$.

Condition (ii) of Proposition 8 requires h_j be larger than a certain weighted average of h_K and h_{M+1} for $j = K+1, \ldots, M$, where the weight on h_K (h_{M+1}) decreases (increases) as j grows. We next consider two special cases under which this condition is satisfied:

- If $h_j \geq \max\{h_K, h_{M+1}\}$ for all $j = K+1, \ldots, M$, then condition (ii) of Proposition 8 is satisfied. In that case, pooling replaces several larger holding costs $h_{K+1}, \ldots, h_M$ by a smaller holding cost h_K.
- If $\alpha_L = 1$ for some $L \in \{K, \ldots, M\}$, then condition (ii) of Proposition 8 is equivalent to having $h_j \geq h_K$ for $j \in \{K+1, \ldots, L\}$ and $h_j \geq h_{M+1}$ for $j \in \{L+1, \ldots, M\}$. Recall that when $\alpha_L = 1$, then condition (iii) of Theorem 1 holds if the pooled station is placed at the position of station L in the original system. Hence, condition (ii) of Proposition 8 guarantees that the unit holding costs at stations $K+1, \ldots, M$ are lowered by pooling (from h_j, where $j = K+1, \ldots, M$, to either h_K or h_{M+1} depending on how j compares with L).

We conclude this section by discussing additional implications of Proposition 7. First, note that Proposition 7 implies that pooling at the end of the line decreases the total holding cost of each job in the system if (1) pooling improves the departure times from the system, (2) the holding costs at the stations being pooled are no smaller than the holding cost at the pooled station (i.e., $h_K \leq h_j$ for $j = K+1,\ldots,N$), and (3) there is an infinite buffer in front of the first station that is pooled (note that $D_\ell(i) = D_\ell^{(K,N)}(i)$ for all $\ell = 1,\ldots,K-1$ and $i \geq 1$ when $b_K = \infty$). This suggests that *complete pooling always decreases the total holding cost as long as it decreases the departure times from the system and the first station is the cheapest place to store jobs.*

Finally, when $K = M$, Proposition 7 leads to a result that compares two tandem lines that have N stations, a common arrival stream, and are not necessarily related by pooling. For $m \in \{1,2\}$, $j \in \{1,\ldots,N\}$, and $i \geq 1$, recall that $D_j^{(m)}(i)$ is the departure time of job i from station j in system m, and let $H^{(m)}(i,t)$ denote the holding cost of job i in system m by time $t \geq 0$. Proposition 7 implies that if $\sum_{\ell=1}^{N} h_\ell'(\min\{t, D_\ell^{(2)}(i)\} - \min\{t, D_\ell^{(1)}(i)\}) \geq 0$ for all $i \geq 1$ and $t \geq 0$, then $H^{(1)}(i,t) \leq H^{(2)}(i,t)$ for all $i \geq 1$ and $t \geq 0$. Consequently, if $h_1 \geq h_2 \geq \cdots \geq h_N$, then speeding up the departures from each station (e.g., by adding buffers or reducing service times) improves the total holding cost. On the other hand, if $h_1 \leq h_2 \leq \cdots \leq h_N$, then increasing the departure times from each station (except for the last station) is beneficial in terms of total holding cost. In other words, *"pushing" jobs toward the end of the line is beneficial when holding costs decrease across stations, and "pulling" jobs from the beginning of the line is preferable when holding costs increase across stations.* These conclusions are intuitively reasonable and are consistent with observations made by Veatch and Wein (1994).

Remark 1 — The sample-path constructions of Propositions 7 and 8 can be extended to usual stochastic and increasing convex orders when there is a common arrival process to both systems and this arrival process is independent of the service time processes.

Remark 2 — Although the above results on holding costs are obtained under the natural assumption that the holding cost per unit time of a job at the pooled station and its input buffer is equal to h_K, these results can be easily generalized to the case where the holding cost $h^{(K,M)}$ at the pooled station and its input buffer is arbitrary (and not necessarily equal to h_K). In particular, for this general case, Propositions 7 and 8 remain intact except that inequality (9.6) is replaced with

$$\sum_{\ell \in \{1,\ldots,K-1,M+1,\ldots,N\}} h_\ell' \left(\min\{t, D_\ell(i)\} - \min\left\{t, D_\ell^{(K,M)}(i)\right\} \right) + \sum_{\ell=K}^{M} h_\ell' \min\{t, D_\ell(i)\}$$

$$- (h^{(K,M)} - h_{M+1}) \min\left\{t, D_M^{(K,M)}(i)\right\} + (h^{(K,M)} - h_K) \min\{t, D_0(i)\} \geq 0,$$

$$\forall i \geq 1 \quad \text{and} \quad t \geq 0;$$

and inequality (9.9) is replaced with

$$\sum_{\ell \in \{1,\ldots,K-1,M+1,\ldots,N\}} h_\ell' \left(D_\ell(i) - D_\ell^{(K,M)}(i) \right) + (h_K - h_{M+1}) \sum_{\ell=K}^{M} \alpha_\ell D_\ell(i)$$

$$- (h^{(K,M)} - h_{M+1}) D_M^{(K,M)}(i) + (h^{(K,M)} - h_K) D_0(i) \geq 0, \quad \forall i \geq 1.$$

9.5 Extensions

In this section, we consider three extensions of the original (tandem) model that we introduced in Section 9.1. In particular, we consider pooling in closed tandem lines in Section 9.5.1, pooling under other blocking mechanisms (besides the production blocking) in Section 9.5.2, and pooling in more general queuing networks than tandem lines in Section 9.5.3.

9.5.1 Closed tandem networks

Consider a closed tandem line of $N \geq 2$ stations in which the jobs that leave station N join the queue of station 1. Assume that no external arrivals occur and that there are $I \geq 1$ jobs in the system at any given time. Such a tandem network can be used for modeling production lines that are operated under the CONWIP (Constant WIP) work release policy (see Spearman et al, 1990). In production lines operated under CONWIP, the release of new jobs is not allowed into the line if the total WIP in the line is at or above a certain limit (this limit is I in our model).

To avoid deadlocks in the pooled and unpooled closed networks (i.e., to ensure that there will not be a case where all stations in the system are blocked), we assume that $I < \min\{\sum_{j=1}^{N} b_j + N, \sum_{j=1}^{K-1} b_j + P + Q + \sum_{j=M+2}^{N} b_j + N - M + K\}$. Then we can show that Theorem 1 and Propositions 1, 3, 4, and 5 also hold for closed tandem lines. For example, results corresponding to Theorem 1 and Proposition 1 for closed tandem lines are given in the following proposition, which is proved in Argon and Andradóttir (2006).

Proposition 9. For a closed tandem line and $1 \leq K \leq M \leq N$, Theorem 1 and Proposition 1 still hold. Moreover, condition (iii) of Theorem 1 and Proposition 1 can be replaced by $P + Q \geq I$ if $K = 1$ and $M = N$ and $P \geq \min\{I - 1, \sum_{k=K}^{L_{max}} b_k + L_{max} - K\}$ and $Q \geq \min\{I - 1, \sum_{k=L_{min}+1}^{M+1} b_k + M - L_{min}\}$ otherwise.

The number of buffer spaces in an open network are often restricted to control the WIP. This is especially important when it is assumed that there is an infinite supply of jobs in front of the first station. However, in closed networks, the WIP level is constant and, hence, the addition of buffers is more justifiable unless there are physical restrictions. With this reasoning, Proposition 9 implies that *pooling in closed tandem lines will always be beneficial when the service times in the pooled system are sufficiently small* (see conditions (i) and (ii) of Theorem 1 and condition (9.2)).

Finally, note that Proposition 9 implies that the pooled system in which the pooled station is placed at the position of station $L \in \{K, \ldots, M\}$ in the unpooled system is superior to the unpooled system as long as condition (9.4) holds and the buffers around the pooled station are sufficiently large. Once this result holds, then it is easy to show that pooling under other buffer allocation structures (such as placing the pooled station at the position of a buffer) is also beneficial in a closed tandem line, using Corollaries 6.4 through 6.7 in Glasserman and Yao (1994). For example, by Corollary 6.4 in Glasserman and Yao, if $I > \sum_{j=1}^{K-1} b_j + Q + \sum_{j=M+2}^{N} b_j + N - M + K$, then decreasing P by one unit and allocating this buffer space to any other station improves the system throughput.

9.5.2   *Other blocking mechanisms*

In this section, we consider the general blocking mechanism of Cheng and Yao (1993), which generalizes and unifies some well-known blocking mechanisms, such as production, communication, and kanban blocking mechanisms. This general blocking mechanism is characterized by a set of three parameters at each station, namely upper limits on the number of raw jobs, the number of finished jobs, and the total number of jobs (raw or finished). Suppose that in the unpooled system, a_j, c_j, and k_j denote these upper limits on the number of raw jobs, finished jobs, and total number of jobs at station j, respectively, where $1 \leq a_j \leq k_j$, $0 \leq c_j \leq k_j$, and $a_j + c_j \geq k_j$ for $j = 1,\ldots,N$. Suppose also that when stations K through M are pooled, a_j, c_j, and k_j for stations $j \in \{1,\ldots,K-1, M+1,\ldots,N\}$ are not decreased, and that $a^{(K,M)}$, $c^{(K,M)}$, and $k^{(K,M)}$ are the corresponding parameters at the pooled station. Let $T_j(i)$ be the service completion time of job i at station $j \in \{1,\ldots,N\}$ in the unpooled line and $T_j^{(K,M)}(i)$ be the service completion time of job i at station $j \in \{1,\ldots,K-1,M,\ldots,N\}$ in the pooled line, for $i \geq 1$. The following result is proved in Argon and Andradóttir (2006).

> *Proposition 10.   If all conditions of Theorem 1, except for condition (iii), hold, and if*

$$a^{(K,M)} \geq \sum_{j=K}^{L_{max}-1} k_j + a_{L_{max}}, \quad c^{(K,M)} \geq \sum_{j=L_{min}+1}^{M} k_j + c_{L_{min}}, \quad \text{and} \quad k^{(K,M)} \geq \sum_{j=K}^{M} k_j, \quad (9.10)$$

where L_{min} and L_{max} are defined in Theorem 1, then we have that $T_j^{(K,M)}(i) \leq T_j(i)$ and $T_M^{(K,M)}(i) \leq T_{L_{max}}(i)$ for all $j \in \{1,\ldots,K-1,M,\ldots,N\}$ and $i \geq 1$.

Note that it is easy to obtain an extension of Proposition 10 to usual stochastic and increasing convex orders similar to Proposition 1. We now consider Proposition 10 under some well-known blocking mechanisms, assuming that a_j, c_j, and k_j for stations $j \in \{1,\ldots,K-1, M+1,\ldots,N\}$ are kept the same in the pooled line:

1. *Kanban blocking* is a blocking mechanism that is motivated by the kanban control mechanism in "just-in-time" production systems (see, e.g., Tayur, 1993). Under this blocking scheme, each station has an upper bound on the number of (raw or finished) jobs that it can hold. The jobs in the station can be all raw jobs, all finished jobs, or a combination of raw and finished jobs. Therefore, the number of blocked jobs at a station can be more than one under the kanban blocking mechanism as opposed to, at most, one blocked job under the production blocking mechanism.

When the blocking mechanism is of kanban type in the original line, then $a_j = c_j = k_j$ for $j = 1, \ldots, N$, and condition (9.10) becomes $a^{(K,M)} \geq \sum_{j=K}^{L_{max}} k_j$, $c^{(K,M)} \geq \sum_{j=L_{min}}^{M} k_j$, and $k^{(K,M)} \geq \sum_{j=K}^{M} k_j$. This condition is satisfied when $a^{(K,M)} = c^{(K,M)} = k^{(K,M)} = \sum_{j=K}^{M} k_j$, corresponding to kanban blocking in the pooled line. This means that when all conditions of Theorem 1 except for condition (*iii*) hold for a line operated under the kanban blocking mechanism, then it is sufficient to keep the total number of storage spaces equal in the pooled and unpooled systems for pooling to be beneficial.

2. *Communication blocking* is a blocking scheme that is commonly observed in communication systems, such as computer networks. Under this blocking mechanism, a server cannot start processing a job at a station unless there is space available in the output buffer. In other words, if the blocking mechanism is a communication type in the original line, then we have $c_j = 0$ and $a_j = k_j = b_j + 1$ for all $j = 1, \ldots, N$. In that case, condition (9.10) becomes

$$a^{(K,M)} \geq \sum_{j=K}^{L_{max}} b_j + L_{max} - K + 1, c^{(K,M)} \geq \sum_{j=L_{min}+1}^{M} b_j + M - L_{min},$$

(9.11)

$$\text{and} \quad k^{(K,M)} \geq \sum_{j=K}^{M} b_j + M - K + 1.$$

If the conditions on $a^{(K,M)}$ and $c^{(K,M)}$ are satisfied in condition (9.11), then $k^{(K,M)} = a^{(K,M)} + c^{(K,M)}$ satisfies the condition on $k^{(K,M)}$. This corresponds to dividing the total number of storage spaces at the pooled station into two groups: $a^{(K,M)}$ buffers (including the space for the server) that can be used only by raw jobs and $c^{(K,M)}$ buffers that can be used only by finished jobs. This is equivalent to having the pooled station operate under communication blocking with $P = a^{(K,M)} - 1$ buffers in front of the pooled station and $Q = c^{(K,M)} + a_{M+1} - 1$ buffers after the pooled station. Since condition (9.11) holds when P and Q satisfy condition (*iii*) of Theorem 1, Theorem 1 also holds under the communication blocking mechanism.

3. When the blocking mechanism is of *production* type in the original line, then $a_j = k_j = b_j + 1$ and $c_j = 1$ for all $j = 1, \ldots, N$, so that condition (9.10) can be rewritten as condition (9.11) except that $c^{(K,M)} \geq \sum_{j=L_{min}+1}^{M} b_j + M - L_{min} + 1$. By letting $k^{(K,M)} = a^{(K,M)} + c^{(K,M)} - 1$, the pooled station operates under the production blocking mechanism with $P = a^{(K,M)} - 1$ and $Q = c^{(K,M)} + a_{M+1} - 2$. Condition (*iii*) of Theorem 1 now implies that condition (9.10) holds.

9.5.3 More general queuing networks

Pooling several stations in queuing networks other than (open or closed) tandem lines may be beneficial in terms of system throughput. For various queuing networks with probabilistic routing, the service completion times of jobs from each station can be shown to be stochastically nondecreasing in the service times (see, e.g., Glasserman and Yao, 1992, Proposition 5.5). For such networks, pooling several adjacent stations in series will decrease the service completion times stochastically if the pooled service time is sufficiently small and the buffer allocation around the pooled station is done properly. More specifically, for such networks it is possible to obtain a result similar to inequality (9.5) for usual stochastic orders if a condition on the service times similar to condition (9.4) holds for usual stochastic orders. For example, we can obtain such a result for pooling a group of stations in series in a branch of a tree-like network, where for each station ℓ, all jobs arriving at station ℓ come from the same station j_ℓ and the input buffer of each station that has an external arrival stream is infinitely large. Note that Theorem 1 and Proposition 1 are not special cases of this result because the condition on the service times in Theorem 1 and Proposition 1 is weaker than the condition on the service times in this new result (compare conditions (9.2) and (9.4)).

9.6   Numerical results

In order to quantify the improvement obtained by pooling and gain better insights into how partial pooling should be performed in tandem lines, we have conducted a number of numerical experiments. In this section, we assume that Equation (9.1) holds with identical workers. We also assume that the service times at each station $j \in \{1,\ldots,N\}$ in the unpooled system (i.e., $X_j(i)$, where $i \geq 1$) are independent and exponentially distributed with rate $\eta_j \geq 0$. The steady-state throughput (THP) and WIP of each system under consideration are then obtained by solving the balance equations of the resulting CTMC.

   In Sections 9.6.1 and 9.6.2, we study balanced and unbalanced tandem lines, respectively, under the assumptions that the total number of storage spaces in the pooled and unpooled systems are equal and that the server rates are additive. Note that most of the studies in the literature on cross-trained and cooperative workers make the latter assumption (see, e.g., Mandelbaum and Reiman, 1998, and Van Oyen et al., 2001). In Section 9.6.3, we look at the effects of pooling when the server rates are additive, but additional buffers are not provided to equalize the total number of storage spaces in the pooled and unpooled systems. Finally, we discuss the effects of pooling when the server rates are not additive in Section 9.6.4.

9.6.1   Pooling in balanced lines

In this section, we observe the effects of pooling on tandem lines with $N \in \{3,4,5\}$ stations, in which the workload allocation is balanced among all stations. We assume that $\eta_j = 1.0$ for $j = 1,\ldots,N$. In Table 9.1 and Table 9.2, we present the steady-state throughput and WIP for different

Table 9.1 Throughput and WIP of Balanced Lines with $N \in \{3,4,5\}$ and $b_j = 0$, for $j = 2,\ldots,N$

System	THP	% Inc. in THP	WIP	% Inc. in WIP	% Dec. in WIP
		$N = 3$			
1-2-3	0.5641	–	2.3590	–	–
(12)-b3	0.7808	38.42	2.2740	–	3.60
1-b(23)	0.7808	38.42	2.2877	–	3.02
(123)	1.0000	77.27	1.0000	–	57.61
		$N = 4$			
1-2-3-4	0.5148	–	3.0646	–	–
(12)-b3-4	0.6329	22.94	3.2966	7.57	–
1-(23)-b4	0.6376	23.85	2.7009	–	11.87
1-b(23)-4	0.6376	23.85	3.2374	5.64	–
1-2-b(34)	0.6329	22.94	2.6534	–	13.42
(123)-bb4	0.8421	63.58	2.8226	–	7.90
(12)-bb(34)	0.8682	68.65	2.8682	–	6.41
1-bb(234)	0.8421	63.58	2.8616	–	6.62
(1234)	1.0000	94.25	1.0000	–	67.37
		$N = 5$			
1-2-3-4-5	0.4858	–	3.7786	–	–
(12)-b3-4-5	0.5553	14.31	4.1346	9.42	–
1-(23)-b4-5	0.5773	18.83	3.7463	–	0.85
1-b(23)-4-5	0.5643	16.16	4.0763	7.88	–
1-2-(34)-b5	0.5643	16.16	3.2566	–	13.81
1-2-b(34)-5	0.5773	18.83	3.6562	–	3.24
1-2-3-b(45)	0.5553	14.31	3.1862	–	15.68
(12)-b3-b(45)	0.7229	48.81	3.6278	–	3.99
(12)-b(34)-b5	0.7273	49.71	3.6856	–	2.46
(12)-bb(34)-5	0.6789	39.75	4.2353	12.09	–
1-(23)-bb(45)	0.6789	39.75	2.8399	–	24.84
1-b(23)-b(45)	0.7273	49.71	3.5389	–	6.34
(123)-bb4-5	0.6565	35.14	4.2524	12.54	–
1-(234)-bb5	0.6731	38.55	2.9400	–	22.19
1-b(234)-b5	0.7099	46.13	3.5768	–	5.34
1-bb(234)-5	0.6731	38.55	4.1015	8.55	–
1-2-bb(345)	0.6565	35.14	2.7811	–	26.40
(1234)-bbb5	0.8781	80.75	3.3475	–	11.41
(123)-bbb(45)	0.9090	87.11	3.4008	–	10.00
(12)-bbb(345)	0.9090	87.11	3.4172	–	9.56
1-bbb(2345)	0.8781	80.75	3.4086	–	9.79
(12345)	1.0000	105.85	1.0000	–	73.54

Source: Argon, N.T., and S. Andradóttir, *Queueing Systems,* **52** (2006), no. 1, 5–30. With permission.

Table 9.2 Throughput and WIP of Balanced Lines with $N = 4$ and $b_j = 3$, for $j = 2, 3, 4$

System	THP	% Inc. in THP	WIP	% Inc. in WIP	% Dec. in WIP
1-*bbb*2-*bbb*3-*bbb*4	0.7477	–	8.0813	–	–
(12)-*bbbbbb*3-*bbb*4	0.8253	10.38	9.3771	16.03	–
1-*bbb*(23)-*bbbbbb*4	0.8418	12.59	6.6562	–	17.63
1-*bbbb*(23)-*bbbbb*4	0.8525	14.02	7.2741	–	9.99
1-*bbbbb*(23)-*bbbb*4	0.8557	14.44	7.8049	–	3.42
1-*bbbbbb*(23)-*bbb*4	0.8525	14.02	8.3266	3.04	–
1-*bbbbbbb*(23)-*bb*4	0.8418	12.59	8.9137	10.30	–
1-*bbb*2-*bbbbbb*(34)	0.8253	10.38	6.1837	–	23.48
(123)-*bbbbbbbbbb*4	0.9455	26.45	6.9025	–	14.59
(12)-*bbbbbbbbbb*(34)	0.9576	28.07	6.9576	–	13.90
1-*bbbbbbbbbb*(234)	0.9455	26.45	6.9884	–	13.52
(1234)	1.0000	33.74	1.0000	–	87.63

Source: Argon, N.T., and S. Andradóttir, *Queueing Systems,* **52** (2006), no. 1, 5–30. With permission.

pooling structures for lines with $N \in \{3, 4, 5\}$ and $b_j = 0$ for $j \in \{2, ..., N\}$ and with $N = 4$ and $b_j = 3$ for $j \in \{2, 3, 4\}$, respectively. We also provide the percentage increase in throughput and percentage increase/decrease in WIP that are obtained over the unpooled system by each pooling structure. To describe the system configurations, we use hyphens to separate the stations, put the pooled stations between parentheses, and denote each buffer space with a small letter "*b*." For example, when $N = 4$ and $b_2 = b_3 = b_4 = 0$, then 1 - 2 - 3 - 4 denotes the original system and 1 - *b* (23) - 4 denotes the system for which stations 2 and 3 are pooled, $P = 1$ and $Q = 0$.

We can summarize our conclusions from Table 9.1 and Table 9.2 as follows:

- For balanced lines, pooling any group of adjacent stations improves the system throughput regardless of the buffer allocation around the pooled station. This is consistent with Proposition 1, part (*i*) of Proposition 2, and Theorem 1.5 in Meester and Shanthikumar (1990).
- In terms of system throughput, pooling provides a substantial benefit in balanced lines (the improvement in the system throughput gained by pooling lies between 10.38 and 105.85% in Table 9.1 and Table 9.2). Moreover, the more stations that are pooled, the better the throughput gets (this conclusion is consistent with Proposition 4).
- Pooling stations near the middle of the line (and allocating the buffers around the pooled station evenly) generally yields better throughput than pooling stations at the beginning or end of the line when systems with the same number of pooled stations are compared. Also, pooling that results in a more balanced buffer allocation provides better throughput. These results are consistent with the recommendations for increasing system throughput via buffer and workload allocation (see, e.g., Carnall and Wild, 1976 and Hillier et al., 1993).

- The amount of benefit that pooling has on the throughput decreases as the buffer sizes increase (compare the results in Table 9.1 and Table 9.2). This is reasonable because, as the number of buffers increases, it becomes less important to use pooling to improve the throughput (note that in a balanced system with infinite buffers, identical workers, and additive rates, pooling will not affect the throughput (see Section 9.2.1)).
- Pooling may increase or decrease the WIP. This does not contradict our discussion in Section 9.4, since in this section we assume that there is an infinite supply of jobs at the beginning of the line. Moreover, pooling several stations at the end of the line usually yields a better WIP when systems with the same number of stations after pooling are compared. We believe that this is due to the fact that in such cases, the jobs are pushed out of the system more effectively and the added buffers are closer to the end of the line (see So, 1997, and Papadopoulos and Vidalis, 2001).

9.6.2   Pooling in unbalanced lines

In this section, we look at the effects of pooling on the steady-state throughput and WIP of unbalanced tandem lines with four stations. We first study lines with zero buffers and present the steady-state throughput of these lines under several pooling structures in Table 9.3. We consider various combinations of service rates: lines with a single bottleneck, lines with two bottlenecks, and completely unbalanced lines. For each set of service rates and number of stations in the pooled system, the maximum throughput is highlighted in Table 9.3 (the table does not include all possible combinations of service rates because of the reversibility property of tandem lines in terms of system throughput (see Yamazaki and Sakasegawa, 1975).

Some of our conclusions from Table 9.3 can be summarized as follows:

- For all combinations of service rates considered in Table 9.3, the system throughput is improved not only by pooling stations at the beginning or end of the line (which is consistent with Theorem 1), but also by pooling the two intermediate stations. The percentage increase in the throughput falls in the ranges 116 to 165%, 4 to 125%, or 0.5 to 73% when the pooled system has one, two, or three stations, respectively.
- The bottleneck station should be one of the pooled stations in order to obtain the most benefit out of pooling (this conclusion is in agreement with Proposition 1). This seems to have a bigger impact on the throughput than any other factor. Moreover, it is desirable to pool the bottleneck station with as many stations as possible (see Proposition 4). When it is only feasible to pool the bottleneck station with one other station, then it is generally better to pool the bottleneck station with its faster neighboring station, except in some cases when

Table 9.3 Throughput of Unbalanced Lines with $N = 4$ and $b_j = 0$, for $j = 2, 3, 4$

η_1	η_2	η_3	η_4	Throughput								
				1-2-3-4	(12)-b3-4	1-(23)-b4	1-b(23)-4	1-2-b(34)	(123)-bb4	(12)-bb(34)	1-bb (234)	(1234)
0.2	1.0	1.0	1.0	0.1916	**0.3250**	0.1946	0.1987	0.1934	**0.4254**	0.3330	0.1999	0.5000
1.0	0.2	1.0	1.0	0.1884	**0.3250**	0.3110	0.3110	0.1933	**0.4254**	0.3330	**0.4254**	0.5000
0.2	0.2	1.0	1.0	0.1328	**0.1996**	0.1658	0.1819	0.1333	0.2726	0.2000	0.1967	0.3333
0.2	1.0	0.2	1.0	0.1452	0.1801	0.1658	0.1819	**0.1855**	0.2726	**0.2786**	0.1967	0.3333
0.2	1.0	1.0	0.2	0.1539	**0.1855**	0.1574	0.1574	**0.1855**	0.1967	**0.2786**	0.1967	0.3333
1.0	0.2	0.2	1.0	0.1324	0.1801	0.1966	0.1966	0.1801	0.2726	**0.2786**	0.2726	0.3333
0.2	0.6	1.0	1.4	0.1832	**0.2968**	0.1912	0.1980	0.1846	0.3911	0.3000	0.1999	0.4773
0.2	0.6	1.4	1.0	0.1838	**0.2980**	0.1925	0.1980	0.1846	0.4043	0.3000	0.1999	0.4773
0.2	1.0	0.6	1.4	0.1881	**0.3095**	0.1912	0.1980	0.1934	0.3911	0.3321	0.1999	0.4773
0.2	1.0	1.4	0.6	0.1910	**0.3174**	0.1945	0.1971	0.1934	0.4133	0.3321	0.1999	0.4773
0.2	1.4	0.6	1.0	0.1904	**0.3166**	0.1925	0.1980	0.1962	0.4043	0.3470	0.1999	0.4773
0.2	1.4	1.0	0.6	0.1926	**0.3227**	0.1945	0.1971	0.1962	0.4133	0.3470	0.1999	0.4773
0.6	0.2	1.0	1.4	0.1816	0.2968	0.2844	**0.3057**	0.1846	0.3911	0.3000	**0.4133**	0.4773
0.6	0.2	1.4	1.0	0.1828	0.2980	0.2921	**0.3086**	0.1846	0.4043	0.3000	**0.4133**	0.4773
1.0	0.2	0.6	1.4	0.1814	**0.3095**	0.2863	0.2917	0.1930	0.3911	0.3321	**0.4043**	0.4773
1.0	0.2	1.4	0.6	0.1876	**0.3174**	0.3086	0.2921	0.1930	0.4133	0.3321	0.4043	0.4773
1.4	0.2	0.6	1.0	0.1823	**0.3166**	0.2917	0.2863	0.1954	0.4043	0.3470	0.3911	0.4773
1.4	0.2	1.0	0.6	0.1870	**0.3227**	0.3057	0.2844	0.1954	0.4133	0.3470	0.3911	0.4773

Source: Argon, N.T., and S. Andradóttir, *Queueing Systems*, **52** (2006), no. 1, 5–30. With permission.

pooling the bottleneck station with its slower neighboring station leads to a line with either a smaller number of bottleneck stations and/or a bowl-shaped workload allocation (see, e.g., Hillier and Boling, 1966, for the bowl phenomenon in workload allocations, which indicates that center stations should be faster than the end stations to achieve higher system throughput).

Next, we study a tandem line of four stations for which the buffer sizes may be nonzero and the service rate η_j at each station $j \in \{1,\ldots,4\}$ is generated independently from a uniform distribution on the range $[0.5, 2.5]$. We consider both lines that have the same amount of buffer spaces between any two stations (i.e., $b_2 = b_3 = b_4 \in \{0,3\}$) and lines for which the buffers between any two stations are generated independently from a discrete uniform distribution on the set $\{0,1,2,3\}$. In Table 9.4, we provide the estimated probability of

Table 9.4 Throughput and WIP of Unbalanced Lines with $N = 4$

System	Prob. of Incr. in THP	% Incr. in THP	% Decr. in THP	Prob. of Incr. in WIP	% Incr. in WIP	% Decr. in WIP
		Common Buffer Size = 0				
(12)-B3-4	1.0000	26.00±0.44	–	0.8882	7.00±0.14	6.98±0.46
1-(23)-B4	1.0000	25.13±0.36	–	0.0862	2.23±0.20	12.62±0.21
1-B(23)-4	1.0000	25.23±0.36	–	0.7688	8.12±0.21	5.47±0.23
1-(23)-4*	1.0000	25.18±0.35	–	0.4144	3.91±0.11	8.83±0.20
1-2-B(34)	1.0000	25.95±0.45	–	0.1144	5.50±0.34	12.60±0.16
(123)-B4	1.0000	67.82±0.67	–	0.3306	6.65±0.27	20.92±0.46
1-B(234)	1.0000	67.57±0.68	–	0.4448	14.71±0.48	17.31±0.40
(1234)	1.0000	110.09±0.34	–	0.0000	–	65.71±0.14
		Common Buffer Size = 3				
(12)-B3-4	1.0000	16.13±0.44	–	0.7746	16.71±0.62	28.77±1.05
1-(23)-B4	0.9996	16.13±0.41	0.02±0.10	0.2682	6.61±0.40	24.45±0.42
1-B(23)-4	0.9986	16.19±0.41	0.01±0.01	0.6676	17.90±0.60	18.83±0.54
1-(23)-4*	1.0000	16.14±0.41	–	0.4634	7.59±0.26	18.97±0.40
1-2-B(34)	1.0000	16.10±0.45	–	0.2036	15.42±0.65	21.84±0.36
(123)-B4	1.0000	37.83±0.66	–	0.4266	25.20±1.31	50.64±0.84
1-B(234)	1.0000	37.63±0.67	–	0.5130	38.89±1.20	41.42±0.75
(1234)	1.0000	63.77±0.57	–	0.0000	–	85.03±0.20
		Buffer Sizes ~ Uniform {0, 1, 2, 3}				
(12)-B3-4	1.0000	20.92±0.48	–	0.8266	14.85±0.44	19.86±1.06
1-(23)-B4	0.9944	20.72±0.42	1.27±0.40	0.1942	6.29±0.45	20.40±0.40
1-B(23)-4	0.9936	20.95±0.43	1.23±0.42	0.7112	17.07±0.64	13.07±0.49
1-(23)-4*	1.0000	20.62±0.41	–	0.4486	7.39±0.28	14.97±0.36
1-2-B(34)	1.0000	20.76±0.49	–	0.1582	15.30±0.98	19.71±0.31
(123)-B4	1.0000	50.52±0.71	–	0.4084	20.09±0.93	38.77±0.78
1-B(234)	1.0000	50.27±0.72	–	0.4876	36.10±1.38	31.36±0.68
(1234)	1.0000	80.57±0.58	–	0.0000	–	78.62±0.23

Source: Argon, N.T., and S. Andradóttir, *Queueing Systems,* **52** (2006), no. 1, 5–30. With permission.

having an increase in the throughput and WIP as a result of pooling, using results from 5000 randomly generated systems. We also tabulate 95% confidence intervals on the percentage increase and decrease in throughput and WIP achieved by each pooled system. The percentage increase in throughput is based on only those combinations of service rates (and buffers) for which an increase in throughput is realized. Percentage decrease in throughput and percentage increase/decrease in WIP are computed similarly.

In Table 9.4, we use a capital letter "B" to indicate the location where the buffers between the pooled stations and the additional buffers are placed. Note that when stations 2 and 3 are pooled, we only consider placing the pooled station at the position of either station 2 or station 3 (and not at the position of a buffer). If we place the pooled station at the position of the slower one of stations 2 and 3, we denote this system by 1-(23)-4* (we have recommended this buffer allocation scheme earlier based on Proposition 1).

Some of our conclusions from Table 9.4 are as follows:

- When systems with the same number of stations after pooling are compared, the choice of pooling strategy has a substantial effect on the average WIP but only a small impact on the average throughput. Also, pooling provides smaller increases in throughput for larger buffer sizes. These results are consistent with our observations for balanced lines (see Section 9.6.1).
- Intermediate pooling may decrease the steady-state throughput when the buffers between at least some of the stations are positive (pooling at the beginning and end of the line always improves the throughput, see Theorem 1). However, if the buffer allocation is performed as suggested in Section 9.2.2, then intermediate pooling always improves the throughput. Also, even if the buffer allocation is not done properly, intermediate pooling decreases the throughput only very rarely and the decrease is not substantial.
- If pooling is done properly, then a greater benefit is realized in an unbalanced line than in a balanced line (see the percentage increase in throughput in Table 9.1, Table 9.2, and Table 9.4).

9.6.3   *Pooling without additional buffers*

In this section, we observe the effects of pooling on the throughput and WIP when it is infeasible or undesirable to add extra buffers to equalize the number of storage spaces in the pooled and unpooled lines. For this purpose, we consider a tandem line of four stations. First, we look at the balanced case with $\eta_j = 1$ for $j = 1, \ldots, 4$. The results, which are presented in Table 9.5, show that both throughput and WIP are improved in the balanced case, even without extra buffers (although the improvement is less than when extra buffers are added (Table 9.1)). One difference between Table 9.1 and Table 9.5 is that all the pooling strategies decrease the WIP when no buffers are added.

Table 9.5 Throughput and WIP of Balanced Lines with $N = 4$, $b_j = 0$, for $j = 2, 3, 4$, and No Extra Buffers

System	Throughput (THP)	% Increase in THP	WIP	% Decrease in WIP
1-2-3-4	0.5148	–	3.0646	–
(12)-3-4	0.5855	13.73	2.4125	21.28
1-(23)-4	0.5850	13.64	2.3857	22.15
1-2-(34)	0.5855	13.73	2.3773	22.43
(123)-4	0.7033	36.62	1.7033	44.42
(12)-(34)	0.7273	41.28	1.7273	43.64
1-(234)	0.7033	36.62	1.7033	44.42
(1234)	1.0000	94.25	1.0000	67.37

Source: Argon, N.T., and S. Andradóttir, *Queueing Systems,* **52** (2006), no. 1, 5–30. With permission.

Next, we study unbalanced tandem lines for which the service rate η_j at each station $j \in \{1, \ldots, 4\}$ is generated from a uniform distribution with range $[0.1, 20.1]$ (we have used a wider range than in Section 9.6.2 since the frequency of observing a decrease in throughput by pooling at either end of the line is quite low unless the stations are highly unbalanced). We consider both lines that have the same number of buffer spaces between any two stations (i.e., $b_2 = b_3 = b_4 \in \{0, 3\}$) and lines for which the buffers between any two stations are generated independently from a discrete uniform distribution on the set $\{0, 1, 2, 3\}$. In Table 9.6, we tabulate the estimated probability of having an increase in the throughput and WIP as a result of pooling using the same notation as in Table 9.4. We also provide 95% confidence intervals on the percentage decrease/increase in throughput and WIP achieved by pooling. These results are obtained from 5000 randomly generated systems, and the confidence intervals are computed as in Table 9.4.

Table 9.6 suggests that the probability of observing an increase (decrease) in the system throughput (WIP) is much larger than that of observing a decrease (increase), and the percentage decrease in the throughput is much smaller than the percentage increase in throughput. Moreover, the probability of observing a decrease in throughput (and also the percentage decrease in throughput) by pooling is higher for intermediate stations than for stations at the beginning or end of the line. This is reasonable since, with added buffers, pooling at the beginning or end of the line is always beneficial. Note, however, that when the buffers around the pooled station are allocated as recommended in Section 9.2.2, then the probability of a decrease in throughput for pooling intermediate stations becomes closer to that for pooling stations at the beginning or end of a line.

The numerical results in this section suggest that even if extra buffers are not added to equalize the total number of storage spaces in the unpooled and pooled systems, pooling will still lead to an increase in throughput for most tandem lines. Moreover, even when pooling decreases the system throughput when extra buffers are not added, it decreases the throughput only slightly.

Table 9.6 Throughput and WIP of Unbalanced Lines with $N = 4$ and No Extra Buffers

System	Prob. of Incr. in THP	% Incr. in THP	% Decr. in THP	Prob. of Incr. in WIP	% Incr. in WIP	% Decr. in WIP
Common Buffer Size = 0						
(12)-3-4	0.9810	25.93 ± 0.75	0.05 ± 0.01	0.0726	6.08 ± 0.50	23.45 ± 0.23
1-(23)-4	0.9008	27.38 ± 0.76	1.16 ± 0.12	0.0136	1.23 ± 0.28	19.95 ± 0.21
1-2-(34)	0.9822	25.92 ± 0.76	0.05 ± 0.01	0.0000	−	17.34 ± 0.22
(123)-4	0.9854	69.68 ± 1.38	0.12 ± 0.03	0.0122	5.23 ± 1.06	42.99 ± 0.33
1-(234)	0.9864	69.44 ± 1.39	0.12 ± 0.03	0.0000	−	37.30 ± 0.33
(1234)	1.0000	148.59 ± 1.35	−	0.0000	−	61.87 ± 0.36
Common Buffer Size = 3						
(12)-*B*3-4	1.0000	25.34 ± 0.83	−	0.3072	24.53 ± 1.50	21.02 ± 0.78
1-(23)-*B*4	0.9584	25.94 ± 0.83	0.05 ± 0.01	0.0834	9.73 ± 1.06	20.02 ± 0.44
1-*B*(23)-4	0.9644	25.79 ± 0.83	0.06 ± 0.02	0.3038	30.15 ± 1.07	12.42 ± 0.34
1-(23)-4*	0.9822	25.30 ± 0.82	0.12 ± 0.03	0.1226	7.38 ± 0.78	12.77 ± 0.31
1-2-*B*(34)	1.0000	25.15 ± 0.84	−	0.1920	20.30 ± 0.65	17.31 ± 0.40
(123)-*B*4	1.0000	64.70 ± 1.50	−	0.1496	40.14 ± 3.90	43.99 ± 0.91
1-*B*(234)	1.0000	64.44 ± 1.52	−	0.4028	44.51 ± 1.62	31.63 ± 0.71
(1234)	1.0000	117.74 ± 1.83	−	0.0000	−	80.21 ± 0.49
Buffer Sizes ~ Uniform {0, 1, 2, 3}						
(12)-*B*3-4	0.9988	27.24 ± 0.82	0.04 ± 0.08	0.2538	18.61 ± 1.05	19.44 ± 0.54
1-(23)-*B*4	0.9512	27.81 ± 0.81	0.47 ± 0.09	0.0618	7.98 ± 1.13	19.72 ± 0.39
1-*B*(23)-4	0.9544	27.77 ± 0.80	0.50 ± 0.11	0.2210	31.04 ± 1.75	13.52 ± 0.25
1-(23)-4*	0.9762	27.06 ± 0.79	0.41 ± 0.12	0.0762	7.30 ± 0.95	14.14 ± 0.24
1-2-*B*(34)	0.9990	27.15 ± 0.83	0.04 ± 0.04	0.1318	22.75 ± 1.29	17.29 ± 0.33
(123)-*B*4	0.9996	71.10 ± 1.45	0.05 ± 0.54	0.1032	27.08 ± 2.76	40.07 ± 0.69
1-*B*(234)	0.9994	70.84 ± 1.47	0.09 ± 0.35	0.2724	44.16 ± 2.38	29.15 ± 0.51
(1234)	1.0000	128.57 ± 1.67	−	0.0000	−	74.05 ± 0.47

Source: Argon, N.T., and S. Andradóttir, *Queueing Systems*, **52** (2006), no. 1, 5–30. With permission.

9.6.4 *Pooling with nonadditive server rates*

In this section, we observe the effects of pooling on the throughput and WIP when the pooled servers are subadditive or superadditive. For this purpose, we consider a tandem line of four stations with $b_j = 0$ for $j \in \{2, 3, 4\}$. When stations K through M are pooled, we assume that the rate at which the pooled team performs task $j \in \{K, \ldots, M\}$ is given by $g_j^{(K,M)}(\mu_{K,j}, \ldots, \mu_{M,j}) = v \sum_{\ell=K}^{M} \mu_{\ell,j}$,

where v is a positive real number. By Equation (9.1) and the discussion following it, $v = 1$ corresponds to the case where the worker rates are additive, $v < 1$ corresponds to the case where the rates are subadditive, and $v > 1$ corresponds to the case where the rates are superadditive. It is easy to see that the throughput of a tandem line with complete pooling is linear in v and the WIP of a completely pooled line is independent of v. In the remainder of this section, we study how v changes the effects of partial pooling on the throughput and WIP of two different tandem lines.

Figure 9.1 presents plots of the percentage change in throughput and WIP obtained by pooling two stations at the beginning or end of a balanced line (with $\eta_j = 1$ for $j = 1, ..., 4$) as a function of $v \in \{0.1, 0.2, ..., 1.9\}$. Similarly, Figure 9.2 provides plots of the percentage change in throughput and WIP by partial pooling as a function of $v \in \{0.1, 0.2, ..., 1.9\}$ for an unbalanced line with bottleneck at station 1 ($\eta_1 = 0.2$ and $\eta_j = 1.0$ for $j \in \{2, 3, 4\}$).

Some of our observations from Figure 9.1 and Figure 9.2 are as follows:

- The benefit of pooling on system throughput is nondecreasing in v. This is an expected result because the departure times from a tandem line are known to be nondecreasing functions of the service times (see, e.g., Glasserman and Yao, 1994, Proposition 4.4.3). Moreover, for the given examples, the WIP increases/decreases with v - when stations at the beginning/end of the line are pooled. This is an intuitive result because speeding up the station at the beginning of the line pulls jobs into the system faster (and, hence, increases the WIP), whereas speeding up the station at the end of the line pushes jobs out of the line more quickly (and, hence, decreases the WIP). When stations in the middle of the line are pooled, the relationship between WIP and v is more complicated. In particular, we have obtained numerical results showing that when stations in the middle of the line are pooled, WIP may not change monotonically as v increases.
- The impact of superadditivity on the throughput and WIP appears to be smaller than that of subadditivity in balanced lines and also in unbalanced lines when nonbottleneck stations are pooled. (We have obtained additional numerical results for these lines that show that the throughput and WIP for $v = 10$ are not much different from their values when $v = 1.9$.)
- In the given examples, pooling is beneficial in terms of system throughput for all $v \geq 0.6$. This shows that partial pooling is beneficial in these examples even when the efficiency of the workers is reduced by 40% when they collaborate.
- In the unbalanced example, the throughput of the pooled line is very sensitive to changes in v when the bottleneck station is among the pooled stations, whereas it is relatively insensitive to changes in v when nonbottleneck stations are pooled. This is an intuitive result because the service rate of the bottleneck station is expected to have the biggest effect on the system throughput among all the stations in the line.

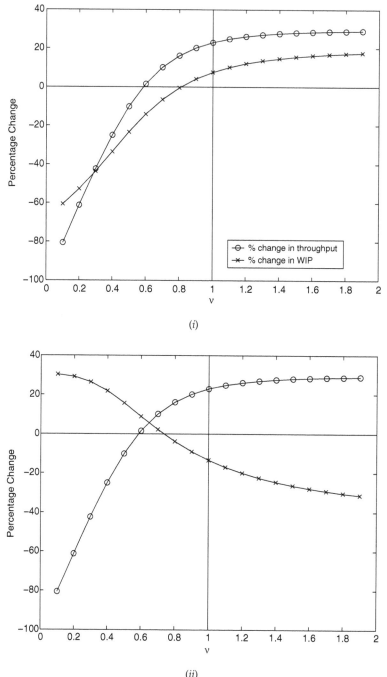

Figure 9.1 Percentage change in throughput and WIP as a function of v when two stations at the beginning (*i*) (12)-b3-4 or end (*ii*) 1-2-b(34) of a balanced line are pooled. (From Argon, N.T., and S. Andradóttir, *Queueing Systems*, **52** (2006), no. 1, 5–30. With permission.)

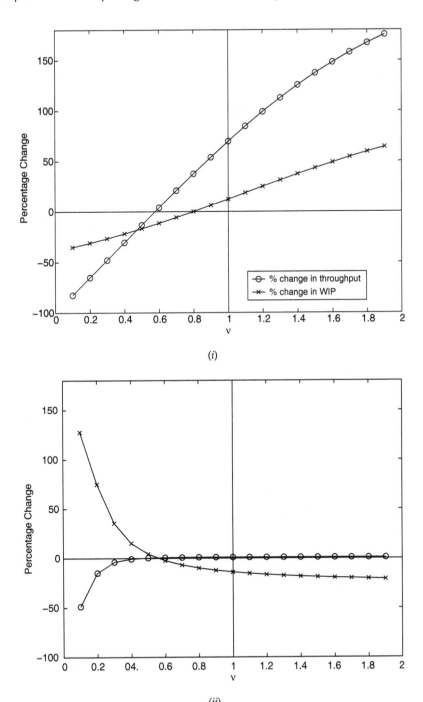

(i)

(ii)

Figure 9.2 Percentage change in throughput and WIP as a function of v when two stations at the beginning (*i*) (12)-b-3-4 or end (*ii*) 1-2-b(34) of an unbalanced line with bottleneck at station 1 are pooled. (From Argon, N.T., and S. Andradóttir, *Queueing Systems*, **52** (2006), no. 1, 5–30. With permission.)

9.7   Conclusions

For a tandem network of single-server queues with finite buffers, general service times, and cross-trained servers, we have provided conditions that guarantee that pooling several stations permanently (to obtain a pooled station with a single team of cooperative servers) decreases the departure time of each job from the system and, hence, increases the steady-state system throughput. Moreover, we have studied the effects of pooling on other performance measures, namely the long-run average sojourn time, WIP, and holding costs, in a tandem line with an external arrival stream at the first station. We have proved that whenever pooling decreases the departure time of each job from the system, then it also decreases the WIP at any given time and the sojourn time of each job in the system. By contrast, we have shown that pooling may increase the average holding cost per job, even when the sojourn time of each job is decreased by pooling, and have provided sufficient conditions under which pooling is guaranteed to decrease the average holding cost per job. Finally, we have extended our results to queuing systems other than open tandem lines operated under the production blocking mechanism, and have performed a numerical study on tandem lines with identical cooperative workers. Based on our numerical and theoretical analyses, we have the following additional insights about partial pooling in a tandem line:

1. When different workers have similar capabilities at the various tasks, then it is usually possible to improve the throughput by pooling these workers into a team if they do not loose much efficiency when they work together. Moreover, the resulting increase in the system throughput can be substantial.
2. If the service times at the pooled station are not too large (e.g., when the workers are identical and their rates are additive), then pooling several stations at the beginning or end of a tandem line is always beneficial. However, pooling intermediate stations needs to be performed with care because pooling may decrease the system throughput if the allocation of buffers around the pooled station is not done well. We recommend placing the pooled station at the position of the slowest station it replaces.
3. In a balanced tandem line with additive server rates, pooling stations near the center of the line results in higher system throughput than pooling stations near either end of the line. In an unbalanced line with additive server rates, the bottleneck station should be among the pooled stations in order to achieve the highest throughput by pooling. In general, the throughput of the (partially) pooled system is highly sensitive to the efficiency of the pooled team if the bottleneck station is among the pooled stations.
4. Increasing the number of pooled stations (either by increasing the number of stations in a group of pooled stations or by increasing the

number of groups of pooled stations) generally results in increased system throughput.

5. Pooling stations near the end of the line and placing buffers close to the end of the line usually decreases WIP if there is an infinite supply of jobs at the first station.

6. Pooling at the end of a line generally reduces holding costs if it decreases the departure times from the system, the holding costs increase across stations, and the buffer in front of the first station to be pooled is large.

7. Whether or not pooling improves the system throughput does not appear to depend heavily on the choice of the blocking mechanism.

Acknowledgments

This work was supported by the National Science Foundation under Grant No. 9523111, Grant No. 0000135, Grant No. 0217860, and Grant No. 0400260.

References

Ahn, H., I. Duenyas, and R.Q. Zhang, Optimal Stochastic Scheduling of a Two-Stage Tandem Queue with Parallel Servers, *Advances in Applied Probability*, 31 (1999), no. 4, 1095–1117.

Andradóttir, S., and H. Ayhan, Throughput Maximization for Tandem Lines with Two Stations and Flexible Servers, *Operations Research*, 53 (2005), no. 3, 516–531.

Andradóttir, S., H. Ayhan, and D.G. Down, Server Assignment Policies for Maximizing the Steady-State Throughput of Finite Queueing Systems, *Management Science*, 47 (2001), no. 10, 1421–1439.

Andradóttir, S., H. Ayhan, and D.G. Down, Dynamic Server Allocation for Queueing Networks with Flexible Servers, *Operations Research*, 51 (2003), no. 6, 952–968.

Argon, N.T., *Performance Enhancements in Tandem Queueing Networks and Confidence Interval Estimation in Steady-State Simulations*, Ph.D. dissertation, School of Industrial and Systems Engineering, Georgia Institute of Technology, Atlanta, 2002.

Argon, N.T., and S. Andradóttir, Partial Pooling in Tandem Lines with Cooperation and Blocking, *Queueing Systems*, 52 (2006), no. 1, 5–30.

Bartholdi, J.J. III, D.D. Eisenstein, and R.D. Foley, Performance of Bucket Brigades when Work is Stochastic, *Operations Research*, 49 (2001), no. 5, 710–719.

Benjaafar, S., Performance Bounds for the Effectiveness of Pooling in Multi-processing Systems, *European Journal of Operations Research*, 87 (1995), no. 2, 375–388.

Bischak, D.P., Performance of a Manufacturing Module with Moving Workers, *IIE Transactions*, 28 (1996), no. 9, 723–733.

Buzacott, J.A., Commonalities in Reengineered Business Processes: Models and Issues, *Management Science*, 42 (1996), no. 5, 768–782.

Buzacott, J.A., and J.G. Shanthikumar, *Stochastic Models of Manufacturing Systems*, Prentice-Hall, Englewood Cliffs, NJ, 1993.

Bramson, M., Instability of FIFO Queueing Networks, *The Annals of Applied Probability*, 4 (1994), no. 2, 414–431.

Calabrese, J.B., Optimal Workload Allocation in Open Networks of Multiserver Queues, *Management Science*, 38 (1992), no. 12, 1792–1802.

Carnall, C.A., and R. Wild, The Location of Variable Work Stations and the Performance of Production Flow Lines, *International Journal of Production Research*, 14 (1976), no. 6, 703–710.

Cheng, D.W., and D.D. Yao, Tandem Queues with General Blocking: A Unified Model and Stochastic Comparisons, *Discrete Event Dynamic Systems: Theory and Applications*, 2 (1993), no. 3–4, 207–234.

Conway, R., W. Maxwell, J.O. McClain, and L.J. Thomas, The Role of Work-in-Process Inventory in Serial Production Lines, *Operations Research*, 36 (1988), no. 2, 229–241.

Glasserman, P., and D.D. Yao, Monotonicity in Generalized Semi-Markov Processes, *Mathematics of Operations Research*, 17 (1992), no. 1, 1–21.

Glasserman, P., and D.D. Yao, A GSMP Framework for the Analysis of Production Lines, in *Stochastic Modeling and Analysis of Manufacturing Systems*, Ed., D.D. Yao, Springer-Verlag, New York, 1994.

Glasserman, P., and D.D. Yao, Structured Buffer-Allocation Problems, *Discrete Event Dynamic Systems: Theory and Applications*, 6 (1996), no. 1, 9–41.

Hillier, F.S., and R.W. Boling, The Effect of Some Design Factors on the Efficiency of Production Lines with Variable Operation Times, *Journal of Industrial Engineering*, 17 (1966), no. 12, 651–658.

Hillier, F.S., K.C. So, and R.W. Boling, Notes: Toward Characterizing the Optimal Allocation of Storage Space in Production Line Systems with Variable Processing Times, *Management Science*, 39 (1993), no. 1, 126–133.

Mandelbaum, A., and M.I. Reiman, On Pooling in Queueing Networks, *Management Science*, 44 (1998), no. 7, 971–981.

Meester, L.E., and J.G. Shanthikumar, Concavity of the Throughput of Tandem Queueing Systems with Finite Buffer Storage Space, *Advances in Applied Probability*, 22 (1990), no. 3, 764–767.

Muth, E.J., The Production Rate of a Series of Work Stations with Variable Service Times, *International Journal of Production Research*, 11 (1973), no. 2, 155–169.

Papadopoulos, H.T., and M. I. Vidalis, Minimizing WIP Inventory in Reliable Production Lines, *International Journal of Production Economics*, 70 (2001), no. 2, 185–197.

Shaked, M., and J.G. Shanthikumar, *Stochastic Orders and Their Applications*, Academic Press, San Diego, CA, 1994.

Smith, D.R., and W. Whitt, Resource Sharing for Efficiency in Traffic Systems, *Bell System Technical Journal*, 60 (1981), no. 1, 39–55.

So, K.C., Optimal Buffer Allocation Strategy for Minimizing Work-in-Process Inventory in Unpaced Production Lines, *IIE Transactions*, 29 (1997), no. 1, 81–88.

Spearman, M., D. Woodruff, and W. Hopp, CONWIP: A Pull Alternative to Kanban, *International Journal of Production Research*, 28 (1990), no. 5, 879–894.

Tayur, S.R., Structural Properties and a Heuristic for Kanban-Controlled Serial Lines, *Management Science*, 39 (1993), no. 11, 1347–1368.

Tembe, S.V., and R.W. Wolff, The Optimal Order of Service in Tandem Queues, *Operations Research*, 22 (1974), no. 4, 824–832.

Van Oyen, M.P., E.G.S. Gel, and W.J. Hopp, Performance Opportunity of Workforce Agility in Collaborative and Noncollaborative Work Systems, *IIE Transactions*, 33 (2001), no. 9, 761–777.

Veatch, M.H., and L.M. Wein, Optimal Control of a Two-Station Production/Inventory System, *Operations Research*, 42 (1994), no. 2, 337–350.

Yamazaki, G., and H. Sakasegawa, Properties of Duality in Tandem Queueing Systems, *Annals of the Institute of Statistical Mathematics*, 27 (1975), no. 2, 201–212.

Zadavlav, E., J.O. McClain, and L.J. Thomas, Self-Buffering, Self-Balancing, Self-Flushing Production Lines, *Management Science*, 42 (1996), no. 8, 1151–1164.

chapter 10

Design of a labor-intensive assembly cell in the clean room of a medical device manufacturer

Viviana I. Cesaní

Contents

10.1   Introduction

This chapter details the design of an assembly cell in the clean room of a medical device manufacturer. The discussion focuses on technical design issues where special emphasis is given to labor cross training and labor allocation using static and dynamic evaluation models. The analysis presented is based on information obtained through on-site interviews with management and employees, through the examination of company documents, and by observing the manufacturing system in place at the plant. The study was conducted over a 6-month period that ended in February 2006.

The company under study (name omitted for confidentiality) is a manufacturing division of a major medical devices manufacturer. This particular division is responsible for producing specialized medical instruments and endoscopy equipment. The company's medical instruments unit manufactures irrigation equipment used to clean injuries in orthopedic surgeries and trauma, cement mix systems used for implants, and post surgery blood recollection systems. The endoscopy unit manufactures high quality cutting accessories used in arthroscopic surgeries to repair tissues, and irrigation and suction systems used in laparoscopic surgeries.

This project focused on the endoscopy unit, specifically in the cutters' area, which generates 65% of the revenue for the unit. The company's corporate strategy includes the introduction of several new products in forthcoming years to respond to a fast evolving business environment, with most of their products in the growth phase of their product lifecycle. Providing the floor space for these new products is particularly important because production space is not only limited but, due to the nature of the products, most of the required process steps need to be performed inside a clean room. Because clean rooms require a controlled environment, their associated costs are significantly higher than in other production areas.

Another important issue in the manufacturing of cutters is its labor-intensive nature. The assembly process is mostly manual and their associated costs are relatively high. Therefore, management wants to evaluate if it is making the best use of its labor resources. Furthermore, one goal in this particular company division is to implement the principles of lean manufacturing in every unit. Last year, the endoscopy unit had one of the area assembly lines in the irrigation and suction area transformed into a continous flow assembly line with a transfer batch of one unit between workstations (i.e. one-piece flow). Management was very satisfied with the results and wanted to apply a similar approach in the cutters' area.

In a labor-intensive process such as the one in the cutters' area, the assignment of operators plays a major role and directly affects its production output. To respond to unexpected variations in the workload and fluctuations in the supply of labor, management in this area has encouraged the development of a multifunctional workforce. The concept of using multifunctional operators is referred to as labor flexibility. Labor flexibility is attained by cross training operators in different tasks. This type of flexibility has been recognized as a tool for enhancing manufacturing performance in cellular manufacturing systems, since it provides several strategic advantages, such as the ability to reduce manufacturing flow times and work-in-process (WIP) inventories, and improve customer service while providing an efficient use of both labor and equipment (see, e.g., Cesaní and Steudel, 2005; Slomp et al., 2005).

Given the company's constraint of limited resources and their objective of implementing lean manufacturing principles, a redesign of the assembly area based on cellular manufacturing principles was considered. Issues such as operators' cross training, operations' staffing levels, and possible labor assignments in order to achieve a desired performance level were examined in detail. The analysis presented is based on a methodology that includes the combination of static workload analysis and stochastic simulation modeling.

This chapter explains the transformation of the cutters' assembly area into an assembly cell and is organized as follows: Section 10.2 presents background information including product characteristics, demand pattern, production process, and labor issues; Section 10.3 describes the technical details of the cell including the static workload analysis, layout configuration, and operational issues and assumptions used in the development of the simulation model. Section 10.4 presents the experimentation phase in which the performance of the system under different labor assignments was examined using simulation modeling; and, finally, Section 10.5 includes conclusions and final remarks.

10.2   Background

10.2.1   Description of the process

The cutters' unit handles 180 to 200 different products, which are currently assigned to 12 categories that are, in turn, grouped into 3 main families. Table 10.1 illustrates the distribution of product categories per family, namely

Table 10.1 Distribution of Product Categories by Part Family

Part family		
Leibinger	BIC	Formula
1. Ent	3. Bur	8. Formula Angle BIC
2. Hummer 4	4. Standard	9. Formula Standard
	5. TMJ	10. Formula Bur
	6. BIC	11. Formula TMJ
	7. Angle BIC	12. Formula BIC

Figure 10.1 Endoscopy product picture — cutter.

Leibinger, BIC, and Formula. The products designated as "Formula" are the new generation of the BIC family because it includes an additional safety feature that prevents its repeated use and, thus, avoids risk for patients. For this purpose, a computer chip that records and controls product usage and instrument speed is programmed and installed in the product. Figure 10.1 shows one of the products manufactured in this area. The cutters are inserted into a power unit referred to as the handpiece, which is part of a modular system used in surgical procedures. The handpiece is a reusable unit, while the cutters are designed as a single-use disposable product.

The demand pattern for the different product categories vary considerably, as shown in Figure 10.2. This product vs. quantity (P-Q) chart was constructed using data from the company's master production schedule (expressed in boxes of 5 units each) for 1 quarter. The figure shows that products within the BIC and Formula BIC categories have the highest volumes of demand representing 37 and 26%, respectively, of the total demand. These are followed by BUR (12%) and Formula BUR (8%), while the remaining eight categories have relatively smaller demands. When considering

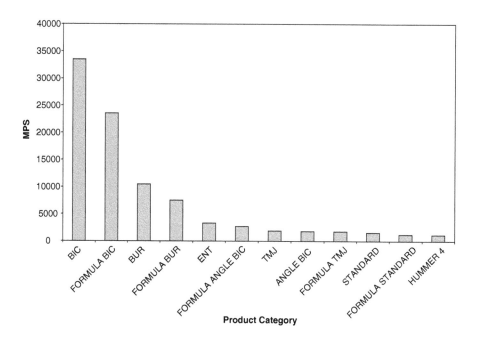

Figure 10.2 P to Q chart by product category.

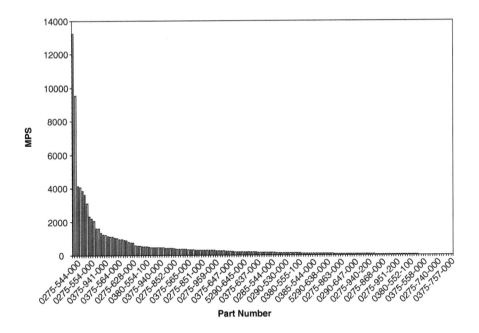

Figure 10.3 Pareto chart.

the distribution of demand by family, BIC has the highest volume (55%), followed by Formula (40%) and Leibinger (5%). In this high-volume, low-mix production environment, most of the products are assembled in a make-to-stock fashion, while a very small fraction is made-to-order.

Figure 10.3 illustrates the demand vs. individual part numbers comprising the cutters' product mix. From this chart, it is evident that only six to seven part numbers generate over 50% of the demand. Further analysis shows that 15% of the part numbers are responsible for approximately 80% of the total demand, thus showing that Pareto's principle applies to this product mix.

10.2.2   Processing steps

The manufacturing of cutters includes production activities that are done in three different production areas: the machine center, the clean room, and a packaging area located outside the clean room. The focus of our analysis is in the assembly area, which is located inside the clean room and is comprised of 1633 square feet. The operational policy for this area includes an 8-hour shift and approximately 240 working days per year. The effective working time per shift is reduced to 7 hours because time is dedicated daily to ergonomic exercises, safety gowning, lunch breaks, and other work allowances.

10.2.2.1   Production control

The material coordinator, together with the planner, daily check the inventory level per item to determine the days of available inventory. The forecasted demand and firm customer orders are subtracted from the inventory to determine what particular products should be replenished. Once an order is issued, the material handler pulls the necessary components from the warehouse and the group leader from the machine center takes the order. The order is then processed in the machine center and, when completed, transferred to the clean room for assembly and packaging. Once in the clean room, orders wait in a transfer room for a material handler or the cutters' supervisor to feed them into the assembly area and allocate them in the cutters' racks until they are ready to be processed.

10.2.2.2   Material flow

A material flow decision-making diagram for each product family is shown in Figure 10.4. As the figure shows, the products are composed of two main parts: housing and cutter or bur. The difference between cutters and burs is that cutters are used to cut tissue while burs are used to cut bone. The basic process steps needed for the assembly of cutters and burs are presented in Figure 10.5. This line flow diagram includes the sequence in which processes are executed including those that can be done simultaneously. The description of each process step is as follows:

> *Glue*: Plastic parts are attached separately to both cutter/bur and housing using a gluing solution. For the housing, this part represents the drive shaft to which a power cable is inserted. For the cutter (bur), the plastic part is the hub and is responsible for keeping the unit in a stable position. A curing time of 2 hours is required for most product categories, while products in the categories ENT, Angle BIC, and Formula Angle BIC require a 6-hour curing time.
>
> *Programming*: A computer chip is programmed and installed in the housing. The purpose of this chip is to prevent product re-use and control device speed. This operation is performed using either a numerical control (NC) machine or a manual process.
>
> *O-ring*: A rubber ring is attached to the housing to facilitate the rotation of the cutter or bur inside the housing. This process is completely manual.
>
> *Alcohol*: After the programming operation is performed, the outer part of the housing is cleaned with alcohol. This process is completely manual.
>
> *Spring*: A metal spring is inserted into the cutter or bur with the purpose of facilitating its removal from the handpiece and to centralize the piece. This process is completely manual.
>
> *Dipping*: A plastic tray with capacity for 300 cutters is submerged into a grease and alcohol mixture for the purpose of lubricating and strengthening them. Cutters (since burs do not have to pass through this step) are submerged for 1 minute followed by a 15-minute

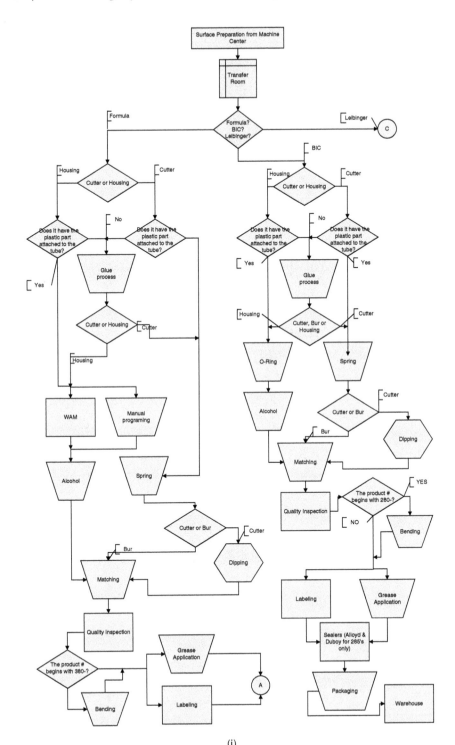

(i)

Figure 10.4 Actual material flow diagram.

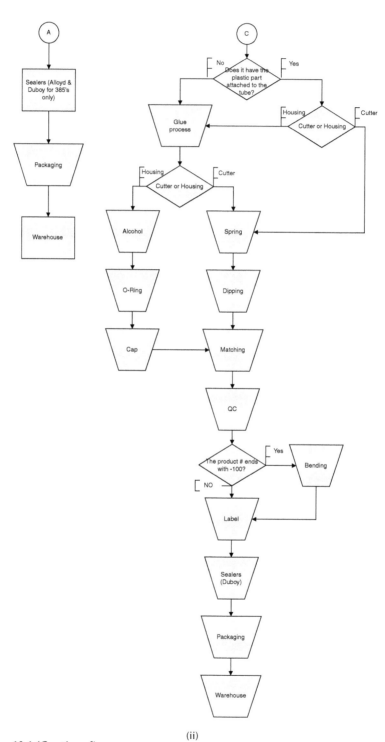

(ii)

Figure 10.4 (Continued).

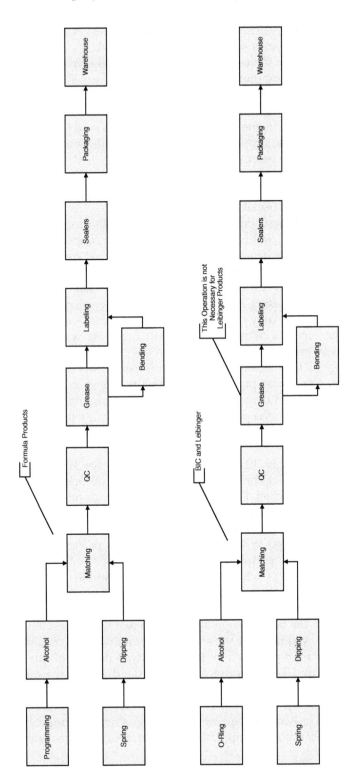

Figure 10.5 Current line flow diagram.

drying process using air blowers. This process is characterized by emission of fumes and is completely manual.

Matching: After the assembly operations for the cutter/bur and the housing are completed, they are assembled as a single piece.

Quality Control: Torque and leakage tests are performed to check cutter's (bur's) functionality. Only 20% of the cutters or burs are inspected using a sampling technique. Specialized equipment is used for this operation.

Grease: A grease coating is applied to the cutter or bur to facilitate rotation. This process is completely manual.

Bending: In this process, specific part numbers that belong to the categories ENT, Hummer 4, Angle BIC, and Angle Formula BIC are bent using a special fixture.

Labeling: In this process, a computer label with product information is printed out and attached to a plastic tray containing the assembled unit. The labels are automatically printed by a machine and attached manually. Operators in this area perform both operations.

Sealing: This process consists of sealing the packages containing the cutters/burs using a wax sheet. It can be performed in two types of machines: Alloy or Duboy. A particular situation in this process is the amount of particles present; products with an excessive amount are recycled.

Packaging: The packaging operation is performed outside the clean room. The sealed products are manually placed in packages of five units each and then placed into a box that contains six packages.

Warehouse: After the packaging operation is completed, boxes are put in storage and then sent for outside sterilization.

As shown in Figure 10.4 and Figure 10.5, the specific process route depends on the product characteristics because not all operations are performed with every product. Furthermore, it also depends on the condition in which parts are received from the machine center (e.g., to accelerate assembly lead time, high-volume part numbers have passed through the glue operation before they are delivered to the clean room). Another particular operation is dipping, which is only performed on the cutters. The "Leibinger" family is only composed of cutters, while the other two main families "BIC" and "Formula BIC" include cutters and burs. Furthermore, only specific part numbers that belong to the categories ENT, Hummer 4, Angle BIC, and Angle Formula BIC pass through the bending operation, which requires a special fixture.

The processing times for each assembly operation were obtained using the value stream map provided by the company. A portion of this diagram is shown in Figure 10.6. As appreciated from the diagram, the majority of the operations consist simply of a cycle time, which is the time per unit needed to perform the operation. This is the case for most manual processes including O-ring, Alcohol, Spring, Matching, Grease Application, and Bending. A summary of the operations and processing times is presented in Table 10.2.

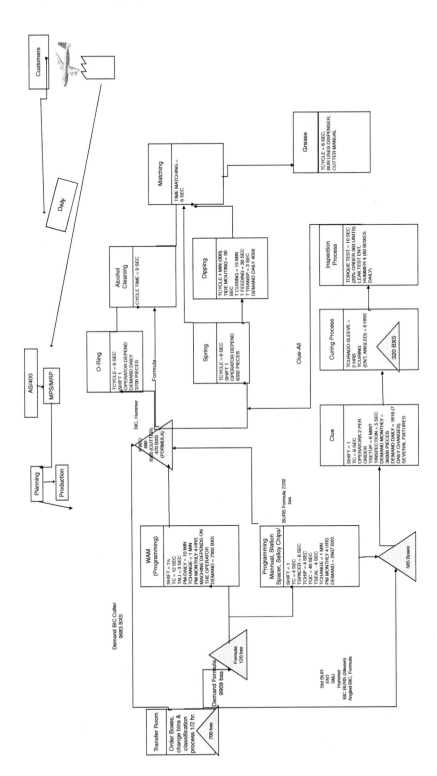

Figure 10.6 Value stream map.

Table 10.2 List of Operations and Processing Times Per Product Family

Process step	Formula	Part family BIC	Leibinger
Glue	CT = 6 sec/order S/T = 20 min CuT = 6 hrs. (Angle) or 2 hrs. (others) Insp = 5 sec	CT = 6 sec/order S/T = 20 min CuT= 6 hrs. (Angle) or 2 hrs. (others) Insp = 5 sec	CT = 6 sec/order S/T = 20 min CuT = 6 hrs. (ENT) or 2 hrs. (others) Insp = 5 sec
Programming (WAM)	CT = 12 sec/unit Tali = 3 sec Daily PM = 10 min T change = 1 min		
O-ring		CT = 6 sec/unit	CT = 6 sec/unit
Spring	CT = 6 sec/unit	CT = 6 sec/unit	CT = 6 sec/unit
Alcohol	CT = 9 sec/unit	CT = 9 sec/unit	CT = 9 sec/unit
Dipping	CT = 1 min/order S/T = 30 sec/order CuT= 15 min/order FT = 30 sec/order TT = 3 sec/unit	CT = 1 min/order S/T = 30 sec/order CuT = 15 min/order FT = 30 sec/order TT = 3 sec/unit	CT = 1 min/order S/T = 30 sec/order CuT = 15 min/order FT = 30 sec/order TT = 3 sec/unit
Matching	CT = 6 sec/unit	CT = 6 sec/unit	CT = 6 sec/unit
QC	Torque and leak tests = 8 min/order	Torque and leak tests = 8 min/order	Torque and leak tests = 8 min/order
Grease	CT = 6 sec/unit	CT = 6 sec/unit	
Bending	CT = 10 sec/unit	CT = 10 sec/unit	CT = 10 sec/unit

Note: CT = cycle time; CuT = curing time; FT = feeding time; S/U = set up; TT = transportation time. Tchange = change over time; Insp = inspection time; PM = preventative maintenance.

The glue operation involves times for setup, cycle, waiting, and inspection. The setup is done before processing the batch in order to prepare the equipment with the required features. The cycle time is the processing time per unit to perform the operation. For this process, there is a required waiting time of 6 hours for part numbers in categories ENT and Angle BIC and 2 hours for the remaining categories. This curing time is important for product quality and may represent a significant delay in the process because a long waiting time is needed to assure that parts are completely dried before continuing with subsequent processes. The only process that could be done while parts are waiting is manual programming, which only applies to the "Formula" products passing through glue.

The programming operation is required for the products in the "Formula" family and is done either by using manual or automatic programming. Both methods contain a cycle time and a changeover time. The cycle time is the time needed to program a single piece, while the changeover time is considered per order and represents the time needed between orders for line clearance. Manual programming is done using a desktop computer. Automatic programming, on the other hand, is performed using a numerical control machine that has a significant smaller lead time than the manual option and results in less ergonomic issues.

The dipping operation, only performed on the products containing a cutter, have time values for the processes of mounting, feeding, and transporting the plastic container used to place the pieces. The container has a capacity of 300 pieces, which are processed simultaneously, representing a limitation in terms of the transfer batch size. The transfer batch is forced to be the same as the batch size in order to keep the units together for the dipping process. Depending on the order size, the container can be used at partial or full capacity.

The quality control inspection process includes torque and leak tests, which are important in assuring product quality and performed only on 20% of the pieces within each order. The cycle time for this process is considered by order.

10.2.2.3   Production batches
The planner selects the production batch sizes depending on product sales volume and process characteristics. For example, the high-volume part numbers are produced in batches of 600 units and packed in 30-unit carton boxes containing 6 smaller packages with 5 units each. The products passing through the glue process are characterized by having smaller batch sizes, including 100, 75, or 50 units because of the long process waiting time. The remaining products (i.e., those that do not need be glued) have batch sizes of either 150 or 300 units, depending on the order size.

10.2.2.4   Current layout
The current layout is presented in Figure 10.7. As previously explained, orders wait in the cutters' rack until they are ready to be processed. The necessary

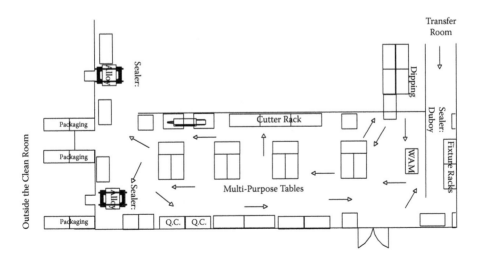

Figure 10.7 Current layout.

process steps to be performed depend on the order's part number. In the layout, there are areas dedicated to specific processes including dipping, programming (WAM), sealing (Alloy and Duboy), and quality control. The remaining process steps are performed on the multipurpose tables, which, based on the required process step, operators set up their workstations as needed. The complete batch is processed in each step before it is transferred to the next in sequence.

10.2.2.5   Material handling

A material-handling operator moves the material containers in and out of storage areas (i.e., machine center, transfer room, and cutters' rack). Because cutters, burs, and housings are lightweight, operators are responsible for material movement within the production area.

10.2.3   Labor issues

There are 27 persons currently assigned to the areas inside (i.e., assembly and labeling and sealing areas) and outside the clean room (i.e., packaging area). During peak season, this number could increase to 30 operators because the company does part-time hiring. The current distribution of personnel (operators and group leaders) per production area is presented in Table 10.3. Group leaders are responsible for administrative duties, keeping the daily work flowing and, during peak seasons, they also serve as operators because they familiar with many processes. Moreover, there is a manufacturing engineer who provides technical support to the area and a manufacturing supervisor who oversees the entire operation.

The area operates under a team structure in which operators and group leaders meet at least once a day to discuss production issues. This meeting is typically held after the daily ergonomic exercises and right before lunch. The manufacturing supervisor holds one-on-one monthly meetings with operators to discuss their progress and to address issues, such as their desire for cross training in other processes. Management has an open-door policy in which operators are encouraged to give their input at anytime.

Table 10.3 Distribution of Personnel Per Production Area

Production area	Number of personnel assigned
Assembly Operations	12 Operators
Quality Control	2 Operators
	1 Group leader
Sealing and Labeling	4 Operators
Packaging	5 Operators
(Outside the clean room)	1 Group leader
Quality Control	2 Operators
(Outside the clean room)	
Total	27

At the beginning of the fiscal year, operators, supervisors, group leaders, and management convene to set production targets. Operators are paid using traditional hourly wages, but their end-of-year compensation is linked to team performance and to whether their targets are reached. In the past, this scheme has been very effective in keeping operators motivated because the monetary amount received has been significant.

10.2.3.1 Labor flexibility

The concept of using a multifunctional workforce is referred to as labor flexibility, which is achieved by cross training operators in different tasks. According to Molleman and Slomp (1999), two important elements of labor flexibility are multifunctionality and redundancy. In this particular assembly area, multifunctionality (MF) is defined as the number of different types of processes an operator is able to perform, and redundancy (RD) is defined as the number of operators that can perform a specific process. By introducing multifunctionality, which is an attribute of the operator, and redundancy, which is an attribute of the process, a variety of labor strategies can be used to fulfill the demand for assembly time.

The operator–process matrix for the assembly area is presented in Table 10.4. According to the table, the multifunctionality of the operators varies from 6 to 9 operations, while the redundancy of each process varies from 3 to 15 operators. Operations, such as O-ring, Alcohol, Spring, Matching, and Grease are of low complexity and, therefore, most of the operators are able to master many of these processes and are considered to be highly multifunctional. Programming is the most challenging operation because it requires the use of a NC and, therefore, only five operators are able to master it. A similar case occurs in the quality control (QC) station where specialized equipment is needed for the leak and torque tests.

10.2.3.2 Cross training

The time needed for cross training in most assembly operations is only 1 day, while cross training in the more technically complex operations, such as WAM and manual programming, can take 2 to 3 days. The glue operation can only be performed by a subset of operators due to the dexterity skills required for the task. The bending operation is another particular case in which cross training is done with products because specific products have specific characteristics and processing requirements.

The cross-training process is conducted on the job and one of the operators guides the person, who is in training, through the different processes. The training operator receives an additional compensation.

10.2.3.3 Job design and responsibilities

Labor assignments are made by the manufacturing supervisor or by the group leaders. Operators work in the process to which they are assigned and remain there during the entire shift. However, in processes, such as dipping and glue, there is job rotation every 4 hours to prevent intoxication

Table 10.4 Operator – Process Matrix

	Blue	Program. WAM	O-Ring	Alcohol	Spring	Dipping	Matching	QC	Bending	Grease	MF
1	X		X	X	X	X	X			X	7
2	X		X	X	X	X	X		X	X	8
3	X	X	X	X	X	X	X		X	X	9
4	X		X	X	X	X	X		X	X	8
5			X	X	X		X		X	X	6
6	X		X	X	X	X	X			X	7
7	X	X	X	X	X	X	X		X	X	9
8	X		X	X	X	X	X			X	7
9	X		X	X	X	X	X			X	7
10	X		X	X	X	X	X			X	7
11	X		X	X	X	X	X			X	7
12	X		X	X	X	X	X			X	7
13	X	X	X	X	X		X	X	X	X	9
14		X	X	X	X		X	X	X	X	8
15	X	X	X	X	X		X	X		X	8
RD	13	5	15	15	15	11	15	3	7	15	

Note: MF = multifunctionality, RD = redundancy.

by fumes. From the above discussion, it is clear that one of the main benefits of cross training in this particular production area is that it makes job rotation feasible. Several empirical studies, including Adler and Cole (1993) and Berggren (1989), have shown that cross training could be motivated by human factors issues, such as reducing the burden of task repetitiveness and preventing physical problems.

10.2.3.4 Value of cross training

Management considers cross-trained operators as valuable assets because it allows them to respond to unexpected variations in the workload. Cross training permits job rotation in critical operations, such as dipping, and to handle operators' scheduled and nonschedule absenteeism. From the operators' perspective, cross training allows them to perform a variety of tasks and relieve the boredom associated with performing the same operation continuously. Furthermore, cross training makes operators more competitive because the degree of operators' multifunctionality is considered in their performance appraisal.

10.3 Design of the assembly cell

10.3.1 Problems and constraints with the current system

There are several opportunities in the assembly area at the company:

1. The production flow is not unidirectional because workstations are not arranged based on product flow.
2. Although operators are highly multifunctional, the company is not taking full advantage of this feature because labor allocation is mainly based on fixed assignments (i.e., operators remain at their assigned workstations for the entire shift).
3. The transfer batch between operations is the complete batch, thus the company is not taking advantage of the possible lead-time reductions that a piece flow can offer.
4. The variability in the batch sizes used for the different part numbers creates confusion in the assembly area and complicates the packaging operation because each box should contain 30 units and the current batch sizes are not multiples of 30.
5. Although the goal of the company is to complete 6000 units per day (i.e., 1200 packages of 5 units each), baseline data showed that this target is not always achieved using regular production time. An average of 12% overtime was reported for the period examined.

The company wanted to address these concerns along with the previously discussed strategic issues of new product introduction, limited clean room space, and efficient labor use in the most effective manner. Therefore, cellular manufacturing was proposed for this assembly area.

Cellular manufacturing systems are small organizational units within the firm designed to exploit similarities in how information is processed, products are made, and customers are served. Manufacturing cells foster continual performance improvements by closely locating people and equipment required for processing families of like product (Hyer and Wemmerlöv, 2002). Companies that adopt cellular manufacturing have reported significant improvements in important performance measures, such as lead times, production space, inventories, quality, and cost (Wemmerlöv and Johnson, 1997; Askin and Estrada, 1999). According to Irani et al. (1999), cells can be implemented in any manufacturing environment where one or more products may require the same combination of processes and resources. They also identified some of the conditions favorable for the introduction of cellular manufacturing including clearly defined and stable families, steady demand pattern and stable delivery schedules for assemblies, assembly equipment with low set-up times, and skilled and flexible workforce, among others. Based on the previous discussion, cellular manufacturing was considered an appropriate approach for the cutters assembly area.

10.3.2   *Layout configuration and operational issues*

Given the company's desire to implement the principles of lean manufacturing in every unit, the design of the assembly area included many of the elements necessary for lean production. Lean manufacturing is a management philosophy focused on waste elimination. Waste is defined as a non-value-added activity and in manufacturing these unnecessary activities include overproduction (producing more than what the market demands), waiting time (people waiting for materials, equipment, or other employees), transportation (moving materials long distances), inefficient processing (poor techniques, technologies, or fixturing), building inventories anywhere in the system (besides the inventories resulting from overproduction), motion (unnecessary walking and hand movements by operators), and scrap (producing defective parts and products).

Cells are implemented to support a lean operating philosophy. A 10-step approach to lean production is presented in Black and Hunter (2003, p. 25) and includes:

1. Re-engineer the manufacturing system
2. Setup reduction/elimination
3. Integrate quality control into the system
4. Integrate preventive maintenance into the system
5. Level, balance, sequence, and synchronize
6. Production control
7. Reduce work-in-process (WIP)
8. Integrate suppliers
9. Automation
10. Computer-integrated manufacturing

The company adopted lean manufacturing as its manufacturing philosophy in 2000 and since then it has already addressed many of these issues in its production areas. Therefore, for the purpose of this study, the analysis focused on core characteristics of cells including efficient layout and the use of multifunctional operators.

According to Mahmoodi et al. (2001), the cell design process can be lengthy, requiring substantial effort to evaluate candidate cells according to a diversity of operational and economical performance measures. They proposed a three-step framework for evaluating cellular manufacturing configurations including: static evaluation using computational schemes, stochastic evaluation using queuing theory techniques, and dynamic evaluation using computer simulation. For the purpose of this study, the static and dynamic evaluations are used. The stochastic evaluation was not considered because a detailed examination of labor allocation was needed and this can only be assessed using simulation modeling.

10.3.2.1 Static evaluation

One of the most important steps in the cell design process is the configuration of the factory floor. A static evaluation model based on the assembly area annual workload was used to assist in this process. Static evaluation models assume a deterministic environment (e.g., predictable schedules, no breakdowns, and part availability) and ignore all dynamics, interactions, and uncertainties typically seen in manufacturing systems (Suri and Diehl, 1986). The company initially proposed a three-cell configuration layout based on product families, but after a static workload analysis, it was evident that the utilization of the different processes within the assembly area was very low and, thus, the possibility of using only one cell was entertained.

A process analysis using the material flow diagrams and current flow line (Figure 10.4 and Figure 10.5) showed that all products pass through the same basic operations. Furthermore, detailed process mapping of the three families using the value stream map (Figure 10.6) confirmed that the routings had only very minor variations. Due to the consistent routing patterns, only one assembly cell was proposed for further analysis.

The annual workload summary for the proposed assembly cell is presented in Table 10.5 and its corresponding workload profile is shown in Figure 10.8. The operation with the highest workload is alcohol and requires three processing stations, followed by programming, glue, matching, and spring, which require two stations each. The remaining stations need only one processing station. Since at the time of the case study the company was considering the acquisition of a new WAM machine for the programming operation for further analysis, it is assumed that this machine will replace manual programming. As previously mentioned, manual programming was associated with ergonomic issues and the company wanted to eliminate these issues.

Several constraints limit the possible layout configurations in this assembly area. The dipping and glue processes need to be located at a reasonable

Table 10.5 Annual Workload Summary (Hours)

	Initial product	Housing	Housing	Housing	Cutter			Housing and cutter	Final product	Final product	Final product
	Glue	Program	O-Ring	Alcohol	Spring	Dipping	Cutter	Matching	QC inspection	Bending	Grease application
LEIBENGER	488	0	111	167	111	30	111	111	119	39	0
BIC	1415	0	1230	1846	1230	93	1230	1230	490	75	1230
FORMULA BIC	549	2529	0	1376	917	75	917	917	301	111	917
TOTAL REQUIREMENTS	2452	2529	1342	3388	2259	198	2259	2259	910	225	2148
AVAILABLE	1640	1640	1640	1640	1640	1640	1640	1640	1640	1640	1640
Required Number of Stations	1.49	1.54	0.82	2.07	1.38	0.12	1.38	1.38	0.55	0.14	1.31
Assigned Number of Stations	2	2	1	3	2	1	2	2	1	1	2
AVERAGE WORKSTATION UTILIZATION	75%	77%	82%	69%	69%	12%	69%	69%	55%	14%	65%

Assumptions:

Average batch sizes: 600 units (part no. considered high-volume items), 75 units (orders that passed through glue), 25 units (remaining part numbers).

Available time per year: 12 months/year × 20 days/month × 1 shift/day × 7 hours/day = 1680 hours/year.

Cleaning time: 10 min/day × 1 hr/60 min × 12 months/year × 20 days/month × 1 shift/day = 40 hours/year.

Available production time = available time − clearance time = 1640 hours/year.

Formulas: (given per operation):

Annual workload per part number = (number of orders/year) × (setup time + order size × cycle time per piece).

Annual workload per product family = $\Sigma_{\text{all part numbers within the family}}$ annual part number workload.

Total required annual workload = annual workload–Leibenger + annual workload–BIC + annual workload–Formula BIC.

Required number of workstations = total required annual workload/available production time.

Utilization = required annual workload/(number of workstation − available production time).

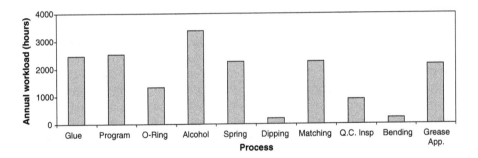

Figure 10.8 Workload profile.

distance from the other operations because they emit harmful fumes that can affect operators if they are exposed for a prolonged period of time. The sealing operations are affected by the presence of particles; an excessive amount can cause the product to be reprocessed and, therefore, the Alloy and Duboy machines also need to be located at a reasonable distance from the assembly operations.

Based on the product flow, static workload analysis, and previously mentioned constraints, several layout configurations were proposed. After evaluating the pros and cons of each proposal, the configuration shown in Figure 10.9 was selected. This configuration includes a combination of a U-shaped cell and a linear layout. In the U-shaped portion, the assembly of the cutters and burs and their housings are processed in parallel. Once these two parts are completed, they are matched and continue as a single product to the QC, bending, and grease operations, respectively.

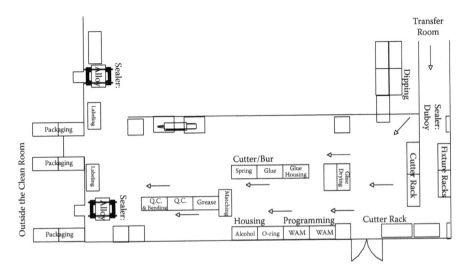

Figure 10.9 Proposed layout.

One of the important features of the proposed layout is flexibility because additional stations can be easily accommodated in the assembly area. This will allow the company to handle increases in demand or the introduction of new products without major configuration changes. Furthermore, when compared to the current layout, the proposed alternative resulted in a 17% reduction of the space occupied by the assembly unit because workstations are placed more closely together and reduced in number. A particular feature of the layout is that due to the proximity of the workstations, operators are able to work on more than one process and, thus, allowing for the consideration of several labor strategies within the area.

10.3.2.2   Production leveling

One of the most important operational aspects in the design of an assembly cell using the concepts of lean manufacturing is production leveling. Leveling is the process of planning and executing an even production scheduling. Leveling production for a given period develops steady material flow, which results in a repetitive manufacturing environment. To perform level production, a mixed-model assembly line must be created by smoothing aggregate production requirements and sequencing final assembly. Detailed computations of the mixed-model assembly schedule for this cell are presented in Appendix 10.1.

10.3.2.3   Dynamic evaluation

In order to examine the performance of the proposed layout and labor strategies, a dynamic evaluation model using stochastic simulation was used. A simulation model is the only methodology that is robust enough to systematically examine the role and impact of key variables on system performance. By using this model, the status of each product and workstation can be traced over time, thereby identifying the actual loading patterns, idle times, systems bottlenecks, operator utilization, and queue lengths.

The simulation model for this assembly area was developed using *ARENA™* software (Kelton et al., 2004). In the model, the assembly cell produces each part type in proportion to its share of overall demand, thus allowing for a level production. Job interarrival times are calculated based on the total available time and number of jobs. The release of jobs depends on the number of batches calculated using the logic presented in Appendix 10.1.

Clean room operations include a combination of semiautomatic and manual processes. The only process step that could be considered as a semiautomatic process is programming, when performed in the WAM machine. However, because the cycle time is relatively low, it requires an attendant operator for the entire process. The cycle time in the WAM machines is modeled as constant because this is an NC machine and little or no variability in processing time is expected.

The processing and setup times in each manual process are modeled using a normal distribution with a mean equal to its average cycle time and a coefficient of variation (CV) of 0.5. According to Hopp and Spearman

(2000), these times are subject to natural variation, which is characterized with low variability (i.e., $0 < CV < 1$).

In order to address the concerns regarding batch sizes and to facilitate packaging, the proposed batch sizes are multiples of 30. For high volume part numbers, the batch size is 600 (i.e., 20 boxes of 30 units), 60 for those part numbers that pass through dipping and 240 for the remaining part numbers.

Based on the assembly area current and proposed operational policies, the following assumptions and constraints were used in constructing the simulation model:

1. The available production time per year is 1640 hours (see Table 10.5 for computations).
2. All operators perform at 100% efficiency at each workstation.
3. Operators are confined to the assembly area and there is no transfer of operators to packing and labeling and sealing areas.
4. Operators are responsible for assembly area part handling; walking time between processes is assumed to be negligible.
5. The simulation will stop when the 240 working days have been completed.
6. Parts are removed from processing queues according to the low-value first rule using the part type as reference value (see Appendix 10.1). This rule was used in order to keep the orders together.
7. There are no breakdowns in the manual processes. In the programming workstation, WAM machines are assumed to have a 5% downtime. These machines undergo daily and monthly preventive maintenance during nonproduction time.
8. The only process in which rework is necessary is the sealing process when an excessive amount of particles are present and, because our focus is on the assembly area, a rework rate is not included in the model.

The specific characteristics used in building the simulation model for this assembly cell are summarized in Table 10.6.

The mean production rate, the percentage of annual production completed, and the mean operator utilization are used to assess the performance of the system under different labor allocation scenarios. Each experimental condition was run for 240 days including a warm-up period.

10.4 Experimental phase

The issues of labor cross training and labor allocation are explored in this section. Several labor strategies are evaluated in order to determine which strategies are more appropriate for this assembly area. Labor assignments are classified from the operator's perspective as fixed or floating assignments. Assignments are fixed when an operator is assigned to only one process. Assignments are floating when the operator is assigned to more

Table 10.6 Assembly Cell: Features and Assumptions

Condition	Description
Number of parts/families	180 parts grouped into 12 categories and 3 part families (i.e., Leibenger, BIC, Formula)
Cell size	15 stations grouped into 10 different workstations
Number of operations per part	6 to 10, depending on part type
Interarrival times	Based on empirical distribution (see Appendix 10.1)
Manual operations	
Cycle	Normally distributed ($\mu = CT_{process}$, $CV = 0.5$)
Setup times	Normally distributed ($\mu = $ Setup time$_{per\ order}$, $CV = 0.5$)
Programming operation	Constant
Queue service discipline	Low-value first rule using the part type as reference value (see Appendix 10.1).
Part flow	One-piece flow
Workstation type	NC machines: Programming Manual: Glue, O-ring, Alcohol, Spring, Dipping, Matching, Grease Specialized equipment: QC, Bending
Operator responsibilities	Setup, processing, part handling
Machine downtime	WAM – 5%
Operator's "where to move" rule	Depends on the type of operator's labor assignment
Average production rate (based on actual demand pattern)	5648 units per day (see Appendix 10.1 for computations, 1,355,390 units/year ÷ 240 days/year)

than one process at a time. The multifunctionality and redundancy of each labor strategy will depend on the type of assignment given the operators.

The first step in our investigation is to determine the appropriate number of operators that should be assigned to the assembly area. For this purpose, the performance of the system is evaluated using floating assignments in which the workload is completely shared among operators (i.e., operators can work on any process as required). Examples of the labor strategies used are provided in Appendix 10.2. The number of operators was systematically increased from 1 to 20 and the performance of the system under the scenarios estimated. The results are shown in Figure 10.10 and Figure 10.11.

Figure 10.10 clearly shows that in order to satisfy the expected annual demand, which requires an average production rate of 5648 units per day, at least 11 fully cross-trained operators need to share the workload. The average utilization of the operators as shown in Figure 10.11 remains close to 100% when 1 to 10 operators are assigned and starts to slowly decrease as additional operators are added.

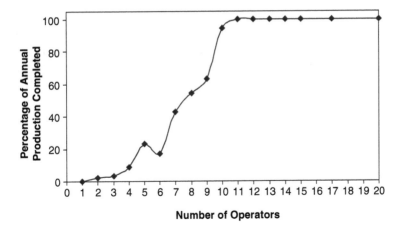

Figure 10.10 Percentage of annual production completed vs. number of floating operators.

The company's current operational policy considers 15 operators per shift (including assembly and QC operators) in the assembly area. However, despite the fact that most operators are highly multifunctional as previously discussed, labor strategies currently used have low values for the operator's multifunctionality. The typical value is 1 because operators are assigned to only one process and remain there for the entire shift. Therefore, in order to explore the benefits of using labor strategies based on floating assignments, several labor strategies were examined. These labor strategies included 13 operators per shift, which was considered an appropriate staffing level to achieve a good physical flow of material through the cell. Furthermore, based on the simulation results presented in Figure 10.11, this number should provide reasonable average operator utilization.

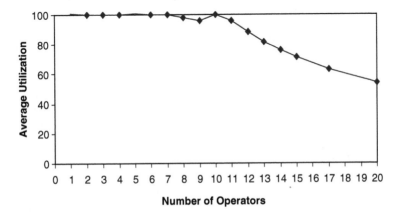

Figure 10.11 Average operator utilization vs. number of floating operators with completely shared assignments.

Table 10.7 Summary of Labor Strategies Based on Primary and Secondary
Operations

Labor strategy	Primary operations	Secondary operations	MF (per operator)	RD (per machine)
1 (Base case)	1	0	1–3	1–2
2	1	1	2–4	1–4
3	1	2	3–5	2–6
4	1	3	4–6	2–8
5	1	4	5–7	4–8
6	1	9	10	13

Note: MF = multifunctionality, RD = redundancy.

Two different sets of experiments were performed. The first set consisted
of labor strategies using the floating technique where some or all of the operators have primary and secondary assignments. When using this technique,
operators work in the primary operation to which they are assigned; whenever
idle and there is work available in their secondary operation(s), they transfer
to that operation to perform the work and return as soon as there is work
available in their primary operation. The proposed labor strategies examined
under this technique are presented in Appendix 10.3 and summarized in
Table 10.7. Labor Strategy 1 represents the base case in which all operators
have fixed assignments (i.e., only primary assignments). In Labor Strategy 2,
operators are assigned a secondary operation. Likewise, in Labor Strategies 3
and 4, operators are assigned to two and three secondary assignments, respectively. Finally, in Labor Strategy 5, operators have one primary assignment
and nine secondary assignments. This labor strategy is added for the purpose
of comparison because it is seldom used in practice. Most secondary operations were selected based on their proximity to the primary operation. In some
cases, the secondary operation was selected based on workload because some
operations have a relatively low workload and the attendant operator can
handle more than one operation without much disruption in the workflow.

The performance of these labor strategies is presented in Table 10.8. In
Labor Strategy 1, the base case, only 75% of the required demand was satisfied.
The average operator utilization was 78% with a standard deviation of 0.15,
resulting in a CV = 0.19. This labor strategy also was characterized by a
significant amount of work in process inventory. In Labor Strategy 2, the
performance was close to 100%, thus showing that adding a secondary operation to each operator resulted in a performance improvement of 33%. The
average operator utilization was 90% with a CV = 0.10. Labor Assignments 3,
4, and 5 were also associated with high performance, average operator utilization of 85%, and a coefficient of variation ranging from 0.09 to 0.10. Finally,
Labor Assignment 6 not only was associated with high production performance but also resulted in the smallest coefficient of variation, CV = 0.04.

The second set of experiments considered labor strategies using fixed
and floating assignments. The workload analysis in Table 10.5 was used to

Table 10.8 Summary of Results: Primary and Secondary Operations

	Labor strategy					
	1	2	3	4	5	6
Percentage of annual production completed	75.14%	99.77%	99.76%	99.80%	99.80%	99.85%
Average production per day	4317.32	5634.35	5634	5636.25	5636.34	5638.75
Operator ID		Operator Utilization				
1	0.68	0.81	0.80	0.67	0.78	0.82
2	0.68	0.98	0.89	0.72	0.72	0.77
3	0.70	0.74	0.73	0.87	0.85	0.85
4	0.70	0.99	0.91	0.83	0.88	0.79
5	0.81	0.95	0.83	0.95	0.94	0.85
6	0.99	1.00	0.96	0.88	0.85	0.85
7	0.99	0.99	0.92	0.88	0.85	0.87
8	0.68	0.80	0.86	0.73	0.82	0.83
9	0.68	0.94	0.91	0.92	0.92	0.86
10	0.93	0.87	0.77	0.93	0.91	0.80
11	0.71	0.87	0.95	0.94	0.90	0.81
12	0.61	0.89	0.77	0.84	0.73	0.88
13	0.99	0.89	0.76	0.86	0.83	0.78
Average utilization	0.78	0.90	0.85	0.85	0.84	0.83
Std. Dev.	0.15	0.09	0.08	0.09	0.07	0.04
CV	0.19	0.10	0.10	0.11	0.09	0.04

Table 10.9 Summary of Labor Strategies Based Fixed and Floating Assignments

Labor strategy	Fixed assignments		Floating assignments		RD (per operation)
	No. operators	MF (per operator)	No. operators	MF (per operator)	
1	8	1	5	2	1–3
2	8	1	5	3	1–3
3	8	1	5	4	1–4
4	8	1	5	5	2–5
5	8	1	5	6	3–5
6	8	1	5	7	3–6
7	8	1	5	8	3–7
8	8	1	5	9	3–7
9	8	1	5	10	5–7

Note: MF = multifunctionality, RD = redundancy.

identify the operations with the highest workloads to which one or two operators were assigned using fixed assignments. The resulting number of operators per process was: Glue (1), Programming (1), O-ring (1), Alcohol (2), Spring (1), Matching (1), and Grease (1). Therefore, the proposed labor strategies have a total of eight operators with fixed labor assignments while the remaining five operators have floating assignments. The number of shared operations in the floating assignments was systematically increased from two to five. The proposed labor strategies examined are presented in Appendix 10.4 and summarized in Table 10.9.

The performance of these labor strategies is presented in Table 10.10. As can be appreciated from the table, production performance in all the strategies is high and overall average operator utilization remains relatively stable. A further analysis of the average operator utilization by the type of labor assignment reveals that as the number of shared operations increases, the average operator utilization in the fixed assignments decreases while there is an increase in floating assignments. This interesting result is depicted in Figure 10.12 and reveals that as the number of shared operations increases, floating operators undertake a higher portion of the workload, thus reducing the workload of operators with fixed assignments.

10.4.1 Implications to the company

The operators' current cross-training levels provide relatively high values for multifunctionality and redundancy. Based on simulation results, from the performance standpoint, the company does not need to further cross train operators. However, from a strategic perspective, cross training can provide several other benefits.

The simulation analysis shows that the company can take further advantage of their cross-trained operators by allowing them to work on more than one process during their shifts. An important outcome of this study is that

Table 10.10 Summary of Results: Fixed and Floating Assignments

	2	3	4	5	6	7	8	9	10
Percentage of Annual Production Completed	99.71%	99.79%	99.77%	99.85%	99.78%	99.82%	99.80%	99.85%	99.84%
Average Production per Day	5631.35	5635.75	5634.50	5638.75	5635.20	5637.58	5636.75	5638.75	5638.75
Operator ID									
Fixed assignments									
1	0.90	0.79	0.82	0.67	0.70	0.67	0.64	0.60	0.72
2	0.97	0.75	0.77	0.75	0.73	0.66	0.67	0.67	0.67
3	0.44	0.54	0.56	0.58	0.55	0.58	0.58	0.57	0.60
4	0.89	0.92	0.86	0.91	0.88	0.85	0.85	0.86	0.86
5	0.80	0.91	0.86	0.90	0.87	0.82	0.82	0.83	0.83
6	0.73	0.75	0.75	0.81	0.70	0.73	0.72	0.69	0.64
7	0.97	0.85	0.86	0.86	0.76	0.80	0.73	0.72	0.68
8	0.77	0.88	0.86	0.91	0.84	0.91	0.81	0.76	0.70
Average utilization	0.81	0.80	0.79	0.80	0.75	0.75	0.73	0.71	0.71
Std. Dev.	0.20	0.14	0.12	0.14	0.13	0.13	0.11	0.11	0.10
CV	0.25	0.18	0.15	0.18	0.17	0.17	0.15	0.16	0.14
Floating assignments									
9	1.00	0.99	1.00	0.97	0.96	0.98	1.00	1.00	1.00
10	0.77	0.92	0.88	0.94	0.99	0.97	0.99	1.00	1.00
11	0.74	0.89	0.90	0.98	0.96	0.97	0.99	1.00	0.99
12	0.89	0.79	0.81	0.95	1.00	1.00	1.00	1.00	1.00
13	0.67	0.90	0.90	0.96	1.00	1.00	1.00	1.00	1.00
Average utilization	0.81	0.90	0.90	0.96	0.98	0.98	1.00	1.00	1.00
Std. Dev.	0.16	0.09	0.08	0.02	0.02	0.02	0.01	0.00	0.00
CV	0.20	0.10	0.09	0.02	0.02	0.02	0.01	0.00	0.00
Overall Results									
Average utilization	0.81	0.84	0.83	0.86	0.84	0.84	0.83	0.82	0.82
Std. Dev.	0.16	0.12	0.11	0.13	0.15	0.15	0.16	0.17	0.17
CV	0.20	0.14	0.13	0.15	0.18	0.18	0.19	0.21	0.20

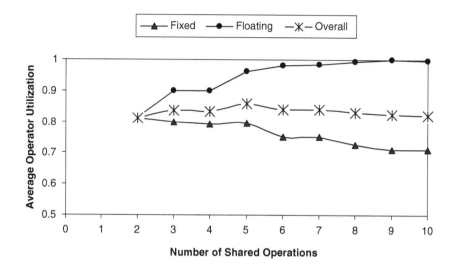

Figure 10.12 Average operator utilization vs. number of shared operations using a combination of fixed and floating labor assignments.

by using labor assignments based on floating techniques, the company can reduce the number of operators needed from 15 to 13 (i.e., a labor reduction of 13%) and completely eliminate overtime. Furthermore, the results reveal that annual demand can be achieved by employing any of the floating techniques examined in the experimentation phase. In the technique where some or all the operators have only primary assignments, significant improvement was achieved by adding one secondary operation. The production rate increased 32% when compared to the base case.

In the floating technique where labor strategies are based on a combination of fixed and floating assignments, the results also revealed that the demand could be achieved using any of the proposed labor assignments. However, it is important to point out that excellent results were obtained even at the lowest levels off multifunctionality where most floating operators work in only two operations. Therefore, it is not necessary to have high levels of multifunctionality in order to achieve good performance.

The next step for the company is to implement the proposed assembly cell system. However, before this process takes place, management should carefully plan the cell implementation project. Issues, such as the change process and possible resistance to the new way of operating, should be addressed *a priori*. Furthermore, preparing the production personnel for cells by training them in a variety of areas, including lean manufacturing, overall cells concepts training, and team work, is critical because appropriate training is one of the most important implementation issues.

In order to have a better workflow in the cell, the company should re-examine the glue and dipping processes. These processes, characterized by emission of fumes, complicate the design of the cell because all required processes should not be closely located within the cells' physical boundaries. For the glue process, which involves a lengthy curing time, the company should perform a viability study in order to decide whether it should be moved to the machine center. As previously mentioned, high volume part numbers have passed through the glue operation before they are delivered to the clean room. Another alternative is changing product design by replacing the glue and curing operations with snap-in components that could be attached in the cell. For the dipping process, the company also should evaluate what other possible alternatives are available.

10.5   Conclusions

This analysis shows that the company can benefit extensively by applying the concepts of lean manufacturing and by implementing assembly cells. Elements such as a layout configuration based on product flow, a cross-trained workforce, workstation proximity to facilitate movement of operators, one-piece flow, and a mixed model assembly schedule are fundamental to this conversion.

Using a two-phase approach that included static workload analysis and stochastic simulation modeling, it was shown that the assembly area can be converted into one assembly cell capable of handling all product categories. Furthermore, by using simulation modeling, fundamental elements in dimensioning the cell were examined. These elements included the number of workstations and the number of operators needed to meet demand. The analysis considered the degree of operator cross training, the number of processes, and the specific processes each operator should be assigned. Key insights from the analysis are that the company can benefit extensively by having cross-trained operators and by using labor assignments based on floating techniques. However, high levels of operator cross training are not necessary in order to achieve good performance.

Regarding the value of cross training, this company considered it as an important element of their corporate strategy. The use of multifunctional operators is perceived by management as a valuable asset because it allows them to respond to unexpected variations in the workload, balance the workload assigned to the operators, have job rotation in critical operations, and handle operators' scheduled and nonscheduled absenteeism. Furthermore, cross training makes operators more competitive because the degree of operators' multifunctionality is considered in their performance appraisal.

Acknowledgments

I would like to thank the NSF–Alliance for Minority Participation for their support in this research and my undergraduate research assistants, Lourdes Medina and Gretchen Torres, for their excellent work. I am particularly grateful to the company for its willingness not only to participate but also for providing data and allocating resources for this investigation.

References

Adler, P.S., and Cole, R.E. (1993). Design for learning: a tale of two auto plants, *Sloan Management Review*, Spring, 85–94.

Askin, R., and Estrada S. (1999). Investigation of cellular manufacturing practices, in *Handbook of Cellular Manufacturing Systems*, S.A. Irani (Ed.), John Wiley & Sons, New York, 25–34.

Berggren, C. (1989). New production concepts in final assembly — the Swedish experience, in *The Transformation of Work? Skill, Flexibility, and the Labour Process*, S. Wood (Ed), Unwin Hyman, London.

Black, J.T., and Hunter S.L. (2003). *Lean Manufacturing Systems and Cell Design*, Society of Manufacturing Engineers, Dearborn, MI.

Bokhorst, J., and Slomp J. (2000), Long-term allocation of operators to machines in manufacturing cells, *Proceedings of the Group Technology Cellular Manufacturing World Symposium*, San Juan, Puerto Rico, 153–158.

Cesaní, V.I., and Steudel H.J. (2005). A study of labor assignment flexibility in cellular manufacturing systems, *Computer and Industrial Engineering*, 48, 571–591.

Hopp, W.J. and Spearman, M.L., (2000). *Factory Physics*, 2d. ed., McGraw Hill, New York

Hyer, N., and Wemmerlöv U. (2002). *Reorganizing the Factory: Competition through Cellular Manufacturing*, Productivity Press, Portland, OR.

Irani, S., Subramanina S., and Allam Y.A. (1999). Introduction to cellular manufacturing systems, in *Handbook of Cellular Manufacturing Systems*, S.A. Irani (Ed.), John Wiley & Sons, New York, 1–23.

Kelton, D., Sadowski R.P., and Storrock D.T. (2004). *Simulation with Arena*, 3rd ed., McGraw-Hill, New York.

Mahmoodi, F., Mosier, T., and Burroughs, A.T. (2001). Framework for cellular manufacturing evaluation process using analytical and simulation techniques, in *Handbook of Cellular Manufacturing Systems*, S.A. Irani (Ed.), John Wiley & Sons, New York, 179–194.

Molleman, E., and Slomp J. (1999), Functional flexibility and team performance, *International Journal of Production Research*, 37(8), 1837–1858.

Slomp, J., Bokhorst, J., and Molleman, E. (2005). Cross-training in a cellular manufacturing environment, *Computer and Industrial Engineering*, 48, 609–624.

Suri, R., and Diehl G.W. (1986). Manuplan: a precursor to simulation for complex manufacturing systems, *Proceedings of the Winter Simulation Conference IEEE*, Piscataway, NJ, 411–420.

Swamidass, P.M. (Ed.) (2000). *Encyclopedia of Production and Manufacturing Management*, Kluwer Academic Publishers, Dordrecht, The Netherlands.

Wallace, H.J., and Spearman M.L. (2000). *Factory Physics*, 2nd ed., McGraw-Hill, New York.

Wemmerlöv, U. and Johnson, D.J. (1997). Cellular manufacturing at 46 user plants: Implementation experiences and performance improvements, *International Journal of Production Research*, 35 (1), 29–49.

Appendix 10.1 Mixed-Model Assembly Dispatching Logic

TOTAL REQUIREMENTS		Create				
Available time calculation:	24600	Interval – # of sec. between days			Total Secs/Day	86400
12 months/year					Secs/Yr	20736000
20 days/year					Nonworking secs/day	61800
1 shift/day	Available time:				Working secs/day	24600
7 hours/day	5904000					
Consider:	10 min/day – cleaning tables					

MODIFICATION OF CREATE SECONDS/YR 20736000

ASSIGN

Group	Classification	Sub-Class	Part Type	NS	Demand/Year	Lot size	No. of Lots	# Lots/Year to Model	Interarrival Time	Demand Modeled	Demand/10 Days	Orders/10 Days
LEIBINGER	Ent	LG	1	1	38482	60	641.37	641	32349.45	38460	1602.5	26.71
LEIBINGER	Ent	LGB	2	2	11024	60	183.73	183	113311.48	10980	457.5	7.63
LEIBINGER	Hummer 4	LG	3	1	10634	60	177.24	177	117152.54	10620	442.5	7.38
LEIBINGER	Hummer 4	LG-all	4	3	3512	60	58.53	59	351457.63	3540	147.5	2.46
LEIBINGER	Hummer 4	LGB	5	2	3067	60	51.12	51	406588.24	3060	127.5	2.13
BIC	Bur	B	6	4	68837	240	286.82	287	72250.87	68880	2870	11.96
BIC	Bur	BG-all	7	5	31651	60	527.52	528	39272.73	31680	1320	22.00
BIC	Bur	BG	8	6	56716	60	945.26	945	21942.86	56700	2362.5	39.38
BIC	Standard	BD	9	7	23141	240	96.42	97	213773.20	23280	970	4.04
BIC	Bic	BD	10	7	185361	240	772.34	772	26860.10	185280	7720	32.17
BIC	Bic	B-hv	11	8	317187	600	528.65	529	39198.49	317400	13225	22.04
BIC	TMJ	BDG-all	12	9	28363	60	472.72	473	43839.32	28380	1182.5	19.71
BIC	Angle Bic	BGDB	13	10	26990	60	449.83	450	46080.00	27000	1125	18.75
FORMULA	F. Standard	FD-W	14	11	17755	240	73.98	74	280216.22	17760	740	3.08
FORMULA	F. Bic	FD-W	15	11	93603	240	390.01	390	53169.23	93600	3900	16.25
FORMULA	F. Bic	F-hv-W	16	12	259590	600	432.65	432	48000.00	259200	10800	18.00
FORMULA	F. TMJ	FG-all-M	17	13	19511	60	325.18	325	63803.08	19500	812.5	13.54
FORMULA	F. TMJ	FD-W	18	11	6879	240	28.66	29	715034.48	6960	290	1.21
FORMULA	F. Bur	F-M	19	14	113020	240	470.92	471	44025.48	113040	4710	19.63
FORMULA	F. Angle Bic	FGDB-M	20	15	40067	60	667.78	668	31041.92	40080	1670	27.83
					LOTS/YEAR 7580.72			7581	DEMAND/YEAR	1355400	Total Orders/ 10 Days:	315.88

DEMAND/YEAR 1355390

Appendix 10.2 Labor Strategies Based on Completely Floating Operator Assignments

Assignment 1

operator	Glue	Programming	O-Ring	Alcohol	Spring	Dipping	Matching	QC	Bending	Grease	MF
1	1	1	1	1	1	1	1	1	1	1	10
Redundancy	1	1	1	1	1	1	1	1	1	1	

Assignment 5

operator	Glue	Programming	O-Ring	Alcohol	Spring	Dipping	Matching	QC	Bending	Grease	MF
1	1	1	1	1	1	1	1	1	1	1	10
2	1	1	1	1	1	1	1	1	1	1	10
3	1	1	1	1	1	1	1	1	1	1	10
4	1	1	1	1	1	1	1	1	1	1	10
5	1	1	1	1	1	1	1	1	1	1	10
Redundancy	5	5	5	5	5	5	5	5	5	5	

Assignment 10

operator	Glue	Programming	O-Ring	Alcohol	Spring	Dipping	Matching	QC	Bending	Grease	MF
1	1	1	1	1	1	1	1	1	1	1	10
2	1	1	1	1	1	1	1	1	1	1	10
3	1	1	1	1	1	1	1	1	1	1	10
4	1	1	1	1	1	1	1	1	1	1	10
5	1	1	1	1	1	1	1	1	1	1	10
6	1	1	1	1	1	1	1	1	1	1	10
7	1	1	1	1	1	1	1	1	1	1	10
8	1	1	1	1	1	1	1	1	1	1	10
9	1	1	1	1	1	1	1	1	1	1	10
10	1	1	1	1	1	1	1	1	1	1	10
11	1	1	1	1	1	1	1	1	1	1	10
12	1	1	1	1	1	1	1	1	1	1	10
13	1	1	1	1	1	1	1	1	1	1	10
Redundancy	13	13	13	13	13	13	13	13	13	13	

Appendix 10.3 Labor Strategies Based on Primary and Secondary Operator Assignments

Assignment 1

operator	Glue	Programming	O-Ring	Alcohol	Spring	Dipping	Matching	QC	Bending	Grease	MF
1	1										1
2	1										1
3		1									1
4		1									1
5			1								1
6				1							1
7				1							1
8					1						1
9					1						1
10						1		1	1		3
11							1				1
12							1				1
13										1	1
Redundancy	2	2	1	2	2	1	2	1	1	1	

Assignment 2

operator	Glue	Programming	O-Ring	Alcohol	Spring	Dipping	Matching	QC	Bending	Grease	MF
1	1					1					2
2	1			1							2
3		1				1					2
4		1		1							2
5			1				1				2
6				1	1					1	2
7				1	1						2
8					1						2
9					1						2
10						1	1	1	1	1	4
11							1	1		1	2
12								1		1	2
13							1				2
Redundancy	2	2	1	4	4	3	4	3	1	4	

Appendix 10.3 Labor Strategies Based on Primary and Secondary Operator Assignments (Cont.)

Assignment 3

operator	Glue	Programming	O-Ring	Alcohol	Spring	Dipping	Matching	QC	Bending	Grease	MF
1	1	1									3
2	1			1							3
3		1	1		1	1					3
4		1	1	1		1					3
5			1			1				1	3
6	1			1						1	3
7			1	1	1						3
8					1		1	1			3
9		1									3
10					1	1	1	1	1	1	5
11				1	1		1	1		1	3
12							1	1	1	1	3
13									1	1	3
Redundancy	3	4	4	5	5	4	4	4	2	6	

Assignment 4

operator	Glue	Programming	O-Ring	Alcohol	Spring	Dipping	Matching	QC	Bending	Grease	MF
1	1	1	1	1	1	1					4
2	1		1	1	1						4
3		1		1		1					4
4	1	1	1		1	1					4
5			1								4
6	1			1				1		1	4
7		1	1	1	1		1			1	4
8	1				1			1			4
9	1	1			1			1			4
10					1	1	1	1	1	1	6
11		1		1	1		1	1		1	4
12	1						1		1	1	4
13										1	4
Redundancy	7	6	5	6	8	4	4	5	2	6	

Appendix 10.3 Labor Strategies Based on Primary and Secondary Operator Assignments (Cont.)

Assignment 5

operator	Glue	Programming	O-Ring	Alcohol	Spring	Dipping	Matching	QC	Bending	Grease	MF
1	1	1	1	1							5
2	1	1		1							5
3		1	1	1	1						5
4		1	1	1	1	1					5
5			1				1	1		1	5
6	1		1	1			1			1	5
7			1	1		1				1	5
8	1	1			1		1	1	1		5
9	1	1			1				1	1	7
10			1		1	1	1	1	1	1	5
11		1		1	1	1	1	1	1		5
12	1							1		1	5
13						1	1	1	1	1	5
Redundancy	7	8	8	7	8	7	7	5	4	6	

Assignment 6

operator	Glue	Programming	O-Ring	Alcohol	Spring	Dipping	Matching	QC	Bending	Grease	MF
1	1	1	1	1	1	1	1	1	1	1	10
2	1	1	1	1	1	1	1	1	1	1	10
3	1	1	1	1	1	1	1	1	1	1	10
4	1	1	1	1	1	1	1	1	1	1	10
5	1	1	1	1	1	1	1	1	1	1	10
6	1	1	1	1	1	1	1	1	1	1	10
7	1	1	1	1	1	1	1	1	1	1	10
8	1	1	1	1	1	1	1	1	1	1	10
9	1	1	1	1	1	1	1	1	1	1	10
10	1	1	1	1	1	1	1	1	1	1	10
11	1	1	1	1	1	1	1	1	1	1	10
12	1	1	1	1	1	1	1	1	1	1	10
13	1	1	1	1	1	1	1	1	1	1	10
Redundancy	13	13	13	13	13	13	13	13	13	13	

Appendix 10.4 Labor Strategies Based on Fixed and Floating Operator Assignments

Assignment 1

operator	Process step										
	Glue	Programming	O-Ring	Alcohol	Spring	Dipping	Matching	QC	Bending	Grease	MF
1	1										1
2											1
3		1	1								1
4				1							1
5				1							1
6					1						1
7							1				1
8										1	2
9	1										2
10		1	1	1							2
11					1	1	1	1			2
12										1	2
13									1		
Redundancy	2	2	2	3	2	1	2	1	1	2	

Assignment 2

operator	Process step										
	Glue	Programming	O-Ring	Alcohol	Spring	Dipping	Matching	QC	Bending	Grease	MF
1	1										1
2											1
3		1	1								1
4				1							1
5				1							1
6					1						1
7							1				1
8										1	3
9	1										3
10		1	1	1		1					3
11		1	1		1	1	1	1			3
12							1	1		1	3
13									1		
Redundancy	2	3	3	3	2	2	3	2	1	2	

Appendix 10.4 Labor Strategies Based on Fixed and Floating Operator Assignments (Cont.)

Assignment 3

operator	Glue	Programming	O-Ring	Alcohol	Spring	Dipping	Matching	QC	Bending	Grease	MF
1	1										1
2		1									1
3			1								1
4				1							1
5				1							1
6					1						1
7							1			1	1
8	1	1	1	1							4
9		1	1	1			1				4
10				1	1		1			1	4
11					1	1	1	1			4
12						1		1	1	1	4
13											
Redundancy	2	3	3	5	3	2	4	2	1	3	

Assignment 4

operator	Glue	Programming	O-Ring	Alcohol	Spring	Dipping	Matching	QC	Bending	Grease	MF
1	1										1
2		1									1
3			1								1
4				1							1
5				1							1
6					1						1
7							1			1	1
8	1	1	1	1		1					5
9	1	1	1	1		1					5
10				1	1	1	1			1	5
11					1	1	1	1	1		5
12					1		1	1	1	1	5
13											
Redundancy	3	3	3	5	4	4	4	2	2	3	

Appendix 10.4 Labor Strategies Based on Fixed and Floating Operator Assignments (Cont.)

Assignment 5

operator	Glue	Programming	O-Ring	Alcohol	Spring	Dipping	Matching	QC	Bending	Grease	MF
1	1										1
2		1									1
3			1								1
4				1							1
5				1							1
6					1						1
7							1				1
8										1	1
9		1	1	1	1	1	1				6
10		1	1	1	1	1	1	1			6
11	1	1		1	1	1	1				6
12	1		1		1	1	1	1	1	1	6
13								1	1	1	6
Redundancy	3	4	4	5	5	4	5	3	2	3	

Process step

Assignment 6

operator	Glue	Programming	O-Ring	Alcohol	Spring	Dipping	Matching	QC	Bending	Grease	MF
1	1										1
2		1									1
3			1								1
4				1							1
5				1							1
6					1						1
7							1				1
8										1	1
9		1	1	1	1	1	1	1			7
10		1	1	1	1	1	1	1			7
11	1	1		1	1	1	1		1		7
12	1		1		1		1	1	1	1	7
13	1	1		1		1		1	1	1	7
Redundancy	4	5	4	6	5	4	5	4	3	3	

Process step

Appendix 10.4 Labor Strategies Based on Fixed and Floating Operator Assignments (Cont.)

Assignment 7

operator	Glue	Programming	O-Ring	Alcohol	Spring	Dipping	Matching	QC	Bending	Grease	MF
1	1										1
2		1									1
3			1								1
4				1							1
5				1							1
6					1						1
7							1				1
8										1	1
9	1	1	1	1	1		1	1		1	8
10	1	1	1	1	1	1	1	1			8
11	1	1	1	1	1	1	1		1		8
12	1			1	1	1	1	1	1	1	8
13		1	1	1		1	1	1	1	1	8
Redundancy	5	5	5	7	5	4	6	4	3	4	

Assignment 8

operator	Glue	Programming	O-Ring	Alcohol	Spring	Dipping	Matching	QC	Bending	Grease	MF
1	1										1
2		1									1
3			1								1
4				1							1
5				1							1
6					1						1
7							1				1
8										1	1
9	1	1	1	1	1	1	1	1		1	9
10	1	1	1	1	1	1	1	1		1	9
11	1	1	1	1	1	1	1	1	1		9
12	1		1	1	1	1	1	1	1	1	9
13		1	1	1	1	1	1	1	1	1	9
Redundancy	5	5	6	7	6	5	6	5	3	5	

Appendix 10.4 Labor Strategies Based on Fixed and Floating Operator Assignments (Cont.)

Assignment 9 operator	Process step										
	Glue	Programming	O-Ring	Alcohol	Spring	Dipping	Matching	QC	Bending	Grease	MF
	0	0	0	0	0						
1	1										1
2		1									1
3			1								1
4				1							1
5				1							1
6					1						1
7							1				1
8										1	1
9	1	1	1	1	1	1	1	1	1	1	10
10	1	1	1	1	1	1	1	1	1	1	10
11	1	1	1	1	1	1	1	1	1	1	10
12	1	1	1	1	1	1	1	1	1	1	10
13	1	1	1	1	1	1	1	1	1	1	10
Redundancy	6	6	6	7	6	5	6	5	5	6	

Index

Related Titles

Human Performance in Planning and Scheduling, Bart MacCarthy, John R. Wilson
ISBN: 0748409297

Increasing Productivity and Profit through Health and Safety: The Financial Returns from a Safe Working Environment, Maurice S. Oxenburgh, Penelope S.P. Marlow, Andrew Oxenburgh
ISBN: 0415243319

The Ergonomics Kit for General Industry With Training Disc, Dan MacLeod
ISBN: 1566703328